Robert Barnes

Lectures on Obstetric Operations

Including the Treatment of Haemorrhage

Robert Barnes

Lectures on Obstetric Operations
Including the Treatment of Haemorrhage

ISBN/EAN: 9783337811365

Printed in Europe, USA, Canada, Australia, Japan

Cover: Foto ©berggeist007 / pixelio.de

More available books at **www.hansebooks.com**

LECTURES ON

OBSTETRIC OPERATIONS.

LECTURES ON

OBSTETRIC OPERATIONS,

INCLUDING

THE TREATMENT OF HÆMORRHAGE,

AND FORMING

A Guide to the Management of Difficult Labour.

BY

ROBERT BARNES, M.D. LOND., F.R.C.P.,

OBSTETRIC PHYSICIAN TO,
AND LECTURER ON MIDWIFERY AND THE DISEASES OF WOMEN AND CHILDREN AT ST. THOMAS'S HOSPITAL;
EXAMINER ON MIDWIFERY TO THE ROYAL COLLEGE OF PHYSICIANS, AND TO THE ROYAL COLLEGE OF SURGEONS,
FORMERLY OBSTETRIC PHYSICIAN TO THE LONDON HOSPITAL;
AND LATE PHYSICIAN TO THE EASTERN DIVISION OF THE ROYAL MATERNITY CHARITY.

LONDON:
JOHN CHURCHILL AND SONS, NEW BURLINGTON STREET.

MDCCCLXX.

LONDON: B. PARDON AND SON, PRINTERS, PATERNOSTER ROW.

TO

RICHARD GRIFFIN, ESQ., J.P.,

OF WEYMOUTH,

IN TESTIMONY OF RESPECT

FOR HIS PUBLIC AND PRIVATE VIRTUES,

AND

IN GRATEFUL ACKNOWLEDGMENT

OF LASTING OBLIGATION,

THIS WORK IS AFFECTIONATELY DEDICATED

BY HIS PUPIL,

THE AUTHOR.

PREFACE.

It has long been my intention to write a systematic work, embracing all the topics usually treated of in Manuals of Obstetrics. When I first formed that intention, the task would have been comparatively easy.

Increased experience, supplying greater store of material, and compelling to more searching judgment of current doctrines and practice, makes it more and more difficult to satisfy myself and to do justice to others.

In the meantime, as occasion prompted, I have discussed in the form of monographs, or lectures published in the Transactions of Societies or in the medical journals, many of the most important questions in obstetrics. Amongst these, the "Lectures on Obstetric Operations," which were published in the *Medical Times and Gazette*, in 1867, 1868, and 1869, on the invitation of the editor, cover a considerable range, and form a fairly comprehensive treatise on a great department of obstetric practice.

I have received so many assurances that these Lectures supply, to a certain extent, a want long felt by practitioners and students, that I gladly accepted the proposal of the publishers to re-issue them at once in the more convenient shape of a separate volume.

To this I was further stimulated by the flattering offer of an American physician to edit the Lectures for me, with notes and additions by himself; and by the enterprise of American publishers, who have announced their intention to republish them. I did not respond to these obliging overtures, because I humbly felt that the author was the most competent person to revise his own work, he being fully conscious that it required revision; and because I hoped that a new edition, corrected and extended by myself, and published in England in the English language, would be not less acceptable to my professional brethren in America than an imperfect reprint done on their side of the Atlantic.

The design of these Lectures was to illustrate the mechanism of the various forms of difficult labour, carefully studying the ways of Nature in dealing with them, so as to deduce from this study indications how to assist Nature when she wanted help; to illustrate and place upon exact foundations the powers and applications of the instruments and operations used in obstetrics; and to give such a view of operative midwifery as would represent more faithfully than was done in any systematic work the practice of those who are most actively engaged in coping with difficult cases.

I believe that those who have had the greatest experience in difficult midwifery will be the most ready to acknowledge that the existing manuals—and I do not limit this proposition to English manuals—are not on a level with the actual practice of the more scientific and skilful operators. Every one who is frequently called upon to act under the sudden emergencies and various difficulties of obstetric practice, if he be endowed with any measure of independent thought and energy, will infallibly work out for himself new principles and methods of action. He will very soon discover the defects left by routine teaching; he will quickly detect that much of that teaching is traditional, conventional; that it shows but faint impressions of having been proved and moulded in the conflict with actual work.

Nothing has astonished me more than to find that the old English straight forceps, variously modified, indeed, but never emancipated from the narrow conditions which cripple the uses of the greatest of all conservative instruments, is still advocated by some range, and form a fairly comprehensive treatise on a great department of obstetric practice.

I have received so many assurances that these Lectures supply, to a certain extent, a want long felt by practitioners and students, that I gladly accepted the proposal of the publishers to re-issue them at once in the more convenient shape of a separate volume.

To this I was further stimulated by the flattering offer of an American physician to edit the Lectures for me, with notes and additions by himself; and by the enterprise of American publishers, who have announced their intention to republish them. I did not respond to these obliging overtures, because I humbly felt that the author was the most competent person to revise his own work, he being fully conscious that it required revision; and because I hoped that a new edition, corrected and extended by myself, and published in England in the English language, would be not less acceptable to my professional brethren in America than an imperfect reprint done on their side of the Atlantic.

The design of these Lectures was to illustrate the mechanism of the various forms of difficult labour, carefully studying the ways of Nature in dealing with them, so as to deduce from this study indications how to assist Nature when she wanted help; to illustrate and place upon exact foundations the powers and applications of the instruments and operations used in obstetrics; and to give such a view of operative midwifery as would represent more faithfully than was done in any systematic work the practice of those who are most actively engaged in coping with difficult cases.

I believe that those who have had the greatest experience in difficult midwifery will be the most ready to acknowledge that the existing manuals—and I do not limit this proposition to English manuals—are not on a level with the actual practice of the more scientific and skilful operators. Every one who is frequently called upon to act under the sudden emergencies and various difficulties of obstetric practice, if he be endowed with any measure of independent thought and energy, will infallibly work out for himself new principles and methods of action. He will very soon discover the defects left by routine teaching; he will quickly detect that much of that teaching is traditional, conventional; that it shows but faint impressions of having been proved and moulded in the conflict with actual work.

Nothing has astonished me more than to find that the old English straight forceps, variously modified, indeed, but never emancipated from the narrow conditions which cripple the uses of the greatest of all conservative instruments, is still advocated by some my views of the physiology and treatment of placenta prævia, I did so on the conviction of their truth, derived from careful observation at the bedside and anxious study and reflection. I now find that these views have received confirmation from so many practitioners of repute, who have put them in practice, that I feel no hesitation in presenting them as containing the true principles of dealing with one of the most formidable of all the dangers attending pregnancy and labour.

I have next sought to place upon its true physiological basis the treatment of the terrible hæmorrhages which occur after the birth of the child. All the ordinary methods of arresting uterine hæmorrhage depend for their efficacy upon their power of directing nerve-force to the uterus to cause its contraction. But if the woman has sunk so low that nerve-force enough for this purpose does not remain, the art of the physician is exhausted; unless, indeed, he can find in the system another power through which other remedies may act. This other power exists in the coagulability of the blood; and there are styptics which will plug the vessels left open by the paralyzed muscular walls. The timely use of all the means we possess for arresting hæmorrhage will almost always keep the loss within limits compatible with recovery; and in those exceptional cases where hæmostatic means have failed, we may still hope to save by the restorative operation of transfusion. Feeling strongly that there is something defective in our art, or in ourselves, if a woman is suffered to bleed to death, I have spent much care upon the exposition of the conditions upon which arrest of hæmorrhage depends, and I have insisted upon the necessity of improving and extending the operation of transfusion.

In seeking illustrations, I have drawn almost exclusively from my own experience and reflection. In so doing I may have failed somewhat in artistic nicety, but I believe that the diagrams are generally clear enough to serve their purpose. I was especially desirous to avoid the errors which ensue from the servile practice of copying; a system fatal to progress, under which errors acquire the accumulating force of tradition, whilst in reality they are often losing what truth was originally in them.

And, moreover, as I had many new ideas to express, I had, in most cases, no choice but to present new illustrations. In a few of the drawings I have had the able assistance of Mr. Dennison, the Librarian of St. Thomas's Hospital.

To my friend and former Resident Accoucheur at the London Hospital, Dr. Woodman, I am indebted for ready and valuable assistance in helping the book through the press.

The guiding principle I have kept steadily before me has been to save the lives, to diminish the suffering, of women in labour, and to rescue their offspring to the utmost extent consistent with that first imperative duty. That the attainment of these great objects may be promoted by diligent study of the principles and skilful execution of the practice set forth in this book, is my most earnest yet confident aspiration.

ROBERT BARNES.

31, GROSVENOR STREET, GROSVENOR SQUARE.
November, 1869.

CONTENTS.

LECTURE I.

LECTURE II.

LECTURE III.

LECTURE IV.

LECTURE V.

LECTURE VI.

LECTURE VII.

LECTURE VIII.

LECTURE IX.

LECTURE X.

LECTURE XI.

LECTURE XII.

LECTURE XIII.

LECTURE XIV.

LECTURE XV.

LECTURE XVI.

LECTURE XVII.

LECTURE XVIII.

LECTURE XIX.

LECTURE XX.

PAGE

LECTURE XXI.

LECTURE XXII.

LECTURE XXIII.

LECTURE XXIV.

LIST OF FIGURES.

OBSTETRIC OPERATIONS.

LECTURE I.

INTRODUCTORY—DESCRIPTION AND SELECTION OF INSTRUMENTS.

Two things have to be considered when attempting to describe the operations in midwifery—

1. What are the emergencies which call upon the Practitioner to operate?

2. What are the means, the instruments at his disposal?

If each accident or difficulty in labour were uniform and constant in all its conditions, it might be possible to apply to its relief the same operation or the same instrument. The history of operative midwifery might be told in an orderly series of simple mechanical formulæ. But how different is the case in practice! How infinite is Nature in her phases and combinations! The dream of Levret will never be realized. In proportion as observation unfolds these combinations, ingenuity is ready to multiply the resources of art. To describe these combinations, and the means of meeting them, is a task of ever-growing difficulty. Partial success only is possible.

The multitudinous array of instruments exhibited at the Obstetrical *conversazione* in 1866, vast as it was, gave but a feeble idea of the luxuriant variety that have been devised. If all these had their individual merits and uses, endless would be the labour of appreciation; the task of describing the operations of midwifery would be hopeless. It is, indeed, true that every instrument, even every modification of an instrument, represents an idea, although sometimes this idea is not easy to understand. Fortunately, it is not always important that the idea should be understood. Many of these instruments are suggested by imperfect observation, by ill-digested experience; many are insignificant variations upon an idea which, in its original expression, was of little value. Huge heaps, then, of instruments may, without loss to science, and to the great comfort of womankind, be cast into the furnace; the ideas of their inventors melted out of them. All that is necessary in relation to them is, to preserve examples in museums, where they may serve as historical records marking the course of obstetric science in its ebb and flow; for, strange to say, obstetric science has its fluctuations of loss as well as of gain, of going back as well as of going forward. These historical specimens will also serve the useful office of warning against the repetition of exploded errors, and of saving men the trouble and vexation of re-inventing.

When Science finds herself in the presence of complicated and disordered facts and ideas, her resource is to classify—that is, to seize a few leading ideas under which the subordinate ones may be grouped. In the first instance, the minor or subsidiary ideas—the epigenetic ideas they may be called—are disregarded. The grand or governing ideas only are studied. Then the process of analysis, the descent to details, to particulars, begins; and again, unless we keep a steady eye upon the governing principles, we are in danger of losing ourselves in the infinitely little, of falling into chaos, of running astray from the parent or guiding truth, in fruitless chase of the multitudinous splinters into which it has been subdivided.

What, then, have we to do? Knowing that, we will see how we can do it. Nature, although always requiring skilful watching, in the majority of cases does not want active assistance. But the cases are many in which pain, agony, may be averted; in which positive danger has to be encountered and thrust aside; in which action must be prompt and skilful.

Labour is a problem in dynamics. Three factors are concerned in the solution:—1st, there is the fœtus, the body to be expelled. 2nd, there is the channel, made up of the bony pelvis and soft parts, through which the body must be propelled; these two together constitute the resisting force, the obstacle to be overcome. 3rd, there is the expelling power, the uterus and voluntary muscles. These factors must be harmoniously balanced to produce a healthy labour. Labour may come to a stand from error in any one of these factors, or from disturbance of correlation. The permutations are almost infinite in kind and degree.

There are many ways in which disturbance may arise. There are not so many ways in which compensation or correction may be made—that is, treatment is more simple than are the causes of disturbance. To take the third factor first. The expelling power may be deficient, the other factors preserving their due relations. This power is a *vis à tergo.* The want of it may be made good in one of two ways. We may, in some cases, spur the uterus and its auxiliary muscles to act. The power may be dormant only; it exists potentially, capable of being roused by appropriate stimulus. This is the case for oxytocics, such as ergot, cinnamon, borax, or cinchona. But the power may not be there, or, if there, it may not be wise to provoke it to action.

An interesting question arises here:—Can we, without resorting to oxytocic medicines, arouse or impart a *vis à tergo?* Can we apply direct mechanical force to push the fœtus out of the uterus, instead of dragging it out? Now, in some cases this seems possible. Von Ritgen,* in a memoir on "Delivery by Pressure instead of Extraction," adverting to the fact that the *natural* mode is by pushing out, said that the *artificial* mode was by dragging out; and asked very pertinently, "Why do we always drag and never push out the fœtus?" Dr. Kristeller † has carried the idea into practice. By means of a dynamometric forceps he has shown that a force of five, six, or eight pounds only is often found sufficient to extract a head that has lain for hours unmoved; so that the force to be administered in the form of pressure need not be very great. Poppel ‡ estimates, from experiments made to

* "Monatsschr. f. Geburtsk," 1856. † *Ibid.*, 1867. ‡ *Ibid.*, 1863.

determine the power necessary to burst the membranes, that the force necessary to effect an easy labour does not much exceed 4 lbs. Dr. Matthews Duncan,* from similar experiments, estimates it at 6 lbs.; whilst the average force to effect labour is only 16 lbs. It is needless to premise that the presentation and the relations of fœtus and pelvis must be normal. The method is as follows:—The patient lying on her back, the operator places his hands spread on the fundus and sides of the uterus, and combining downward pressure with the palms on the fundus with lateral pressure by means of the fingers, the uterus being brought into correct relation with the pelvic axis, its contents are forced down into the cavity. The pressure is so ordered as to resemble the course and periodicity of the natural contractions. Of course the pressure will often excite uterine contraction to aid or even supplant the operator. But it seems that pressure alone is sometimes sufficient. As an adjuvant to extraction, pressure is, I know, of great value. I never use the forceps or any extracting means without getting an assistant to compress the uterus firmly, to maintain it in its proper relation to the axis of the brim, and to help in the extrusion of the fœtus. This resource, then, should not be lost sight of. In certain cases it may obviate the necessity of using the forceps; or it may stand you in good stead when instruments are not at hand. When a *vis à tergo* cannot be had, we have the alternative of supplying power by importing a *vis à fronte*. In the case we are supposing, the means of doing this reside chiefly in two instruments: the lever and the forceps.

* "Researches in Obstetrics," 1868.

In the second order of cases there is a want of correlation between the body to be expelled and the channel which the body must traverse. There are many varieties of this kind of disturbance. The progress of the head may be opposed by rigidity of the soft parts, especially of the cervix uteri. Patience is one great remedy for this. A dose of opium and a few hours' sleep will sometimes accomplish all that is desired. But patience may be carried too far. If the pulse rise and the patient show signs of distress, it is proper to help. I have no faith in belladonna. To excite vomiting by tartar emetic is to add to the distress of the patient without the certainty of relieving her. To bleed is also to indulge in a speculation that will certainly cost the patient strength she will need, and it promises only a doubtful gain. We have two mechanical resources to meet this strictly mechanical difficulty. There is the hydrostatic dilator, which I have contrived for the express purpose of expanding the cervix. In the case of a cervix free from disease, dilatation will commonly proceed rapidly and smoothly under the eccentric pressure of these dilating water bags, which closely imitate in their action the hydrostatic pressure of the liquor amnii. In the case of rigidity from morbid tissue, as from hypertrophy or cicatrices, something more may be necessary. The timely use of the knife will save from rupture, from exhaustion, or from sloughing. I have contrived a very convenient bistoury for this purpose. It is carried by the finger into the os uteri; multiple small nicks are made in its circumference; and by alternate distension with the water bags the cervix may be safely and sufficiently dilated.

The fœtus and the channel may be duly proportioned, but *the position of the child is unpropitious.* In this case all there is to do is to restore the lost relation of position. The hand, the lever, and the forceps are the instruments.

There is disproportion. This may be of various kinds and degrees. The varieties will be more conveniently unfolded hereafter. It is sufficient to say here that all resolve themselves, in practice, into three classes—

1. Disproportion that can be overcome without injury to the mother, and with probable safety to the child.

2. Disproportion that can be overcome without injury to the mother, but with necessary sacrifice of the child.

3. Disproportion that can be overcome with possible or probable safety to both mother and child.

The first class of cases may be relieved by the hands, or by the forceps. The second by reducing the bulk of the child to such dimensions as will permit it to pass through the contracted channel. The perforator, the crotchet, the craniotomy forceps or cranioclast, the cephalotribe, the forceps-saw, and the wire-écraseur are the principal instruments for bringing the bulk of the child down to the capacity of the pelvis. In the third class of cases we cannot insure the mother's safety by sacrificing her child. We therefore seek her *probable* safety by an operation—the Cæsarian section—which evades the difficulty of restoring the relation of bulk and capacity between fœtus and pelvis, by extracting the fœtus through an artificial opening in the mother's abdomen. The instruments required for this purpose are not specially obstetrical. But a bistoury, scissors, needles, and sutures, silk or silver, take but little room, and as they may at any unforseen moment be wanted, they should always be found in the obstetric bag. And we shall have to put into it a few other instruments and accessories in order to be prepared for all emergencies. Let us enumerate all in order.

OBSTETRIC INSTRUMENTS—THE OBSTETRIC BAG.

To save the child.

1. A lever.
2. A long double-curved forceps.
3. Roberton's apparatus for returning the prolapsed funis.

To reduce bulk of child.

4. A craniotome or perforator.
5. A crotchet.
6. A craniotomy-forceps.
7. A cephalotribe.
8. A wire-écraseur and an embryotomy-scissors.
9. Ramsbotham's decapitating hook.

To induce or accelerate labour.

10. A blunt-ended straight bistoury, with a cutting edge of three-quarters of an inch, to incise the os uteri in cases of extreme contraction or cicatrisation. A hernia-knife answers very well.

11. A Higginson's syringe, fitted, on my plan, with a flexible uterine tube nine inches long, which serves for the injection of iced water or perchloride of iron, to arrest hæmorrhage, and also serves to expand.

12. A set of my caoutchouc hydrostatic uterine dilators.

13. Threeor four elastic male bougies (No. 8 or 9).

14. A porcupine quill to rupture the membranes.

15. A flexible male catheter. The short silver female catheter is often useless, and is generally less convenient than the flexible male catheter.

16. A pair of scissors and thread.

For the Cæsarian section.

17. A bistoury.

18. Sutures, silk and silver.

MEDICINES.

1. Chloroform and inhaler.
2. Laudanum.
3. Hofmann's anodyne.
4. Ergot of rye.
5. Solution of perchloride of iron. The liquor ferri perchloridi fortior (Brit. Pharm., 1867). Two ounces of this diluted with six ounces of water is an efficient hæmostatic.

The most convenient mode of packing these instruments is to adapt a travelling leather bag. There is always spare room for anything likely to be wanted, besides its ordinary furniture, or for bringing away a pathological specimen; and by turning out the obstetric furniture you have a travelling bag again.

I will now say a few words in explanation of the instruments recommended.

1. *The Lever.*—The form adopted is that of Mr. Symonds, of Oxford. The blade or fenestra is rather

strongly arched; and there is a joint in the shank, enabling the instrument to be doubled up for greater convenience of carrying.

2. *The Forceps.*—There are several excellent models. I am not bigoted in favour of my own. The best are Simpson's and Roberton's. The essential conditions to be contended for are:—That the blades have a moderate pelvic curve; a head-curve also moderate; an extreme divergence between the fenestræ of three inches; the length of the arc of the cranial bow of about seven inches, to adapt it to the elongation of the fœtal head during protracted labour. There should be between the springing of the bows and the lock a straight shank to lie parallel with its fellow, to carry the lock clear of the vulva and to save the perinæum. In my forceps the shank is further lengthened by a semicircular bow, which forms a ring with its fellow when locked. The use of this is to give a hold for the finger of one hand whilst the other grasps the handles. In Simpson's instrument there is a hollowed shoulder at the head of each handle which answers a similar purpose, and perhaps better. The lock should be easy—a little loose. The English lock is not, I think, surpassed for convenience; but the French lock is a good one. The handles should not be less than five inches long. They should afford a good grasp. Unless they are strong and of fair length, they cannot exert any compressive force for want of leverage, for the fulcrum is at the lock. I think all forceps that have very short handles, especially if not provided with some means, such as the ring or projecting shoulders, which will enable the operator to use both hands,

ought to be rejected. A two-handed instrument can be worked with the utmost nicety and economy of muscular force. A single-handed instrument is necessarily a weak one. The absurd dread of possessing powerful instruments has long been the bugbear of English Midwifery. It has been sought to make an instrument safe by making it weak. There can be no greater fallacy. In the first place, a weak instrument is, by the mere fact of its weakness, restricted to a very limited class of cases. In the second place, if the instrument is weak, it calls for more muscular force on the part of the operator. Now, it is sometimes necessary to keep up a considerable degree of force for some time, and not seldom in a constrained position. Fatigue follows; the operator's muscles become unsteady; the hand loses its delicacy of diagnostic touch, and that exactly balanced control over its movements which it is all-important to preserve. Under these circumstances he is apt to come to a premature conclusion that he has used all the force that is justifiable, that the case is not fitted for the forceps, and takes up the horrid perforator; or he runs the risk of doing that mischief to avoid which his forceps was made weak. The faculty of accurate graduation of power depends upon having a reserve of power. Violence is the result of struggling feebleness, not of conscious power. Moderation must emanate from the will of the operator; it must not be looked for in the imperfection of his instruments. The true use of a two-handed forceps is to enable one hand to assist, to relieve, to steady the other. By alternate action the hands get rest, the muscles preserve their tone, and the accurate sense of resistance which tells him the minimum degree of force that is necessary, and warns him when to desist. A similar reasoning applies to the perforator and the craniotomy-forceps.

Roberton's Tube for Prolapsed Funis.—There are many contrivances for returning the prolapsed cord. Braun's is an excellent one; one by Hyernaux, of Brussels, is also very ingenious and useful, but Roberton's appears to me the most simple. By the knee-elbow posture, indeed, all instruments may occasionally be dispensed with; but still it is well to be provided with this very simple apparatus.

The Perforator.—The instruments designed to open the skull are classed in the Obstetrical Society's Catalogue under four types:—1. The wedge-shaped scissors, having blades cutting on the outer sides. 2. The spear-head. 3. The conical screw. 4. The trepan. Mr. Roberton uses a spear-head. The form most in use in this country is some modification of Smellie's wedge-shaped scissors; but many of these instruments are very clumsy and inefficient. It requires sometimes considerable force to penetrate the cranium. A weak instrument is here especially dangerous: it is apt to slip, to glide off the globe ot the head at a tangent, and to tear the uterus. The conditions of efficiency are these:—The perforating blades must be strong and *straight.* The curve sometimes given is of no use whatever, as it throws the force out of the perpendicular. The shanks must be long, eight inches at least, so as to reach the pelvic brim without interfering with the working of the handles. There should be a broad rest for the hand to give a powerful and steady hold. Almost all the instruments in use fail in this point. The best of all those I have tried and seen is the modification of Holmes' and Naegele's by Dr. Oldham; it fulfils every indication. On the continent, especially in Germany, the trepan, first introduced by Assalini, and variously modified, is chiefly used. To use a trepan, the crown of which can hardly be less than an inch in diameter, you must have at least an equal amount of surface of the cranium accessible, and the crown must be applied quite perpendicularly to the cranium. Now, these conditions are not always present. I have been much pleased in some cases with the trepan of Professor Ed. Martin, of Berlin. But in others, where the pelvic deformity was great, and especially where it was necessary to perforate after the body was born, there was no room for the passage or application of the instrument. I found no difficulty with Dr. Oldham's perforator; it will run up through the merest fissure wherever the finger will go to guide it, and will readily penetrate any part of the skull. This, then, is the perforator to be preferred.*

The Crotchet.—The design of the crotchet was to seize and extract, by taking a hold inside the cranium, after perforation. For this purpose the best crotchet is the one used in the Dublin Lying-in Hospital. It has a curve in the shank, which is set in a transverse bar of wood for a handle. This gives an excellent hold for traction, that does not fatigue or cramp the operator. The crotchet, however, as an extracting instrument, has been

* It is figured in the Obstetrical Society's Catal gue of Instruments, p. 167.

greatly displaced by the craniotomy-forceps. The use to which I now almost restrict the crotchet is to break up the brain and tentoria, so as to facilitate the evacuation and collapse of the skull.

The Craniotomy-Forceps.—The use of this instrument is twofold. It should be able to break up and pick away the bones of the cranial vault, and to grasp firmly the skull to serve as an extractor. In the majority of cases the latter action alone is necessary. For extraction, the essential condition is to have the blades so made that when grasping they shall be perfectly parallel. Unless this be obtained, the blades will only pinch at one point, and the effect will be to break through the bone, to tear through the scalp, and to come away. Each time the attempt is renewed, if ever so little traction is necessary, you are exposed to the same mishap, until you may find no place left that will afford a hold. To remedy this defect, many instruments are armed with horrent teeth and spikes, which only add to the evil. Whereas, if the blades are parallel, they gripe firmly over a wide surface, and do not break away. The hold is obtained by compression, by accurate apposition, not by teeth or spikes. To secure the grasp without fatiguing the hands by compressing the handles, I have adapted a screw to bind the handles together. It is also important that the blades should be distinct, so as to admit of being introduced separately, like the ordinary forceps. These principles are fairly carried out in my craniotomy-forceps, and others as well as myself are well satisfied with this instrument. But recently these principles have been more conveniently adapted in an instrument devised by my friend Dr. Matthews. He makes the female blade to revolve on a central pivot, so that the moment it touches by either end the included scalp and cranium, it adjusts itself in accurate parallelism, and the grip is secured by a sliding ring on the shanks. With such an instrument it is very rarely necessary to take a second hold.

The Cephalotribe.—In cases of great, but not of extreme, distortion of the pelvis, the cephalotribe is an instrument capable of materially accelerating delivery. After perforation, the powerful blades applied to the head, crush in and flatten it, so that it can be drawn through a comparatively narrow passage. The unwieldy bulk and formidable appearance of most of the continental cephalotribes, requiring, as they do, an assistant in their use, must preclude their extensive adoption. Almost every objection is removed in Sir James Simpson's instrument. Dr. Braxton Hicks' modification of Simpson's cephalotribe is, in my experience, a most effective and convenient instrument. Dr. Kidd, of Dublin, has also contrived an excellent instrument. All these can be worked by the operator unaided; and can be carried in the bag with the other pieces of the obstetric armamentarium. Sufficient power to crush down the base of the cranium after perforation is combined with a minimum of size and weight.

The Wire-écraseur.—This instrument I propose to add in order to execute the new method of Embryotomy I have designed, to effect delivery in extreme cases of pelvic deformity. Its use is to make sections of the head *in utero*, so as to reduce it to a bulk easy for extraction. The instrument must be a powerful one; and it should have an endless or Archimedean screw-movement, so as to work a loop of wire large enough to take the head in its equator. Weiss's instrument answers well; but it is very expensive. In connection with this operation it is also necessary to be provided with a pair of *Embryotomy-scissors* to cut up the trunk. The wire can also be used as a decapitator, instead of Ramsbotham's hook. It further forms an excellent snare to catch a foot when too remote to be seized by a loop of tape carried by the fingers.

The Decapitating Hook.—This instrument will be very rarely required, but when the occasion arises, the service it renders is very great. In a protracted transverse presentation, when the child is dead from compression, the uterus spasmodically and closely contracted upon the child, turning cannot be accomplished without subjecting the mother to much suffering and some danger. In such a case it is obviously preferable—the child being past help—to save the mother to the utmost. This hook can be carried over the child's neck, and by a movement of sawing and traction, the head can be severed in a few seconds. Then the body is extracted by pulling on the prolapsed arm. The head remaining alone *in utero* can be easily extracted by the craniotomy-forceps. Thus delivery can be effected, with little cost to the mother, in a few minutes. The same object can be attained by a pair of strong scissors (Dubois) which is made to divide the cervical spine. An excellent decapitator was exhibited* by Jacquemier; in general form

* See Obstetrical Society's Catalogue, p. 47.

it resembles Ramsbotham's, but it has a concealed or sheathed decapitator, the cutting being effected by moveable blades and saw-links. Pajot, again, decapitates by carrying a strong cord round the neck.

The Syringe, Uterine Tube, and Caoutchouc Dilators.—These are perhaps the most frequently useful of all the instruments enumerated. A Higginson's syringe is fitted with a mount, to which the flexible uterine tube or any one of my dilators can be adapted. Three sizes of the dilators are sufficient. They are now in extensive use at home and abroad.

The *elastic male bougies* are useful as the best means of inducing labour, that is, of provoking labour.

The *porcupine-quill* is a most convenient instrument for piercing the membranes; and, although a common quill or steel-pen will answer the purpose, these are not always at hand. The special instruments, as stilets, etc., are really superfluous.

The remaining instruments require no special description.

And, lastly, let me add a few words concerning the obstetric hand, as the master instrument of all, not only as guiding all the rest, but as performing many most important operations unarmed. In ordinary labour it is the only instrument required. It is also the only instrument called for in many of the greatest difficulties. In malpresentations, in placenta prævia, in many cases of contracted pelvis, in not a few cases where, after perforation, the crotchet and craniotomy-forceps have failed to deliver, the bare hand affords a safe and ready extrication. One cannot help seeing that practice is often determined by the accidental perfection of, or familiarity with, particular instruments. Thus, a man who has only reached that stage of obstetric development which is content with a short or single-curved forceps, will be armed with a good perforator and crotchet. He cannot fail to acquire skill and confidence in embryotomy, and greatly to restrict the application of the forceps. Again, the preference generally given on the continent to cephalotripsy over craniotomy and extraction by the crotchet or craniotomy-forceps is the result of the great study directed to the perfecting of the cephalotribe. At the present day we may boast of having good and effective instruments of all kinds, each capable of doing excellent work in its own peculiar sphere, and moreover endowed with a certain capacity for supplanting its rival instruments. For example, the long double-curved forceps is adapted to supplant craniotomy in a certain range of cases of minor disproportion. Hence it follows that it is of more importance to have a good forceps which can save life than it is to have a good perforator and crotchet which destroy life. At the same time, it is eminently desirable to possess the most perfect means of bringing a fœtus through a very narrow pelvis, in order to exclude or to minimise the necessity of resorting to the Cæsarian section. Our aim should then be to get the most out of all our instruments, to make each one as good of its kind as possible. And admirable is the perseverance, marvellous and fertile the ingenuity, that have been brought to this task. I will not say that it has all been misdirected; but certainly the cultivation of the hand, the study of what it can do in the way of displacing cold iron, has been much neglected. It would be not less instructive than curious to carry our minds back to the days when the forceps and other instruments now in use were unknown, and to confront the problem which our predecessors, Ambroise Paré, Guillemeau, and others, had to solve —namely, how to deliver a woman with deformed pelvis without instruments. That they did successfully accomplish in many instances with the unarmed hand what we now do by the aid of various weapons, there can be no doubt. If this implies greater poverty of resources on their part, it not the less implies also greater manual skill. I am confident that the possession of instruments, especially of the craniotomy instruments, has led, within the last century, to a neglect of the proper uses of the hands, which is much to be deplored. We are only now recovering some of the lost skill of our ancestors.

Obstetric Surgery has this peculiarity: its operations are carried on in the dark, our only guide being the information conveyed by the sense of touch. The mind's eye travels to the fingers' ends. The hand thus possesses an inestimable superiority over all other instruments. Its every movement is regulated by consciousness. It is right, then, to ascend a little the stream of knowledge, and to endeavour to recover from the experience of our forefathers their secret of *chirurgery;* to regain, to extend, our power over that great instrument from which the Surgeon derives his name.

LECTURE II.

THE POWERS OF THE FORCEPS—THE FORCE BY WHICH IT HOLDS THE HEAD—THE COMPRESSIBILITY OF THE CHILD'S HEAD—THE LEVER—DEMONSTRATION THAT THE LEVER IS A LEVER, NOT A TRACTOR; ALSO THAT THE FORCEPS IS A LEVER.

To arrive at a just idea as to the application of instruments in difficult midwifery, it is first of all necessary to study carefully what these instruments can do. What, for example, are the powers of the forceps, the lever, of the crotchet and craniotomy-forceps, and of the cephalotribe? When we know these, and have formed a correct idea of the nature of the labour—that is, of the difficulty to be overcome—we shall know which instrument to select, and how to use it. The powers of an instrument must obviously depend upon its construction; but this is true to an extent not often thoroughly appreciated. Take, for example, the noblest of all, the forceps. It is difficult to exaggerate the importance of developing to the fullest extent the powers of this instrument. The more perfect we make it, the more lives we shall save, and the more we throw back into reserve those terrible weapons which only rescue the mother at the sacrifice of her offspring.

Three distinct powers or forces can be developed in the forceps. First, by simply grasping the head and drawing upon the handles, it is a *tractor*, supplementing a *vis à fronte* for the defective *vis à tergo*. Secondly, the forceps consisting of two blades having a common fulcrum at the joint or lock, we can by a certain manipulation use it as a *double lever*. Thirdly, if the blades and handles are long enough and strong enough, and otherwise duly shaped, the forceps becomes a *compressive power* capable of diminishing certain diameters of the child's head, so as to overcome minor degrees of disproportion.

Now, all these powers may be brought into use, and all may be in great measure lost, according to our choice of a good or a bad model. Thus, if we rest satisfied with the short forceps of Denman, we shall only have a feeble tractor, a feeble lever, and an instrument having almost absolutely no compressive force. It is obvious that such a forceps can have but a restricted application. It can only serve to deliver the child when the head is in the pelvis, when very little tractile power is required. Ask yourselves what this means. What is the consequence in practice? Simply this: you are driven in a multitude of cases to perforate, to destroy the child. Such an alternative may well make us reflect whether we cannot extend the powers and the application of the forceps. By simply lengthening the blades and shanks and giving the blades an additional curve adapted to the curved sacrum, we can reach the head detained on the brim of the pelvis. By moderately lengthening the handles and making the instrument stronger, we increase the leverage and tractile power, and we gain a moderate compressive power. Thus we bring within the saving help of the forceps a further number of children that must otherwise be given up to the perforator, or run the risk of turning. You ask, Why hesitate to endow the forceps with this great privilege? Why has the feeble forceps of Denman so long held its sway in this country? The reason is that there are limits beyond which we cannot push the saving powers of the forceps. If we pass beyond these limits, we run into danger of injuring the mother and of losing the child. Now, the great contest in all matters of strife is about boundary lines; and it is concerning these limits that authorities have differed. Some men are afraid of giving power, lest it should be abused. They are so terrified at the possible mischief which great power may work, that they would rather abandon the good which great power is equally capable of working. They tremble lest we should be unable to acquire the skill and the discretion necessary to direct that greater power. Such men virtually say, You shall not apply the forceps where the head has not descended into the pelvic cavity—an arbitrary limit dictated by fear, and fixed by ignorance, that the forceps is just as capable of safely delivering a child whose head is arrested at the brim. For here, as is continually the case in Medicine, experience, arbitrarily limited, excludes progress in knowledge and bars improvement in practice. For example, how can a man acquire a just knowledge of the power of the forceps to deliver a head delayed, by slight disproportion, at the brim, if he always delivers under this difficulty by perforating? Clearly, he bars himself from acquiring that knowledge; and, giving up his intelligence to the delusive dictates of his wilfully limited experience, he refuses even to accept the evidence of those whose experience is greater, because it is directed by a freer spirit of research, by greater confidence in the resources of art.

Let us, then, go back to the study of the powers of the forceps, unshackled by any preconceived opinions as to what the instrument can do or can be permitted to do. First, as to its *tractile* powers. In order to draw, the instrument must take hold. How does it take hold? You may at first sight suppose that this is accomplished by grasping the handles. But in the case of the ordinary forceps, especially the short-handled forceps, there is little or no compressive power, so that the hold cannot be due to the handles. The hold is really due to the curvature of the blades, which fit more or less accurately upon the globular head, and to the compression of the bows of the blades against the soft parts of the mother, supported by the bony ring of the pelvis. This may be made clear by a simple experiment. Take an india-rubber ball, slightly larger in diameter than a solid ring; place the ball upon the ring. Then seize the ball through the ring by the forceps. The blades will be opened out by the ball. Then drawing upon the handles, even without squeezing them together, you will see the blades pressed firmly upon the ball by gradual wedging, as the greatest diameter or equator of the ball comes down into the ring. Just so is it with the child's head and the pelvic brim and canal. The blades are held in close apposition to the head by the soft parts and the pelvis of the mother. The effect of pressure upon the bows of the blades in maintaining the hold, is again proved by the readiness with which the blades slip off as soon as the equator of the head has cleared the outlet. In many cases, this outward pressure upon the bows of the blades is enough to serve for traction. It is not necessary to tie the handles of the forceps. You may even do without handles altogether. Thus, one of the earliest attempts, stimulated by the desire to realize the concealed discovery of the Chamberlens—that of Palfyn—consisted in applying two opposed levers, which did not cross, and therefore could not exert any compressive force. Assalini's forceps was constructed on this principle. It is essentially a tractor, with slight leverage power. Professor Lazarewitch, of Charkoff, brought to the Obstetrical Exhibition a beautiful forceps constructed on Assalini's principle. This instrument I applied in two cases. It held admirably; but all its holding power is due to the pressure exerted by the mother's parts upon the blades.

Mattéi, of Paris, has made another instrument, whose blades do not cross, whose parallel shanks are set in a cross bar of wood to serve for traction; and quite recently Dr. Inglis, of Aberdeen, has proposed a forceps in which the handles are done away with altogether, there being nothing but a short curve of the shank, representing the shoulders on the handles of Simpson's forceps, to serve for traction. I think this sacrifice of all compressive and leverage power, reducing the instrument to a weak tractor, is a retrograde movement. But it proves the proposition that the hold upon the child's head is the result of the adaptation of the curved blades and the outward wedge-pressure of the mother's parts upon the bows of the blades. Now, the strength of the hold depends mainly upon the degree of curvature of the blades and the width of the fenestræ. If the curve is one of large radius, so that the two blades, when in opposition, approach parallelism, and especially if the fenestræ be narrow, the hold will be feeble, and moderate traction will cause the forceps to slip, and this in spite of any compression you can exert upon short handles. But increase the curve so that the blades in opposition form nearly a circle, and the instrument will not slip. This increased head-curve is one feature of the French or continental forceps. The hold is further strengthened by making the points approach nearer together. In the English patterns the points are generally distant from each other an inch or more. In the foreign forceps the distance is often much less than an inch. There is some danger from this proximity of pinching or abrading the skin of the face. So much for the grip and traction.

Let us now study the *compressive power*. This is inconsiderable in almost all the English forceps, but is an important feature in most of the foreign long forceps. The essential condition for compression is, indeed, present in English and foreign. This consists in the crossing of the blades, and in the greatest divergence of the blades, when the handles are brought together, being less than the greatest transverse diameter of the child's head. This diameter is normally from 3¾ to 4 in.; the greatest divergence of the blades is rarely more than 3 in. Therefore, when the blades are sitting loosely on the head, the handles diverge. Practically, the head is rarely grasped exactly in its transverse diameter, but generally in one more or less oblique—something between the transverse and the longitudinal diameter. This, of course, is even longer than the transverse. Now, if we are to exert any direct compression upon the head, we can only do it by squeezing the handles together. For this purpose, the handles must be long and strong on one side of the lock, and the blades must be strong, but not much longer, on the other side of the lock, than are the handles.

FIG. 1.

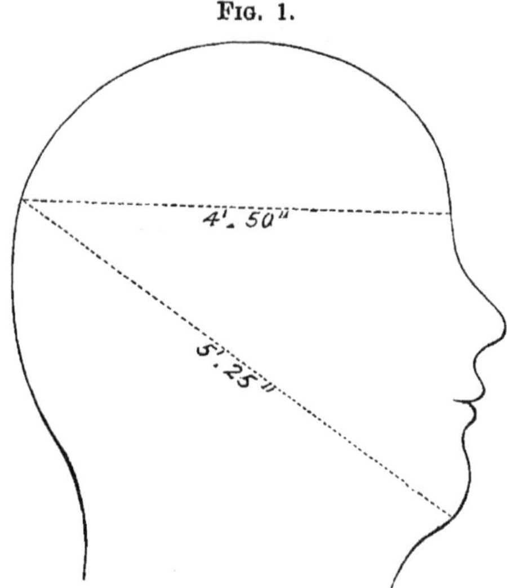

THE ORIGINAL FORM OF THE HEAD BEFORE BEING AFFECTED BY LABOUR.

It would be useless to provide this compressing power if the head were not compressible. That the head is compressible—that is, that we may diminish some of its diameters by lengthening others—is easily proved.

Firstly. It is known that in normal pelves the head in passing, if the labour be protracted, undergoes elongation; from round it becomes conical; the greatest transverse diameter—the interparietal—becomes merged in the lesser or interauricular, whilst the longitudinal diameters are correspondingly increased.

These changes I have demonstrated by actual measurements and outlines.* Diagrams 1 and 2 may be taken as types of the normal head and of the form impressed in protracted labour.

FIG. 2.

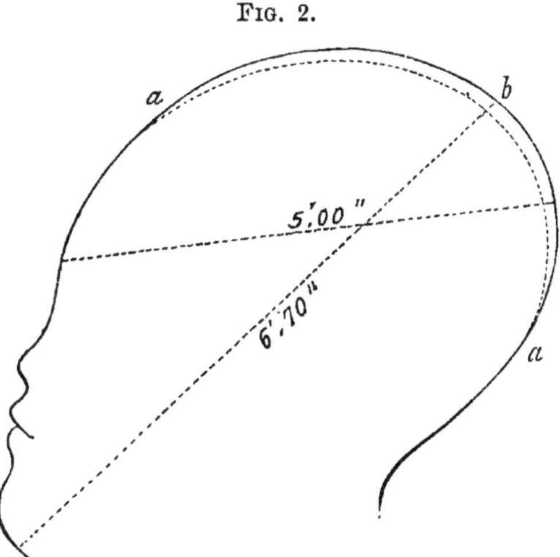

HEAD MOULDED BY PROTRACTED LABOUR.

Thus, just as the pressure of the soft parts and the pelvis is a main agent in fixing the forceps upon the head, so it is in moulding the head to allow of its passing. Indeed, I think this pressure almost entirely accounts for the alteration of form the head undergoes when the English forceps is applied. I can show outlines of heads as strongly

* "Obstetrical Transactions," vol. vii.

altered under the natural forces of labour as they often are under forceps delivery.

Secondly. Numerous experiments have been made with strong forceps upon dead children to determine this point. Baudelocque found that he could lessen the transverse diameter by a quarter to a third of an inch. Siebold gained half an inch. Osiander and Velpeau claim quite as much. More conclusive are the observations of M. Joulin and of M. Chassagny. These gentlemen, in experiments designed to demonstrate the utility of continuous compression and traction by powerful forceps upon the head in difficult labour, have completely proved that a degree of moulding may be effected much beyond that commonly observed. This moulding consists in the elongation of the head, the elongation being gained by the lessening of the equatorial diameters. The process resembles that of reducing wire by drawing it through holes in an iron plate.

Now, another question arises: the head is indeed compressible, but to what extent is it compressible without sacrificing the child's life? For if the maximum of plasticity compatible with life be represented by that degree which is common in severe first labours, then we ought to give the mother all the ease in our power by lessening the diameter of the child's head by perforating. It is very difficult to fix this limit with accuracy. Baudelocque thought compression to the extent of a quarter or a third of an inch was compatible with the safety of the child. The important fact is, that in many cases the child survives, although its head has undergone very great compression and moulding.

The following conditions influence the result:—The degree of development of the head as to size and ossification; and the mode in which the compressing force is applied. If this force be applied *gradually* and *continuously*, a much greater extent of moulding with less injury to the child may be obtained than what Baudelocque thought possible.

At one time it was the practice—more probably with the view of securing the hold than of compressing the head—to tie the handles together; and even now that tying is generally abandoned and condemned, the old custom asserts itself in the preservation of the grooves near the extremities of the handles made to receive the ligature. The objection to tying is this—the continuous compression is opposed to the course of Nature, which intermits the expulsive act, giving periods of rest during which it is presumed that the brain may better adapt itself, and its circulation be maintained. Hence the law that we ought in forceps' labours, and, generally, in all operative labours, to imitate this intermitting action, by interposing intervals of rest, endeavouring so to time our efforts as to be simultaneous with, and in aid of, the natural expulsive efforts. The argument is good both in logic and in physiology. It is not wise to disregard it. But experience proves that there are cases where the moulding of the head can be accomplished more quickly, and without endangering the child, by continuous pressure. Some practitioners, therefore, have recurred to the old practice. Dr. Gayton has adapted a clip to the handles of the forceps, which answers much better than the ligature. Whatever the mechanism resorted to, it is essential that it admit of being instantaneously removed, in order to allow the blades to be taken off. Delore,* who has made many dynamometric observations, concludes that pressure exerted either by the forceps or by the genital organs, may be harmless to the head if spread over a large surface. It is limited and angular pressure that is dangerous. He has also shown that *the greater the traction the greater is the pressure.* The pressure is equal to about half the traction. Thus, if you exert a traction force of fifty pounds, the pressure upon the head is about twenty-five pounds.

To economise traction, then, is to economise pressure. How do we economise traction?

There are three principal rules.

First.—Take sufficient time to allow the head to mould.

Secondly.—Take care to draw in the axis of the brim—that is, traction must be perpendicular to the plane of the brim. If this is neglected, additional force is required, increasing with every degree of angular difference.

Thirdly.—To use slight movements of laterality or oscillation.

This uncertainty and inconstancy in the degree to which compression may be carried with safety to the child, is a justification for tentative or experimental efforts with the forceps. It is the reason why in doubtful cases, where the disproportion in size between pelvis and head is not very decided, we are called upon to make a reasonable trial of the forceps before resorting to craniotomy. It appears to me quite certain that in this country we are yet far from having utilised the powers of the forceps to the

* "Gazette Hebdomadaire," 1865.

highest legitimate extent. I might go further, and say that during Denman's time and until quite recently, we had actually lost ground in this respect, and had reverted to the use of instruments scarcely better than the original rude forceps of the Chamberlens. More than one hundred years ago Smellie contrived and used the long forceps. Perfect used it, and it seems that in his time the long forceps was better known in England than it was during the first half of the present century.

Knowing what the forceps can do, and having an approximate idea of the extent of compressibility of the child's head compatible with the preservation of life, we may now study the various cases in which the instrument may be used, and the modes of applying it.

It is well to begin with the simplest case. This occurs when the head, presenting in the first position, has descended into the cavity of a well-formed pelvis, and is arrested on the perinæum from want of expelling power. In such a case very moderate leverage and tractile power—a force of a few pounds, perhaps—is all that is required. Often the lever or tractor will be quite sufficient. The moulding or diminution of the equatorial diameters will be effected by the sole compression of the mother's structures. The occiput lying behind the left foramen ovale, the tractor may be slipped over it, and the head drawn down towards the pubic arch, using your fingers as a fulcrum. This may be enough, for often, when the head is once started, expulsive action returns. If not, then the tractor may be shifted to the opposite side, so as to lie over the child's face and chin in the hollow of the sacrum. Then drawing down, you give the extension movement to the head, and the cranium soon emerges through the outlet. Several skilful practitioners, who frequently resort to this instrument, contend that it is a true tractor, and point, in confirmation of this view, to the great curve of the blade. But I think reflection will show that it is essentially a lever. It does not directly draw down the head, but by pressing upon one side or point of the head-globe, it causes the globe to revolve upon its centre, its axis representing another lever. If the point opposite to that seized by the lever be moveable, of course, when leverage is applied, the head will roll up on one side as it comes down on the other; but if the opposite point be more or less fixed, as the occiput generally is, against the foramen ovale or left ramus of the pubes, then leverage on the face and chin will effect rotation on that fixed point as a centre, and the bulk of the head will have descended.

The following series of diagrams will illustrate the

FIG. 3.

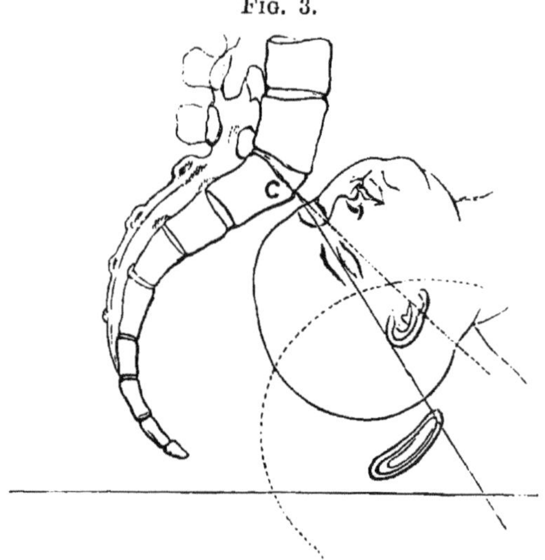

action of the lever bringing the head down by alternate flexion and extension. The lever is supposed to

Fig. 4.

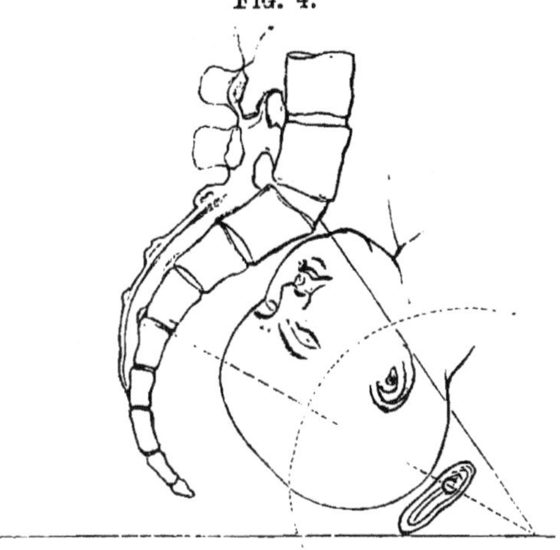

be applied alternately over the occiput and the face. In Fig. 3, c represents the centre of rotation. The

Fig. 5.

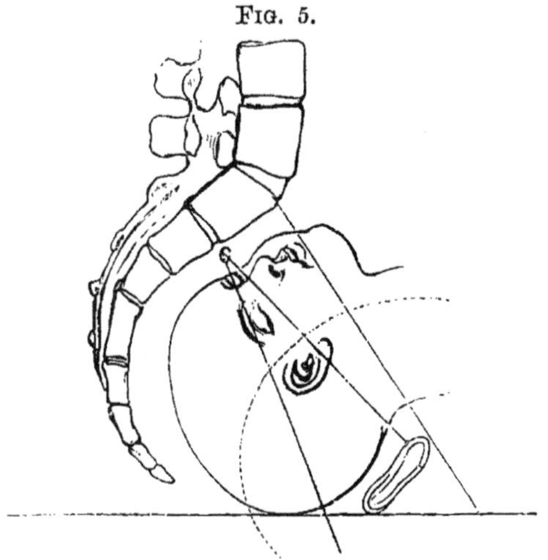

lever applied to the occiput will bring down the pubic

hemisphere of the head-globe, the forehead remaining nearly fixed against the sacrum at C. Flexion is preserved.

In Fig. 4 the lever is reversed. The centre, C, is at the pubes; the facial or sacral hemisphere descends with extension.

In Fig. 5 the lever is shifted back to the occiput, which is made to descend by flexion, the face resting in the sacrum, but at a lower point than in Fig. 3.

FIG. 6.

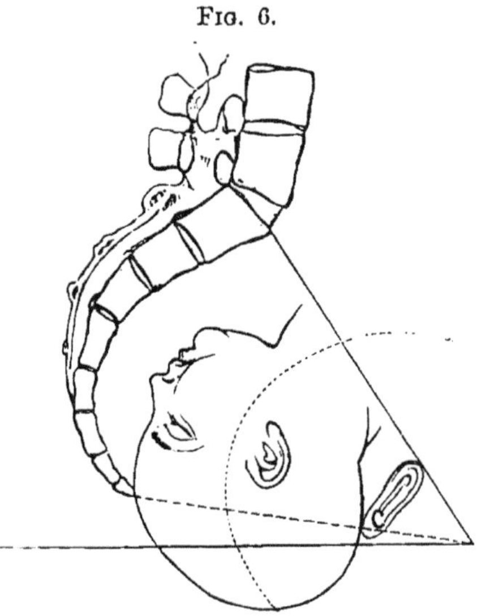

In Fig. 6 the lever shifts to the face. The centre, C, is again at the pubic arch. The facial hemisphere is now made to sweep down over the perinæum, performing the extension-movement of delivery.

Figs. 7 and 8 further illustrate the same points. In Fig. 7 the lever is seen applied to the occiput, bringing down the pubic hemisphere, whilst the opposite point is fixed in flexion at C in the sacrum.

In Fig. 8 the lever is applied over the face, which is brought down in extension, the occiput resting against the pubes.

Fig. 7.

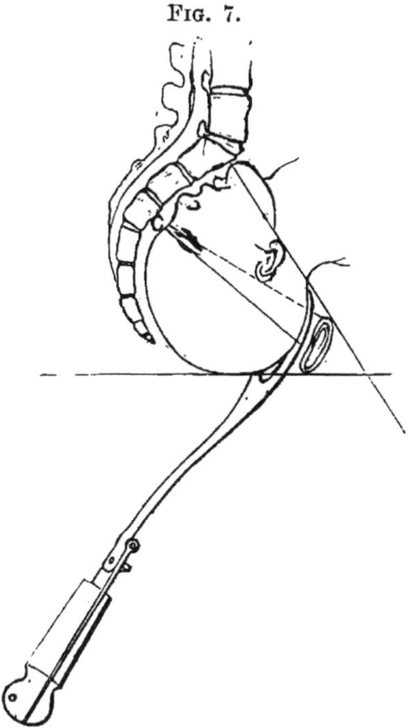

An instrument which claims to be considered with the lever is the whalebone fillet or loop. Its action is entirely similar. If the loop of the fillet be supposed to be substituted in Figs. 7 and 8 for the lever, the demonstration will equally apply. The fillet has lost its place in scientific works; but it is, I believe, largely used by some practitioners, and with great success. The instrument deserves to be remembered for this reason: it can be extemporised. Under exceptional circumstances, when no instruments are to be had, a bit of whalebone can be bent and applied over the occiput or face, in a case of arrest of the head in the pelvis, thus enabling the practitioner to rescue his patient and himself from a trying situation.

Fig. 8.

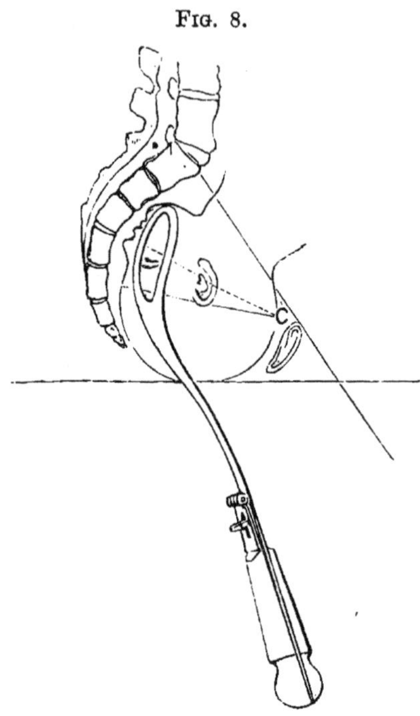

Fig. 9 represents an excellent form of this instrument. It is that of Dr. Westmacott.

I have not used the fillet, having been accustomed to rely upon the forceps, which is undoubtedly a superior instrument.

A similar principle of leverage may be applied by the two blades of the short forceps. But in this case the leverage is applied to the transverse diameter of the head. The lever can in like manner be applied to the side of the head if necessary. When the

FIG. 9.

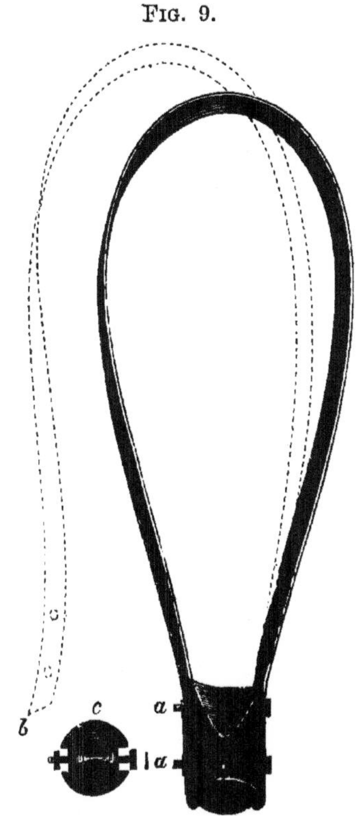

a a, screw-pins and nuts to fix *b*. By removing the nuts, the end *b* of loop is released; *c*, top of handle.

blades are crossed and locked, the common fulcrum is at the lock. Then by gently bearing upon either handle alternately, swaying the instrument backwards and forwards, avoiding all pressure against the pelvic walls, you cause the head-globe to rotate to a small extent alternately in opposite directions upon its own centre. At each partial rotation a little descent is gained, owing to the point opposite to the lever in action being partially fixed by the other blade; and by gentle traction upon the handles. In very many cases this gentle double leverage is enough to effect delivery. Traction is hardly called for at all. The alternate action of the forceps is illustrated in Figs. 10 and 11. In Fig. 10, the head grasped transversely, the handles are first carried to the left. The right or pubic hemisphere descends. The forceps is at right angles with the transverse line which cuts the pelvis obliquely.

Fig. 10.

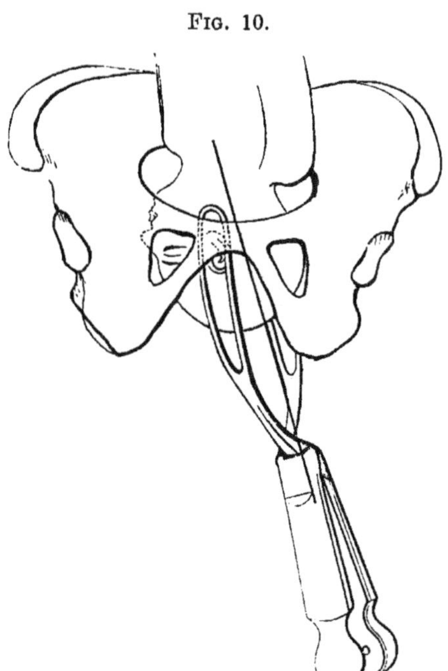

In Fig. 11 the handles are carried across to the right. The left or sacral hemisphere descends.

It is easy to demonstrate this simple leverage action

on the phantom. Thus, if I take each blade of the forceps alternately, unlocked, and use it as a lever, the head advances by a series of alternate side-movements, until it is actually extracted by this power alone. Is it reasonable to throw away a power by means of which we can safely economise the more hazardous traction-force? It is, however, disapproved of by

FIG. 11.

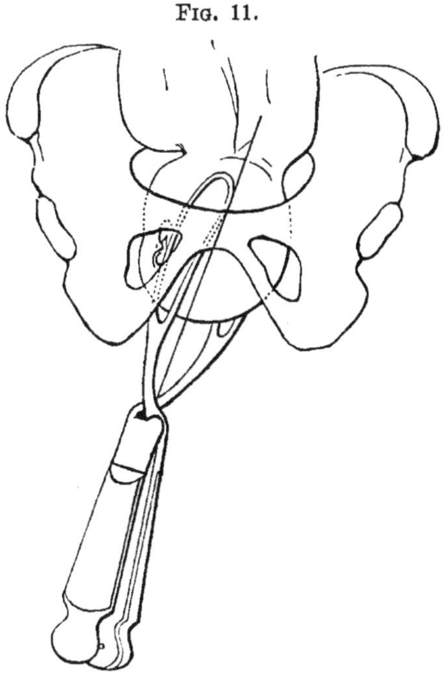

some authorities, who enjoin traction alone. But I believe that pure traction is almost impossible, and I am equally certain that a gentle and careful leverage will enable you to deliver with a great economy of force and time, which means, of course, greater safety to the mother.

LECTURE III.

THE APPLICATION OF THE SHORT FORCEPS—HEAD IN FIRST POSITION—HEAD IN SECOND POSITION—OBJECTIONS TO SHORT FORCEPS—THE APPLICATION OF THE LONG OR DOUBLE-CURVED FORCEPS—INTRODUCTION OF THE BLADES—USE OF LONG FORCEPS—LOCKING—CAUSES OF FAILURE IN LOCKING—EXTRACTION—HOW TO MEASURE THE ADVANCE OF THE HEAD—THE MANŒUVRE OF "SHELLING-OUT" THE HEAD DELAYED AT THE OUTLET—RE-LOCKING—THE HEAD IS SEIZED OBLIQUELY BY THE FORCEPS — TIME REQUIRED FOR EXTRACTION.

WE now come to the mode of applying the short forceps. The head we assume to be in the pelvis, lying in the right oblique diameter, occiput forwards. The child's right ear will be a little to the right of, and above, the symphysis pubis. We have first to consider certain conditions, some of which are necessary to the proper use of the forceps; some which are not necessary, but favourable. 1. The membranes must be ruptured. 2. The cervix uteri must be fairly dilated. 3. The bladder should be empty. 4. The patient must be in a convenient position. Abroad, the patient is usually placed in lithotomy position, on the edge of the bed. With us the pelvis is simply drawn to the edge of the bed, the patient lying on her left side. I think it needless to enter into controversy upon the relative advantages of the two positions. We shall probably adhere to custom. The English method involves much less disturbance of the patient; it involves no exposure; it requires no second assistant; and is in many respects most convenient in home practice. But in cases of convulsions, where the patient is unconscious or unmanageable, it is at times necessary to apply the forceps in the dorsal position. If we use the long French forceps, there is, indeed, little choice. The patient must be in lithotomy position, or if on her side, the pelvis must overhang the edge of the bed to an inconvenient extent. The conditions rendering the dorsal position preferable will be pointed out as the occasions arise.

The operation may be divided into four stages or acts. 1. Introduction of the blades; 2. Locking; 3. Traction and leverage; 4. Removal of the instrument.

1. *Which blade do you pass first?*—In the case of the single-curved forceps, both blades being alike, you cannot take up the wrong one. Seizing, then, either blade, you have to pass it between the head and the sacrum, and feeling the pubic ear, you know the sacral ear is exactly opposite. This blade becomes the posterior or sacral blade. Holding the blade lightly in the right hand, the handle raised and directed forwards, so that the blade shall cross the mother's right thigh obliquely, the point will be guided over the perinæum by two fingers of the left hand, which are

passed up carefully between the child's head and the cervix uteri. The all-essential point is to make out clearly the edge of the os uteri, to pass your fingers inside this edge, and to touch the head itself; then slipping the point of the blade along the inside of your

FIG. 12.

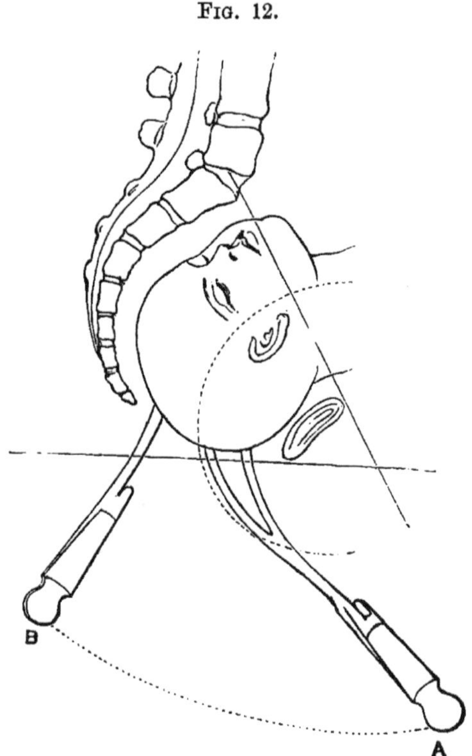

SHOWING APPLICATION OF FIRST OR SACRAL BLADE OF THE SHORT FORCEPS.

A, first stage, blade being guided on to the head. The handle A is then carried slightly downwards and backwards, to get the point of the blade round the head and up into side of the pelvis in the line A B. At B the blade is *in situ*.

fingers, the os uteri resting on the outside of your fingers, the blade will strike the head. This done, you have to adapt the blade to the convexity of the head. The point, therefore, must follow this convexity. This is done by lowering the handle and drawing it backwards, the point being still guided by the fingers of the left hand. When the convexity is well grasped, the handle is further pushed well back against the perinæum, to give room for the manipulation of the second or pubic blade.

FIG. 13.

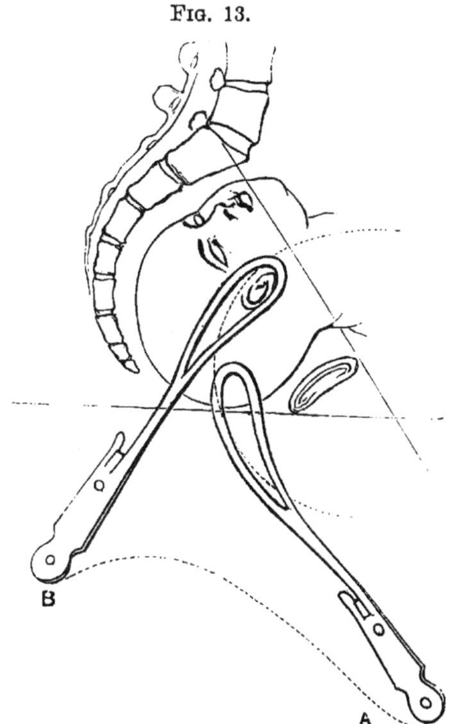

INTRODUCTION OF THE SHORT FORCEPS—THE SECOND OR PUBIC BLADE.

A, the first stage. As the blade passes up the pelvis and round the head, the handle travels in the direction of the line A B. At B the blade is *in situ*. The two blades of Figs. 12 and 13, therefore, correspond at B, and will lock.

Introduction of the Second Blade.—The fingers of the left hand are shifted forward, so as to raise the os uteri from the pubic side of the head. . The handle is held very low, and slightly forwards, crossing the mother's left thigh obliquely. Running the point along the palmar aspect of the fingers behind the right pubic ramus, when the point strikes the head, the handle is raised and carried backwards, so as to take the blade over the convexity of the head. Here you must proceed with the utmost gentleness. It is not by force that you will succeed in passing the blade. Force is quite out of place. You may take this as an axiom: If you are met by resistance that only force can overcome, you are going wrong; and, *vice versâ*, if the blades are slipping in easily, the probability is that you are going right. The rule, then, is this—hold the blade lightly; let it feel its way, as it were; let it insinuate itself into position. It will be sure to slide into the space where there is most room—that is, one blade will go nearly opposite the sacro-iliac synchondrosis, the other will go opposite the foramen ovale.*

The blades introduced, the left hand is withdrawn from the vagina, and the *second* act, or *locking*, is to be done. You seize lightly a handle with each hand, draw them into opposition, and if they have been correctly introduced they will readily lock. A smooth lock is generally an indication that the head is properly grasped. During locking be careful to pass your finger round the lock, in order to remove any hair or skin that might otherwise get pinched. This

* The adherents of the short forceps generally recommend to pass *the upper or anterior blade first.* It would not be easy to prove any advantage in this method. I believe the most skilful practitioners in London and Edinburgh now follow the method recommended in the text—namely, of passing the lower or sacral blade first.

is especially necessary in using the single-curved forceps.

Then come, thirdly, *traction* and *leverage*. These must be exerted in the direction of the pelvic axis. Thus at first the traction will be backwards in a line drawn from the umbilicus to the coccyx. Gradually, as the head descends, the handles will come more forwards, and the face turning a little backwards into the hollow of the sacrum, the handles will also rotate, so that the instrument will approach the transverse diameter of the pelvis. As the head emerges, the vertex appearing under the pubic arch in the genital fissure, the handles, following the extension movement of the head, will describe a circle around the symphysis as a centre, and will therefore at the moment of exit be applied nearly to the mother's abdomen. At this moment, and even earlier if active uterine action have set in, the fourth act—*removal* of the instrument—must be effected. This often requires some smartness. You abandon the grasp of the handles, seize the handle of the pubic blade, draw it downwards and backwards off the head; then, taking the handle of the sacral blade, you draw it upwards and a little backwards.

Head in the Second Position, or occiput to right foramen ovale. In this case you still feel for the pubic ear, which will guide you to the other ear opposite the right sacro-iliac synchondrosis. As the rule is to apply the short forceps over the ears, the introduction of the blades must be governed by the position of the head. You must first then determine the position of the head. So say most, if not all, our systematic authors. So many positions of the head, so many varying modes of applying the forceps! Now listen to the voice of Experience—Experience that so often sets at nought the refinements of theory, and clears out for herself a straight and simple path through the intricacies woven in the closet. Dr. Ramsbotham says:* "In employing the short forceps I lay it down as a rule that the blades should be passed over the ears: the head is more under command when embraced laterally, and there is less danger of injuring the soft parts during extraction. *But I confess that I have for many years been accustomed, however low the head may be, to introduce the blades within each ilium, because they usually pass up more easily in that direction.*" I think I am not wrong in believing that many others do the same thing, some not knowing it, and even imagining that they are following the ancient rule. It is a habit of mine to examine the head in every case of delivery. I have thus many times seen the stamp of the fenestræ on the brow and side of the occiput. This is as clear to read as the impression of a seal on wax. It says, unmistakeably, that the blades found their way into the sides of the pelvis with at most a slight deviation towards an oblique diameter.

All this suggests the question, whether it be really so necessary to "feel the ear" before applying the forceps as has been imagined. If the blades *will* find their way to the sides of the pelvis, clearly it is not necessary to know where the ears are. To feel an ear must in most cases put the patient to much suffering. You can, however, scarcely feel an ear unless the os uteri be well expanded. This being

* "Medical Times and Gazette," 1862.

so, we have an argument in favour of the old rule—not of much worth, it is true, for we can have the assurance that the cervix is properly expanded by other means. With the long forceps, the ancient rule is clearly superfluous.

There is one case in which the short forceps is of especial value. It is when the head descends into the pelvis, its long diameter keeping nearly in the transverse diameter of the pelvis, until it is arrested on the shelf formed by the sacro-sciatic ligaments. At this point, from want of propelling power, the head does not take its screw-movement of rotation on its axis so as to bring the occiput forwards. If the short forceps be now applied in the transverse diameter of the head, by a slight rotatory movement, the axial turn is given, the occiput comes forward, the face goes to the sacrum, and the head is released. In two cases of this kind I thus easily succeeded in delivering after failing with my long forceps. These are the only two cases in which I have ever found the short forceps preferable to the long. And the simple lever would have answered as well. One merit the short forceps has, not without importance to the novice: it is easier to use than the long forceps.

There are *objections* to the single-curved forceps, short or long:—

1. One objection is in the introduction; others in the injuries likely to be inflicted on mother or child. To introduce the second or upper blade, the handle must be much depressed, nearly at right angles with the mother's left thigh, which is flexed upon her abdomen. Now, to do this, the patient's nates must

be dragged over the edge of the bed. To procure and to maintain this position is often a matter of great difficulty and inconvenience.

You may facilitate the introduction of the second blade by introducing a joint into the shank, so as to allow the handle to be doubled up out of the way. Dr. Giles showed at the Obstetrical Exhibition an instrument so modified.

Fig. 14.

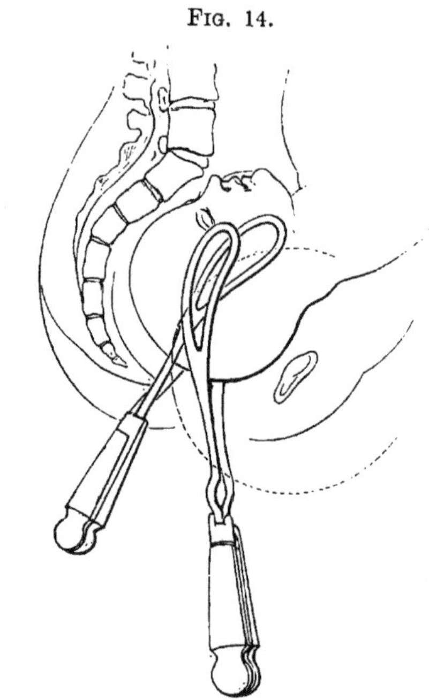

SHOWING THE SINGLE AND DOUBLE-CURVED FORCEPS *in situ*.

The single-curved forceps presses back upon the perinæum, putting this structure on the stretch. The shanks of the double-curved forceps keep clear of the perinæum, the whole instrument approximating to Carus' curve.

2. In extraction, the handles, nearly to the last moment, must be directed more backwards than is necessary with the double-curved forceps, and owing to the bows springing directly from the lock, the

perinæum is wedged open, and not seldom unavoidably torn. In some cases, this injury may be avoided by taking off the blades before the greatest diameter of the head passes. But then the work is not always done, and you may have to put them on again. I may perhaps be told that to suffer the short forceps to tear the perinæum implies want of skill. I reply that men of the highest skill and the largest experience with this instrument have confessed to me that this objection is a real one.

The best single-curved forceps is that of Dr. Beatty, of Dublin. I used it for some time, but have given it up because of these two faults, and of its inadequacy to cope with a large range of cases which come within the power of the long double-curved forceps.

3. The posterior or sacral blade is apt to bruise by one of its edges the sciatic nerve.

The effect is the crushing of some fibres and more or less protracted paralysis of the leg.

4. If the blade be applied as usually taught—*i.e.*, nearly in the transverse diameter of the head—an edge is very likely to press upon the portio dura as it emerges from the temporal bone. The result is paralysis of the facial muscles to which the branches are distributed. The child cannot shut the eye; it cannot suck. I have known a child die of starvation from this cause.

5. If applied as usually inculcated, the anterior blade will often go directly behind the pubis. The edge will bruise the urethra. Hence vesico-vaginal fistula.

From all these objections the long forceps I recommend is nearly altogether free.

THE LONG OR DOUBLE-CURVED FORCEPS.

The application of this instrument is governed by a different law from that which governs the use of the short forceps. The short forceps, according to the recognised rule, must be applied with the blades quite or *nearly* over the transverse diameter of the head. The head determines the manner of applying it. But with the long forceps it is the pelvis that rules the application. The position of the head may be practically disregarded. The pelvic curve of the blades indicates that these must be adapted to the curve of the sacrum in order to reach the brim. They must therefore be passed as nearly as may be in the transverse diameter of the pelvis. One blade will be in each ilium, and the head, whatever its position in relation to the pelvic diameter, will be grasped between them. The universal force of this rule much simplifies and facilitates the use of the instrument. Not only does it apply to the position of the head in relation to the pelvic diameters, but also to all stages of progress of the head from that where it lies above the brim down to its arrest at the outlet.

It has been contended that the short forceps should be preferred in cases where the head is arrested in the cavity, and as a corollary it is urged that in cases of arrest at the brim, where the head has been brought into the cavity by the long forceps, this instrument, after serving so far, should be discarded and replaced by the short forceps. I do not concur in this view. I doubt whether any one who has had any considerable practice with the long forceps has found it worth while to change instruments in the course of delivery. The long forceps possesses a more scientific adaptation to the pelvis throughout the whole canal than the short forceps. And if the long forceps is found in practice capable of taking the head through the pelvis from brim to outlet, it follows that, since the whole contains the parts, the long forceps is qualified to take up the head at any point below the brim.

The pelvis has been compared to a screw. I think a better idea may be formed of its mechanical properties by comparing it to a rifled gun, and the child's head to a conical bullet. But even then the comparison is not complete, for the pelvis, unlike a gun, is a curved tube. Now, just as the head must traverse the pelvis in a helicine course, determined by the relation of form between pelvis and head, so is it natural that an instrument designed to grasp the head should be so modelled as to be fitted to follow this helicine course during introduction and extraction. This indication a well-modelled double-curved forceps fulfils; no single-curved forceps can fulfil it.

First, as to the application when the head is delayed at the brim.

Mode of Applying.—Position of the Patient.—The patient should lie on her left side, the knees drawn up towards the abdomen; the head should be only slightly raised. She should lie across the bed, with the nates near the right edge, about midway between the head and foot. This will facilitate the introduction of the blades, and give room for the sweep of the handles round the pubes at the end of the operation. I do not find it necessary to bring the nates to hang over the edge of the bed. I have often passed both blades when the patient has been lying in the middle of the bed. Sometimes it is very desirable to move her as little as possible.

FIRST ACT.—INTRODUCTION OF THE BLADES.

Selection of the Blades.—First dip them in warm water, wipe dry, and lubricate with oil, or lard, or cold cream. Join them, and, holding the instrument with the concavity of the pelvic curve forwards, and the blades in the position which they are to occupy in the pelvis, you take that one first which is to lie in the left or lower side.

Fig. 15.

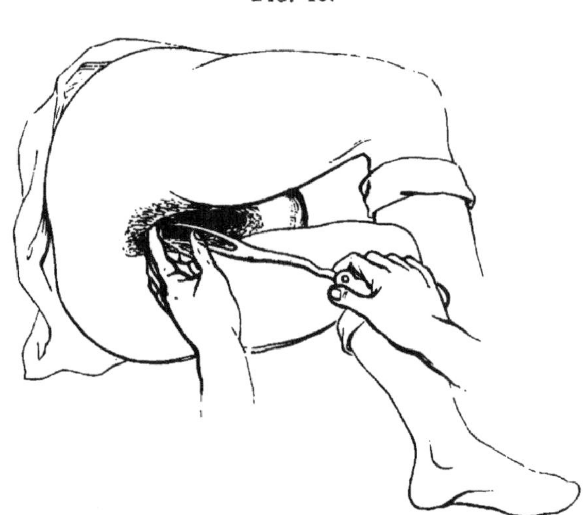

SHOWING THE FIRST STAGE OF INTRODUCTION OF THE FIRST BLADE.

First Stage.—One or two fingers of the left hand are passed in at the perinæum and between the cervix uteri and the head. Then, bearing in mind the relative forms of the instrument, the head, and the pelvic canal, the point of the blade is passed along the palmar aspect of the fingers at first nearly directly backwards towards the hollow of the sacrum.

Second Stage.—The handle is now raised so as to throw the point downwards upon the left side of the head. As the point of the blade must describe a double or compound curve—a segment of a helix—in order to travel round the head-globe, and at the same time to ascend forwards in the direction of Carus' curve, so as to reach the brim of the pelvis, the handle rises, goes backwards, and partly rotates on its axis.

FIG. 16.

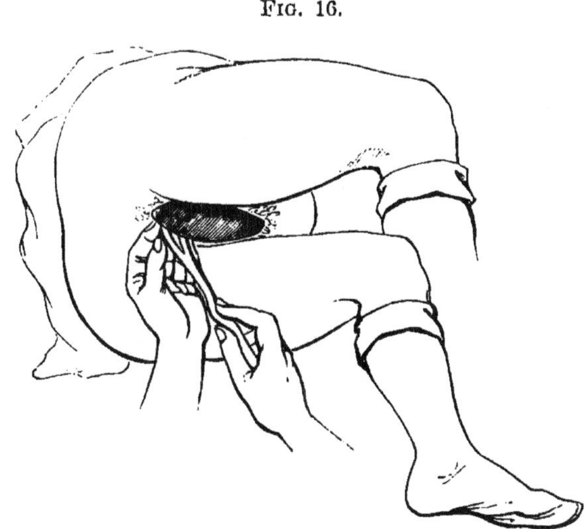

SHOWING SECOND STAGE OF INTRODUCTION OF FIRST BLADE.
The point is running up round the head and into left side of the pelvis.

Third Stage.—The handle is now carried backwards and downwards, to complete the course of the point around the head-globe and into the left ilium. Slight pressure upon the handle ought to suffice. This will impart *movement* to the blade; the *right direction* will be given by the relation of the sacrum and head. The blade is now *in situ;* the shank is to be pressed against the coccyx by the back of the operator's left hand whilst he is introducing the second blade. Its weight aids in maintaining it *in situ.*

Fig. 17.

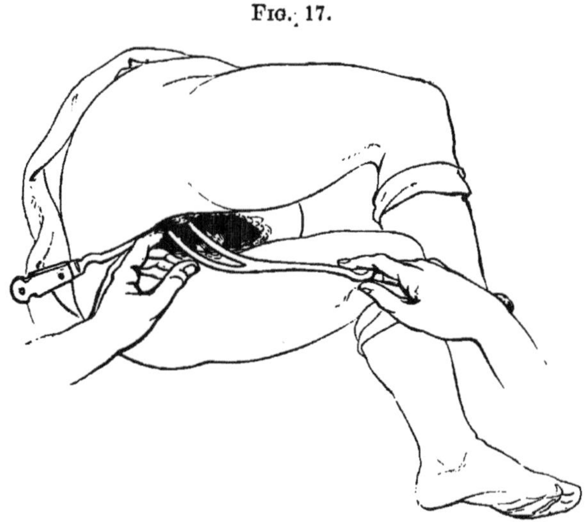

SHOWING LAST STAGE OF INTRODUCTION OF THE FIRST BLADE AND THE CROSSING THE SHANK OF THE FIRST BLADE BY THE SECOND BLADE IN THE FIRST STAGE OF ITS INTRODUCTION.

INTRODUCTION OF THE SECOND BLADE.

First Stage.—Two fingers of the left hand, the back of which is supporting the first blade against the perinæum, are passed into the pelvis between the os uteri and the side of the head which lies nearest to the right ilium. The instrument held in the right hand lies nearly parallel with the mother's left thigh, or crossing it with only a slight angle. The point of the blade is slipped along the palmar aspect of the fingers in the vagina, across the shank of the first blade *in situ*, inside the perinæum towards the hollow of the sacrum.

Second Stage.—As the point has to describe a helicine curve to get round the head-globe and forwards in the direction of Carus' curve, the handle is now depressed and carried backwards until the blade lies in the right ilium. When it has reached this position the handle will be found near the coccyx, nearly in opposition to the first blade.

The application of the long forceps is further illustrated in the following diagrams (Figs. 18, 19).

The Locking.—This is effected by a slight movement of adaptation. A handle is seized in each hand. The handle of the first blade is brought a little forwards over the handle of the second blade. If one blade is a little deeper in the pelvis than the other, it is either brought out, or the other is carried in until the lock is adjusted. This is commonly facilitated by pressing both handles backwards against the coccyx. This movement, by throwing the blades well into the ilia, where there is room, allows the handles to be rotated a little, so as to fall into accurate relation.

Accurate locking is generally evidence that the blades are properly adjusted to the head, and that the pelvis admits of the successful use of the instrument. On the other hand, their not locking is proof of their not being properly introduced, or *of the pelvis not admitting of their application.* In the first case, that of improper introduction, the failure is generally due to neglect in passing the blades exactly in the same diameter of the pelvis—that is, in passing the second blade exactly opposite to the first, so

Fig. 18.

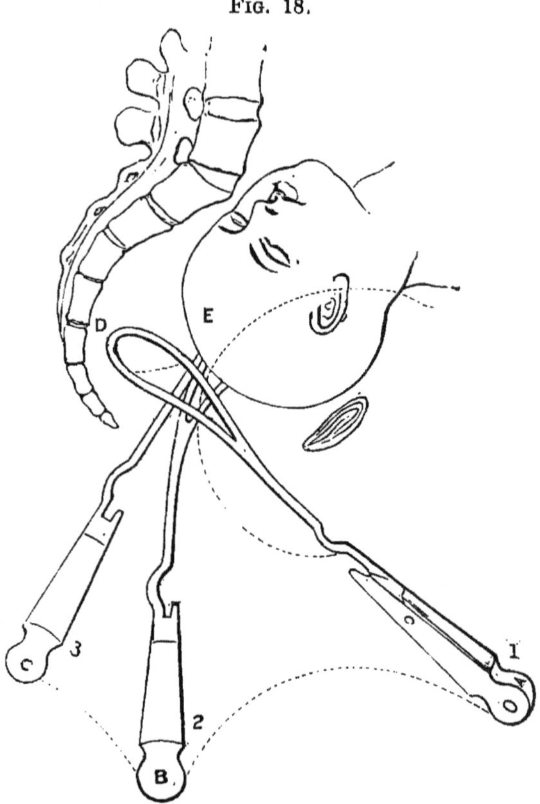

INTRODUCTION OF THE FIRST OR LEFT BLADE OF THE LONG FORCEPS.

1. First stage, or introduction of point of blade in the hollow of the sacrum: A, the handle, is then raised, and at the same time carried across, rotating partly on its axis to B, so that the point D, turning round in the hollow of the sacrum to E, strikes the head, and rises towards the left side of the pelvis. 2. The second stage, or advance of the blade round the head and up in the left ilium. 3. Third stage: The handle B has travelled in the direction B C, still rotating slightly, until at C it is at rest *in situ*, the shank near the coccyx, where it is held by the back of the operator's left hand, whilst the point of the second blade is passed over and across it inside the perinæum, as seen in the next figure.

that if the first blade is applied in the left ilium, opposite one end of the transverse diameter, the right does not lie at the opposite end of that diameter. To remedy this error, the blade must be partly or wholly withdrawn and re-adjusted.

In the second case, that of pelvic unfitness, the locking is prevented by the projecting promontory or

Fig. 19.

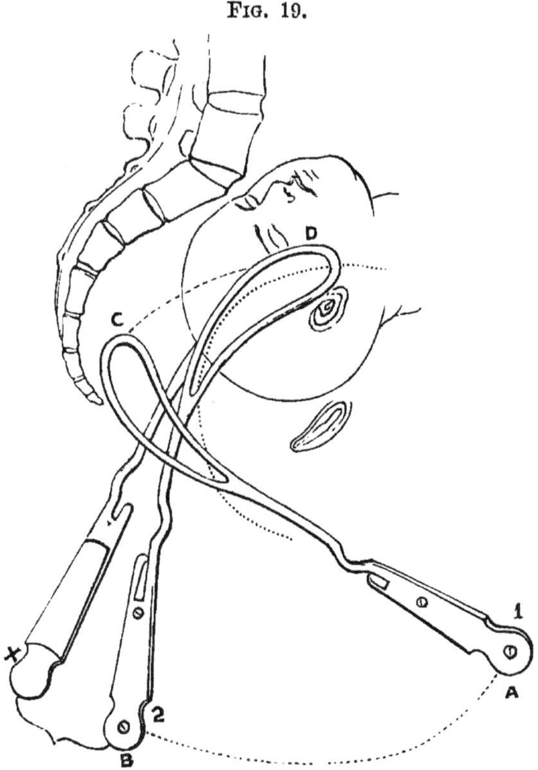

INTRODUCTION OF THE SECOND OR RIGHT BLADE OF THE LONG FORCEPS.

1. First stage of second blade; 2. Second stage of second blade; x, the first blade *in situ*. A, the handle, at the moment of passing C, the point, inside the perinæum into the hollow of the sacrum, across x, the first blade: the handle then drops and goes backwards to B, the point C travelling round the head, and advancing into the right ilium in the direction of the axis of the brim to D; when it has reached this position, it will be found nearly opposed to the first blade, x; the locking is effected by bringing the handle x over the handle B.

other deformity, so distorting the pelvic diameters that the two blades cannot find room to lie in the same diameter opposite to each other. It will commonly be found that the blades will pass one on each side of the promontory, the inside of the blade not looking towards its fellow, but towards the opposite foramen ovale, where you cannot get a blade to lie. When you find this happen you must give up the attempt to use the forceps. Pass the hand into the

Fig. 20.

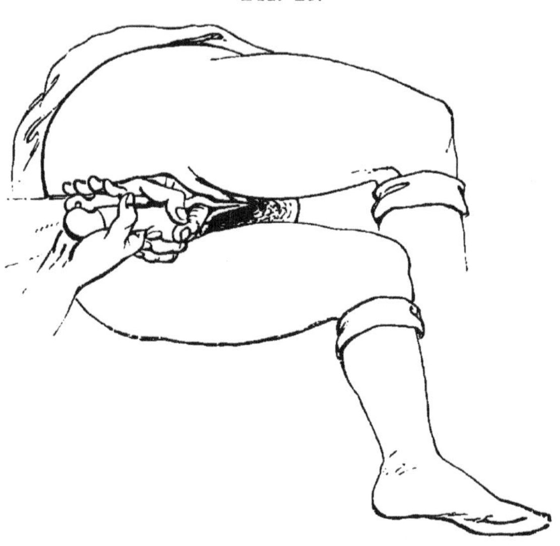

SHOWING THE LONG FORCEPS LOCKED, AND GRASPED BY THE TWO HANDS.
The head being at the brim, traction is backwards.

pelvis, if necessary; explore its dimensions and form carefully; and determine between turning and craniotomy. A correlative proposition may here be stated:—*Wherever the long forceps will lock without force, it may be reasonably concluded that the case is a fit one for the trial of this instrument; and a reasonable attempt should be made to deliver by its aid before passing on to turning or perforation.*

3. *The Extraction.*—Get the nurse to press upon the right hip and support the back. Grasp the

handles with one hand, and apply the fingers of the other hand to the ring or shoulders at the lock. Draw at first backwards in the axis of the brim, during the pains, if any be present, and at intervals of a minute or so if there be none. Concurrently with traction, alternate slight leverage movements

FIG. 21.

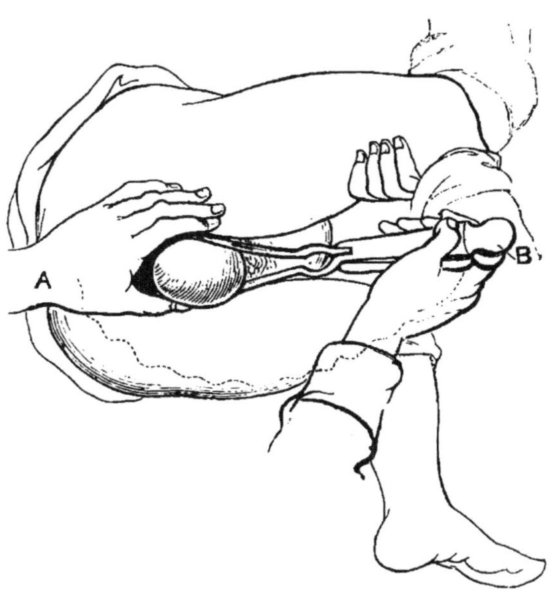

REPRESENTING THE LAST STAGE OF EXTRACTION.

The handles have travelled from A to B, so as at last to touch the abdomen. The dotted line shows the course of the handles, and the slight oscillations practised during the descent of the head.

may be executed, by swaying the handles gently from side to side, always taking care not to press the shanks against the pelvic walls. Each blade is the fulcrum to its fellow. The finger which is used in the ring from time to time gauges the advance of the head.

It is further extremely useful, if you have a competent assistant, to get him to support and press upon the fundus uteri during extraction. This helps to keep the axis of the uterus and of the child in proper relation to the axis of the pelvis; and in proportion to the aid thus given *à tergo*, you lessen, *pro tanto*, the amount of extractive force required.

The advance of the head is measured by the following standards:—First, you feel if the occiput approaches the pubic arch by passing a finger below and behind the pubic bones. Secondly, you sweep your finger round the circumference of the brim, and thus feel if the equator of the head-globe is pressing lower down through the brim. Thirdly, by feeling the direction of the saggital suture: if you find that it is approaching parallelism with the conjugate diameter, you may be certain that the head is descending. Further evidence is found in the rotation of the forceps. As the head can hardly turn upon its cervico-vertical axis without at the same time descending in the pelvis, if the handles of the forceps are observed to rotate, this rotation being imparted by the head, is evidence of advance. Again, as the head descends, of course more and more of the shanks and blades will become visible. This, indeed, is open to a fallacy. Allowance must be made for some degree of slipping, which takes place with all the English instruments whose blades have only a moderate bow. And further, when the head is fairly in the pelvic cavity, the blades lose something of that external support which, as explained in Lecture II., is the chief force in maintaining the grasp upon the head. This is still more marked when the head has partly emerged from the vulva. At this time the blades will be apt to slip away altogether, and it will be necessary to increase the compression on the handles in order to keep your hold. Fourthly, by two or more fingers you measure the space or degree of tightness between the vertex and the floor of the pelvis. At first the fingers find free space; gradually the vertex leaves no room for the fingers. Then the soft floor of the pelvis, the perinæum, is distended by the advancing vertex; it bulges out; it puts the perinæum tightly on the stretch. The anus is protruded. Fæces are often squeezed out. Indeed, the pressure upon the sphincter ani at this stage sets up reflex action. The call to strain or bear down to expel the pelvic contents, whether uterine or rectal, is uncontrollable. Turbulent expulsive action, then, and defecation, constitute certain signs that the head is advancing. To some extent, the increasing scalp-swelling, or caput succedaneum, may give a false impression that the cranium itself is descending. But a little practice and attention will correct this error. When the vertex has reached the floor of the pelvis, the handles of the forceps are found to have turned a little upon their axis, to lie more nearly in the transverse diameter of the pelvis. This is the result and the indication of the screw-rotation of the head. You have no hand in producing it. It is effected by the descending head adapting itself to the cavity of the pelvis.

The handles may now be directed more forwards during traction. The shanks thus avoid stretching the perinæum, and the traction is in the axis of the outlet. An assistant is now useful in holding up the right knee, so as to leave room for the operator to carry the handles well round the pubes in Carus' curve. Here it is often convenient to push the handles forwards rather than to pull. This action is seen in Fig. 21.

During extraction it occasionally happens that the blades will lose their hold, that the handles will twist in opposite directions, and thus unlock. This is generally owing to the operator carrying the handles forward too early. The effect of this is to throw the blades off the head-globe over the face. It is another illustration of the law that the position of the forceps is determined by the relation of the head to the pelvis, and that if you reverse the order by attempting to make the forceps alter this relation you are immediately at fault. The remedy is to carry each handle well back again towards the perinæum, when they will re-lock.

If the head is in the genital fissure, and there is sufficient uterine energy, you may proceed to the

4*th Act. The Removal of the Blades.*—If the head should not be propelled, you may often assist it by a manœuvre which it is well to understand. You apply the palms of both hands one on either side and behind to the perinæum distended by the head; and bearing upon this structure so as to press it a little backwards, whilst the head is pushed forwards towards the pubic arch, the head is, as it were, shelled out by being made to complete its movement of extension. Steady pressure by the hands of an assistant or by a binder upon the fundus uteri will much assist the extension of the head. In this manner I once extricated myself and my patient from an awkward predicament. I had been summoned into the country without knowing the nature of the case, and had no instruments. I found a lady who had been many hours in labour, the head on the perinæum, and no pains. The lever or the forceps would have delivered her in a minute. Neither was to be had. But the manœuvre I have described perfectly succeeded, and put an end to a state of extreme anxiety, and even danger.

Another manœuvre is occasionally serviceable. This is to pass a finger into the rectum, so as to get a point of pressure upon the forehead. In this way it is sometimes possible to bring the face downwards, to start the extension movement, and thus to extricate the head delayed at the outlet. And if at the same time firm downward pressure be made upon the breech through the fundus, as described in the first Lecture, the force propagated through the spine will aid materially in giving the extension movement. This combination of the principles of "pushing," of leverage, and of "shelling-out," may in certain cases enable you to deliver without resorting to the forceps or lever.

When the blades are adjusted, they will not lie exactly in the transverse diameter of the pelvis. The head, lying between the transverse and right oblique diameter, will tend to throw off the blades towards the opposite or left oblique diameter. The head then will be seized obliquely, one blade grasping the right brow, the other the left occiput. This is clearly demonstrated by the impressions of the fenestræ made on the scalp. The blades naturally find their way into this position if they are introduced gently. One tendency of this oblique seizure is to assist the head in its axial rotation, face sacrumwards, as it descends into the pelvis. It is also an answer to an objection urged against the use of the long forceps at the brim —namely, that by seizing the head in its long or fronto-occipital diameter, compression in this direction makes the opposite or bi-parietal diameter bulge out, thus increasing the difficulty of passing the small or conjugate diameter of the pelvis. In most cases the objection is theoretical only—it is mainly based upon experiments made on the dead fœtus on the table.[1]

Elongation or moulding, we have seen, is the result of gradual compression of the equatorial zone. Now the pelvis and the forceps together constitute the compressing ring. Pressure, then, upon the transverse diameter of the head by the opposing points of the sacrum and pubes, simultaneously with pressure upon the longitudinal diameter between the blades of the forceps, tends to *diminish both diameters* by lengthening out the head. Of course it must be understood that the pelvic contraction is of moderate degree only —in short, that the case is a proper one for the forceps. If the conjugate diameter be less than 3·25 inches, the prospect of effecting the desired elongation within a reasonable time is greatly diminished.

I have said that the head is very rarely seized exactly in its longitudinal diameter. An exception occurs in the case of the very flat pelvis, in which there is conjugate contraction with very little projection of the promontory. In this case the head will lie very nearly in the transverse diameter. If, in presumed contraction of the brim, the marks of the blades are on the brow and side of the occiput, the projection of the promontory is not great.

The Time required for Extraction.—If the head be delayed in the cavity of the pelvis for want of expulsive action, or because it rests upon the ischia, maintaining a too near approach to the transverse diameter, and there is no marked hindrance on the part of either the anterior or posterior valve, it is generally sufficient to use slight traction and oscillation for a few minutes. As soon as the head is started by the forceps, the uterus takes up its work, helps the operator, and the labour is quickly over.

If the uterine and perinæal valves obstruct the passage of the head, a little more time and caution are required. (*See* Lecture IV., Figs. 22, 23, pp. 69, 70.)

If the head has to be seized at the brim on account of delay from want of uterine action, time may often be saved by placing the patient on her back, and supporting the uterus against the spine by the hands of an assistant or a binder. This proceeding, by adjusting the axis of the uterus to that of the brim, and getting the aid of gravitation, will greatly facilitate the entry of the head and encourage the action of the uterus. If there is no obstacle from narrowing of the pelvis or want of dilatation of the soft parts, gentle traction and oscillation during ten minutes will generally complete the labour.

In the event, however, of arrest from pelvic contraction or from want of dilatability of the soft parts, time is a necessary element. The process of moulding, of elongation of the head, can only be effected gradually. Here oscillation or leverage must be used with great care. What is wanted is steady compression and traction extended, with moderate intervals of rest, over thirty minutes, or even an hour. Should the head be found to make no advance in entering the brim in that time, the question whether the forceps must not be laid aside for turning or perforation will have to be considered.

LECTURE IV.

CAUSES OF ARREST IN FIRST LABOURS—DISTURBED OR DIVERTED NERVE-FORCE—THE UTERINE AND PERINÆAL VALVES—THE PONDING-UP OF LIQUOR AMNII—THE FORCEPS TO DELIVER THE AFTER-COMING HEAD—THE LONG FORCEPS (*continued*)—APPLICATION IN FRONTO-ANTERIOR POSITIONS OF THE HEAD—THE MECHANISM BY WHICH FRONTO-ANTERIOR, FOREHEAD, AND FACE POSITIONS GENERALLY, ARE PRODUCED—THE MANAGEMENT OF THESE CASES.

A VERY large proportion of cases that call for the forceps are *first labours*. It is therefore well to take a survey of the conditions which lead to this necessity. Disproportion as a cause of arrest we will put aside for the present. In the great majority of first labours the difficulty does not arise from disproportion. The frequency of an easy second labour proves this. The difficulty, then, lies in the soft parts of the parturient canal. And this may be either from want of contractile energy of the uterus or from excessive resistance of the os uteri, vagina, or vulva. I will endeavour to explain the nature of these cases. First, the suspension of uterine and other muscular force. This may be the result of exhaustion from fatigue, or of the discharge of the *vis nervosa* in other directions—metastatic labour, as Dr. Power calls it. Emotion, fear, the shrinking before pain, will frequently cause such a derivation of nerve-force that all labour is suspended. It is in such cases that chloroform finds one of its happiest offices. By removing the sense of pain and of fear, the emotional disturbance is eliminated, the nerve-force responds to the natural call, and labour is frequently resumed and carried out to a successful termination. It is not a figure of speech to say that here chloroform acts like a charm. It may even save the necessity of resorting to instruments.

But not seldom, combined with more or less emotional disturbance, the expelling force gives way before a real mechanical obstacle. It is this:—In primiparæ the cervix dilates slowly. The vertex partly enters the pelvis, capped by the cervix. The anterior portion of the cervix, especially, is carried down before the head, much below the brim. It even gets jammed between the head and the symphysis, and becomes, perhaps, more unyielding from œdema. Now this anterior segment of the uterus forms a valve or plane which guides the head backwards into the sacral hollow in the direction of the axis of the brim. So far it fulfils a useful function, but, having done this, it ought to retire. In pluriparæ it commonly does so, and then the head encounters the second valve formed by the perinæum, which is exactly opposed to the first or uterine valve. The function of this is to guide the head forwards under the pubic arch in the direction of the outlet. Now, it frequently happens in primiparæ that these valves maintain their resistance too long.

The uterine valve may still cap the head when it is propelled to the very floor of the pelvis. In this case the head is prevented from receiving the full impact from the inclined planes of the ischia; it is impeded in its half-quarter axial turn, occiput forwards, and also in its movement of extension. Hence a double difficulty: there is the opposing valve, there is mal-position. Clearly the valve must be got out of the

Fig. 22.

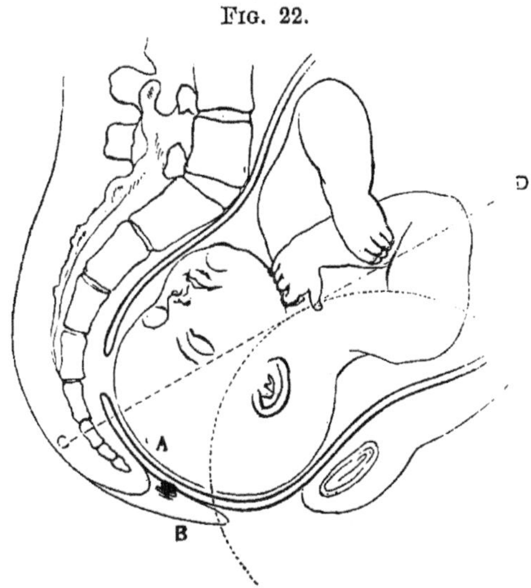

SHOWING THE HEAD ARRESTED IN THE PELVIS BY THE ANTERIOR OR UTERINE VALVE A, WHICH IS CARRIED DOWN INTO CONTACT WITH THE POSTERIOR OR PERINÆAL VALVE B.
The uterine valve A helps to guide the head into the pelvis, in the axis of the inlet C D.

way. How to do it? Sometimes patience will do it; but as patience on the part of the physician may involve agony and danger to the woman, this should not be overstrained. Sometimes one or two fingers may be insinuated between the valve and the head in the intervals of pains, and then the valve may be held back so that the equator of the head may pass it. But you must be careful lest by over-meddling you cause more swelling and rigidity. You may pass up the lever or one blade of the forceps, and bearing upon the occiput, just as you use a shoehorn, the valve, like the heel of the shoe, is held back whilst the head descends upon the inclined plane of the instrument. And here you often get another beneficial result. The head-globe has been lying closely fitting to the ring of the cervix uteri like a ball-valve, ponding up the liquor

FIG. 23.

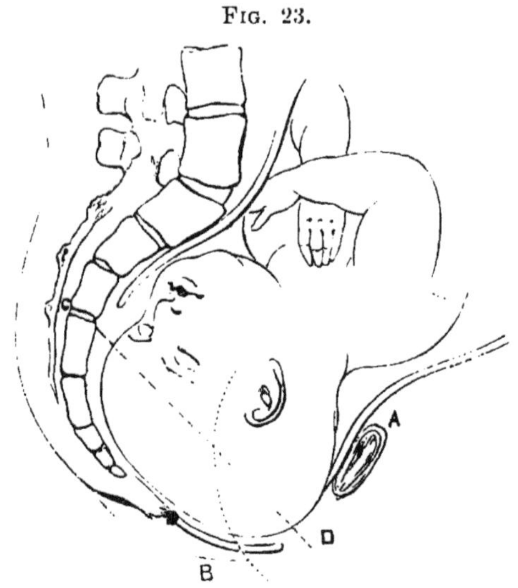

SHOWS THE HEAD ARRESTED AT THE OUTLET BY THE POSTERIOR OR PERINÆAL VALVE B; THE ANTERIOR OR UTERINE VALVE A HAS SLIPPED UP ABOVE THE EQUATOR OF THE HEAD. The posterior valve guides the head out of the pelvis in the axis of the outlet C D.

amnii behind, and impeding the full action of the uterus by over-distending it. The lever or forceps opens a channel for the escape of the pent-up fluid. The uterus then acts immediately, and the labour proceeds. I have often used the forceps successfully for no other purpose than this.

Well, we have now disposed of the uterine valve.

The perinæal valve and the vulva oppose another barrier, all the more troublesome because it has to be encountered by diminished forces. Arrest on the floor of the pelvis, nothing but this valve obstructing, is very common. The lever applied alternately over the occiput and face, or over the sides of the head, answers perfectly in this case. But many will prefer the forceps.

The delay at the vulva is often further increased by intense emotional and sensational nervous disturbance. The uterus seems instinctively to hesitate to contract, lest, by forcing the head upon the acutely sensitive structures of the vulva, it cause intolerable pain. The consequence of this protracted shrinking before pain is twofold:—There is, first, exhaustion of nerve-force; there is, secondly, a condition which I can best describe as a kind of shock, producing prostration, if not collapse, which supervenes whenever an urgent function is suspended or remains unfulfilled.

Another cause of delay may of course reside in the mechanical condition of the resisting structures; rigidity of the cervix uteri, of the perinæum or vulva, may be added to other unfavourable conditions. Rigidity may be due to thickening, œdema, hypertrophy of the tissues, or there may be rigidity without discoverable alteration of texture; and there are the more serious cases of partial or complete occlusion from cicatricial tissue, the result of previous injury or disease. Obstruction from these causes demands special treatment, which will be discussed hereafter.

Most authors describe the application of the *forceps to the after-coming head*—that is, when the head is delayed after the birth of the trunk in breech, footling, or turning labours. The position of the child with its head delayed at the brim, probably compressing the cord, is indeed perilous. Prompt delivery alone can rescue it from asphyxia. How shall we best reconcile the two conditions of promptitude and the minimum of force? It is a point of extreme interest to know what is the greatest time a child can endure being cut off from placental and aërial respiration, and yet recover; for within that time the head must be generally extricated in order to save life. The time is certainly very brief. Here it may truly be said that "horæ momento cita mors venit, aut victoria læta." The data are necessarily wanting in precision. Hugh Carmichael * in two cases removed the fœtus from the uterus within fifteen minutes from the death of the mother. In both cases the fœtus was quite dead, although, on the mother's evidence, it was living just before her dissolution. A similar case has occurred to me. Dr. Ireland was called to a woman who had died suddenly from a blow received from her husband. The Cæsarian section was performed, and a live child was extracted. The interval here was estimated at eight or ten minutes. The following case occurred at St. Thomas's †:—A woman in her ninth month was run over in St. Thomas-street at 7.35, and carried to the Hospital. She died at 7.55. Mr. Green opened the abdomen at 8.8, and the child was withdrawn by Dr. Blundell asphyxiated. Its lungs were inflated, and it survived thirty-four hours.‡ Here, then,

* "Dublin Journal of Medicine," vol. xiv.

† "Med. Chir. Trans," 1822.

‡ It is worthy of remark that in the history of this case Mr. Green especially calls attention to the depressing effect of the warm-bath—a point since enforced by Milne Edwards and Marshall Hall.

we have an instance of partial recovery at the end of thirteen minutes from the mother's death. Cases of extraction of live children within ten minutes are not very rare. But perhaps examples of this kind are not exactly in point. They are not quite analogous to the case of compression of the cord during labour. Numerous observations lead me to conclude that the child will be asphyxiated beyond recovery if aërial respiration do not begin within three, or at most five, minutes after the stoppage of the placental respiration. I think it must be accepted as a general law that if the head compress the cord, the child must be extracted within three minutes. Even if this be done, there will commonly be considerable asphyxia and cerebral congestion, and restorative means will be required.

Now the practical question arises—What is the readiest way of delivering the after-coming head? We can extract by the hands or by the forceps. Which is to be preferred? In many cases undoubtedly the hands are the best instrument. Where the cervix is fully expanded, and the brim of the pelvis is roomy, well-directed manipulation will deliver in a few seconds. And again, if there be any marked contraction of the conjugate diameter, the forceps will probably fail, whereas the hands may extricate the head very quickly. But still some cases may occur in which the forceps will be useful. How to apply it? In the first place, draw down the cord gently, so as to take off any dragging upon the umbilicus, and lay the part which traverses the brim in that side in which the face is found; there is most room for it there. The head is engaged with its long axis more or less nearly in the transverse diameter of the brim. The blades should grasp it in an oblique diameter approaching the antero-posterior. To be able to effect this, the trunk must be carried well forwards over the symphysis in the direction of Carus' curve, and held there by an assistant, so as to leave the outlet clear for manipulation. Then passing your left hand into the vagina, you carry the fingers to the left side of the pelvis, between the cervix uteri and the head. The blade is slipped up along the palmar aspect of the fingers to its place. The like proceeding is then repeated on the right side of the pelvis, and the blades are locked. The assistant supporting the child's body, you then draw the head into the pelvis in the axis of the brim. As soon as this is cleared, you may take off the blades and finish the extraction by the hands. This is done by hooking two fingers of the right hand over the back of the neck, on the shoulders, whilst the left hand seizes the feet above the ankles, a napkin interposed. You then draw in the axis of the outlet. It is the work of a few seconds. Some will prefer completing the extraction with the forceps. If you select this mode, you find the face turned towards the sacral hollow when the head has cleared the brim; the forceps, following this, rotates a little in your hands. When the occiput is appearing under the pubic arch, carry the handles well forwards, so as to bring the face over the perinæum with the least possible strain upon this structure. The face and forehead sweep the perinæum, describing a curve around the occiput resting upon the pubes.

The use of the forceps in this case was strongly inculcated by Busch, of Berlin, who attributes to this practice the extraordinary success of turning in his hands. Of forty-four cases of turning, only three children are stated to have been lost from the effects of the operation. The late Dr. E. Rigby and Dr. Meigs insist also upon the advantage of the practice.

Fig. 24.

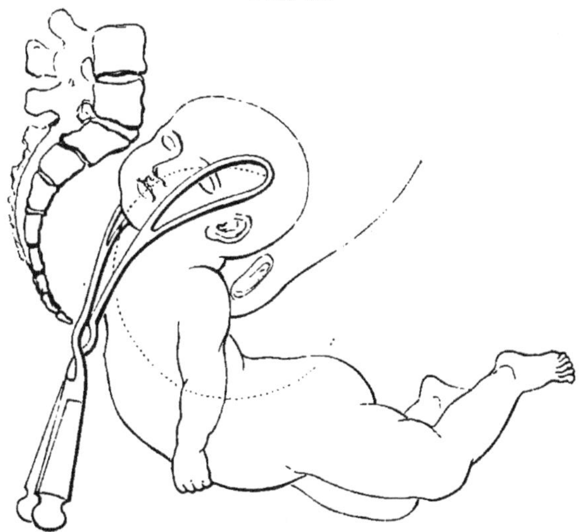

ILLUSTRATES THE APPLICATION OF THE LONG FORCEPS TO THE AFTER-COMING HEAD.

Such, then, is the story of the long forceps applied to the head in the first position at the brim. If the head be in the second position, the blades will seize it—one on the left brow, the other on the right occiput. The occiput will emerge under the right pubic ramus. Here special care is necessary. When the head emerges occiput to the right, if the shoulders are so large as to demand extraction in aid of expulsion, be very careful to direct the face downwards—*i.e.*, to the mother's left thigh—for if through inadvertence you turn the face upwards to the right thigh, you may give a fatal twist to the child's neck, or impede the turn of the shoulders into the antero-posterior diameter of the outlet.

In the case of the third and fourth positions of Naegele, the head will still be seized obliquely; and as it enters the pelvic cavity, it will generally make a quarter axial rotation, face backwards, so as to bring the occiput under a pubic ramus. In the case of fronto-pubic position, the head will be grasped more nearly in its transverse diameter. As it descends into the pelvis, this position may be preserved; and it becomes a question whether delivery should be completed with the forehead forwards, or an attempt made to turn it back into the hollow of the sacrum.

The cause of arrest of labour, of difficulty, when the position is occipito-posterior, is, I believe, this: The head imprisoned in the pelvis is not able to take its normal extension movement. In occipito-anterior positions, the propelling force propagated through the spinal column causes the head to roll up from the floor of the pelvis out by the open space under the pubic arch. But in occipito-posterior positions, the propelling force acts against the escape of the head by driving it against the floor of the pelvis, the occiput naturally rolling back into the hollow under the promontory. If extension-movement then takes place, this, by throwing the occiput against the back, rather increases the difficulty. Release can only be obtained by a movement of flexion. Now, flexion may be useful under two circumstances—first, as already explained, by supplying the essential condition for the spontaneous turn of the face into the sacrum;

secondly, by taking the symphysis as the centre of rotation, and the point against which the root of the nose or the forehead is fixed, whilst the vault of the cranium is made to roll over the floor of the pelvis and through the outlet.

The first question that arises in the presence of an occipito-posterior position is, whether we can hope for the change, spontaneously or by art, to an occipito-anterior position.

Dr. R. U. West* has proved the practicability of procuring the rotation face backwards by artificial means. He applied his fingers to the frontal bones, turning this part backwards, and at the same time raising it up until he felt the posterior fontanelle come down. In another case he brought the occiput down by the lever. As soon as the occiput came down, the rotation seems to have been effected by Nature. This, indeed, is the essential thing to do —to get the occiput down, to restore flexion.

On the other hand, I am persuaded that the head often turns of its own accord, when we think we are helping it. The evidence of Dr. Millar is quite to the purpose. "I met," he says, "a good many cases of occipito-posterior positions in which anterior rotation was effected; but the efficiency, I believed, belonged to me, and not to Nature, because I laboured assiduously to promote it after the manner recommended by Baudelocque and Dewees. . . . I have since experimentally allowed Nature to take her course in a considerable number of such cases, and *I find that the desired mutation is generally accomplished about as well without as with my assistance.*"

* "Glasgow Medical Journal," 1856.

Dr. Leishman, whose excellent book* is full of instruction, says:—"My impression is that rotation can only be effected by artificial means when the head is free above the brim, or when it has quite descended to the floor of the pelvis." If the forehead has come down, Dr. Leishman says:—"No mere rotation can bring about the desired change. Rotation must be so managed that it is combined with a *descent of the occiput* and a corresponding retreat of the forehead."

I have found that the occiput must be brought down below the edge of the sacro-sciatic ligament in order to permit of the rotation face backwards.

It is judicious, I think, to make a reasonable attempt, after the methods of Dr. R. U. West, to bring the occiput down and forwards. This is entirely an affair of leverage. You may act upon either end of the lever represented by the long diameter of the head; or, better still, upon both ends simultaneously. You may apply the blade of the lever as nearly as possible over the occiput, on that side which is most remote from the pubes; draw downwards and forwards at the same time that, with the point of the finger resting on the frontal bone, you press the forehead upwards and backwards. By this manœuvre, under favourable circumstances, the desired change to an occipito-anterior position may be effected.

The leverage may be applied by the forceps. The head being grasped in its transverse diameter, or with only moderate obliquity, a movement of rotation of the instrument on its axis will turn the face backwards into the sacrum. But the forceps cannot at the

* "The Mechanism of Parturition," 1864.

same time so well bring down the occiput as the lever combined with the fingers can do. Professor Elliot, in his admirable practical work, "Obstetric Clinic, New York, 1868," gives cases in which he rotated the occiput forwards by the forceps with success.

But I cannot give more than a qualified assent to the propriety of attempting to rectify the position. It is only exceptionally useful; still more rarely is it necessary, and it is not free from danger. The head can be born very well preserving the occipito-posterior position throughout. Indeed, I think this occurs more frequently than Naegele represents. Nor does the case call for any amount of force. By aid of the forceps the delivery is nearly as easy as when this instrument is applied to an occipito-anterior position. In the event of delay, I therefore advise resort to the long forceps.

The blades should be applied in the sides of the pelvis; they will be guided by the head into the most suitable position. *Extraction*, then, *simply*, without troubling yourselves about rotation, is all that is necessary. If Nature prefer or insist upon rotation, your business is to consent. As the head advances, the occiput may come forwards, and you will feel the handles of the forceps turn upon their axis. But in a large proportion of cases Nature will not insist upon bringing the occiput forwards; and here again your part is simply that of a minister of Nature. The forehead will emerge under the pubes; the cranium will sweep the sacrum and perinæum.

As the blades of the forceps preserve their original position, the handles will turn with the head. It is labour lost—it is encumbering Nature with superfluous help—it is a sin against that most excellent maxim, "*ne quid nimis*," to attempt to promote this turn by twisting with the forceps.

In this latter case there are two things to be observed: first, the perinæum is put more upon the stretch, and therefore requires more care; second, if the handles of the forceps are carried forwards towards the mother's abdomen too soon, the blades will be apt to slip off. The superiority of the long forceps in saving the perinæum is very marked.

The propriety of not attempting to turn the face backwards is even more decided in those more marked cases of fronto-anterior positions in which *the forehead looks nearly directly forwards*. It appears to me that this position is due to unusual flatness of the promontory—a very slight projection of this part. A pronounced projection of the promontory will scarcely permit the head to occupy the antero-posterior diameter; it will throw the occiput to one or the other iliac hollow, so that the moment the head dips into the pelvis the anterior pole is turned into the hollow of the sacrum, or to one side of it.

Upon this point I am glad to quote the authority of Dr. Ramsbotham, whose experience is unsurpassed:—"I prefer extracting it, if possible, with the face under the arch of the pubes, because, as the rotation is made over only one quarter of the half-pelvis, there is less chance of injuring the soft parts. Besides, should the child's body be strongly embraced by the uterine parietes while we are acting, and should it not follow the turn which we are forcing the head to take, we should twist the child's neck, perhaps fatally." In truth, there is no very serious difficulty in extracting with the face forwards.* In this case the fillet or lever seizing the occiput would find its most scientific application.

FIG. 25.

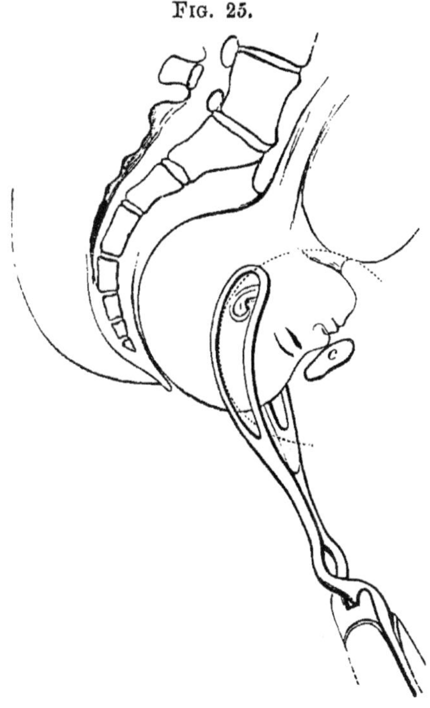

SHOWING THE APPLICATION OF THE FORCEPS TO THE HEAD IN FRONTO-ANTERIOR POSITION

The promontory of sacrum shows very little projection. The head is seized nearly in its transverse diameter. The symphysis, c, is the centre of rotation. The vertex and occiput sweep the perinæum, producing a movement of flexion.

The case is, however, more severe if it is a complete *face-presentation*. You can hardly, by aid of the forceps, so far modify the position of the head as to

* The practitioner or student who wishes to gather instruction from Dr. Ramsbotham is advised to study his clinical reports ("Med. Times and Gaz.," 1862). These give the matured conclusions of this eminent teacher, and show that *his practice*, elaborated out of, and gradually formed in, his encounters with difficult cases, was even superior to the formal and more conventional *doctrines* in his systematic work.

render its course through the pelvis easy; and when you have succeeded in dragging it into the cavity, you may find yourself left with no alternative but to perforate. It is very true that a large proportion of face-labours end happily without assistance. It is equally true that face-presentations supply some of the most difficult cases in practice.

It is convenient in this place to examine *how brow-presentations and face-presentations are produced.* Brow-presentations may be regarded as transitional between vertex and face-presentations; and by analysing the mode in which brow- and face-presentations arise, we shall have the best indications for prevention and treatment. Consider the head as a lever of the third order, the power acting about the middle. The fronto-occipital diameter or axis represents the lever; the atlanto-occipital articulation is the seat of the power. Riding upon this point, the head moves in seesaw backwards and forwards. A force which is generally unnoticed in obstetrics is *friction;* and if friction were uniform at all points of the circumference of the head, it would be unimportant, from a purely dynamic point of view, to regard it. But it is not always so. Friction at one point of the head may be so much greater than elsewhere, that the head at the point of greatest resistance is retarded, whilst at the opposite point the head will advance to a greater extent; or resistance at one point may quite arrest the head at that point. In either case the head must change its position in relation to the pelvis.

Let us, then, take the case where excess of friction bears upon the occiput directed to the left foramen ovale. This point will be more or less fixed, whilst the opposite point or forehead, receiving the full impact of the force propagated through the spine to the atlanto-occipital hinge, will descend—that is, the forehead will take the place of the vertex, and be the presenting part. If this process be continued, the head rotating back more and more upon its transverse axis, the face succeeds to the forehead.

Now, if we can transpose the greatest friction or resistance to the forehead, and still maintain the propelling force, it is clear that the occiput must descend, and that the normal condition may be restored. In practice this is actually done. When at an early stage of labour we find the forehead presenting, we can, by applying the tips of two fingers to the forehead, during a pain, retard its descent, and the occiput comes down. This effected, the rest will probably go on naturally, because, the atlanto-occipital joint being somewhat nearer the occipital than the frontal end of the lever, the shorter or occipital arm of the lever will keep lowest. But if there should still be excess of resistance at the occipital end, we have only to add so much resistance to the frontal end as will maintain the lever in equilibrium. This manœuvre is illustrated in diagrams 26, 27, p. 84.

The face may enter the pelvis, take its turn forwards, and then be arrested, just as the head in cranial presentation may be arrested. In such a case the forceps may be useful. The application is as follows. Assume that it is the first face-position; remember that the object to be accomplished is to make the vault of the cranium and the occiput roll over the floor of the pelvis around the symphysis as a centre, so as to restore flexion. The blades should

Fig. 26.

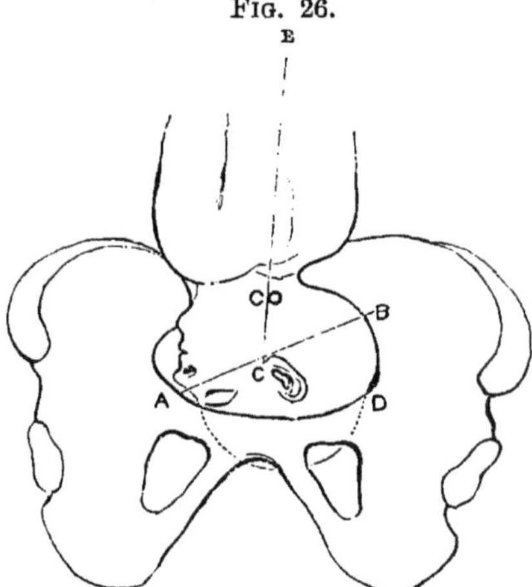

REPRESENTS A CHANGE IN PROGRESS FROM AN ORIGINAL VERTEX PRESENTATION TO A FOREHEAD.

C is the atlanto-occipital joint, or point where the force propagated through the spine E C impinges upon the lever A B C. D is the point of greatest resistance. Therefore the arm A of the lever descends. C′ E, the force, forms an obtuse angle with the arm A.

Fig. 27.

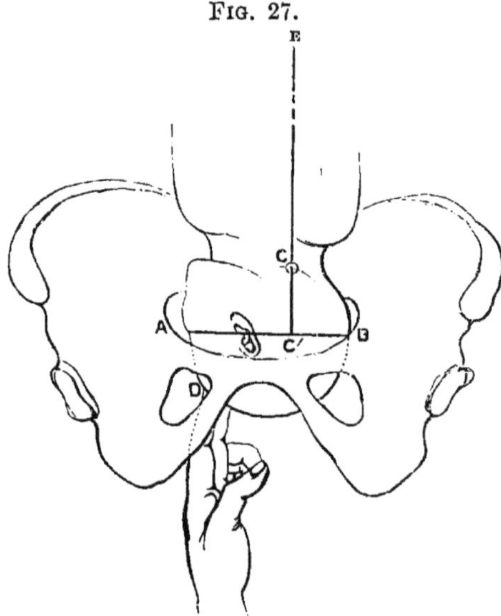

THE FINGERS APPLIED TO THE FOREHEAD AT D TRANSPOSE THE GREATEST RESISTANCE TO THIS POINT.

The force propagated from E to C will therefore drive down B, the occipital or shorter arm. The force E C C′ will form an acute angle with the long arm A, and the tendency will thus be greater to keep the occipital arm B lowest in the pelvis. Or we may help to overcome the resistance at the occipital end of the lever by applying the palm of the right hand externally and pressing the occiput downwards.

seize the head nearly in its transverse diameter. Now, the face presents some degree of obliquity in relation to the pelvis. The first or sacral blade, therefore, must pass up the left side of the pelvis, somewhere between the sacro-iliac joint and the left extremity of the transverse diameter. The second or pubic blade will pass in the opposite point of the pelvis, that is,

Fig. 28.

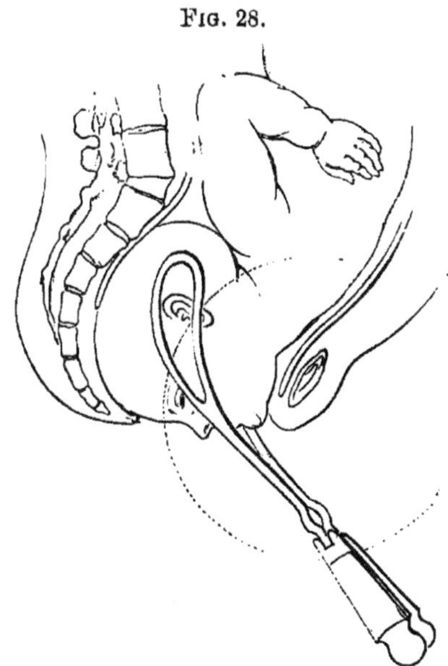

SHOWS THE LONG FORCEPS APPLIED TO THE HEAD IN FACE-PRESENTATION DELAYED IN THE PELVIS.

The curve of Carus—the dotted circle—indicates the direction of traction, restoring flexion.

between the foramen ovale and the right extremity of the transverse diameter. When locked, traction is at first directed downwards, to get the chin fairly under the pubic arch. Then the traction is directed gradually more and more forwards and upwards, so as to bring the vault of the cranium out of the pelvis. The posterior part of the head puts the perinæum greatly on the stretch. It requires great care to extract. Give time for the perinæum to dilate. Carry the forceps well forwards, so that the shanks are out of the way; but not too soon, lest the blades slip off. Extract gently.

But we shall not always be so fortunate even as this. Several of the most difficult cases in which my assistance has been sought have been face-presentations. In some, *the face will not enter the brim.* This is the first order. What shall we do here? If we apply the forceps, one blade is likely to seize beyond the jaw and compress the neck, bruising the trachea. If the attempt be made to seize the head by applying the blades in the oblique diameter, they must be passed very high, and even then may slip. If firmly grasped and traction be made, the faulty extension of the head is increased; the compression of the vessels of the neck and the danger of apoplexy are augmented; and, after all, extraction may have to be completed by perforation. Turning can be effected with infinitely less trouble, and with a better prospect for the child. In the second order of cases, the *face has descended into the cavity.* The birth of a full-grown living or recently dead child, with the forehead maintaining its direction forwards, is almost impossible. The extension of the neck is extreme, the head being doubled back upon the nucha. The face represents the apex of a wedge, the base of which is formed by the forehead, the entire length of the head, and the thickness of the neck and chest. This must be equal to at least seven inches. The bregma and occiput become flattened in, it is true, but much is not to be expected from moulding.

Compression, bearing upon the neck, if great and long-continued, is almost necessarily fatal to the child. Hence arrest or impaction. The turn of the chin forwards under the pubic arch, so as to release the head by permitting flexion round the symphysis, cannot take place. Aid becomes necessary. We have to consider the following points:—

1. Can the head be rotated on its transverse axis, restoring flexion, and so bring the cranium down? This may be accomplished whilst the head is above the brim, but scarcely when it is squeezed into the cavity.

2. Can the turn of the chin forwards be effected by the hand, the lever, or the forceps? This is sometimes possible, and should be tried. It is thus described by Smellie: "After applying the short or long-curved forceps along the ears, push the head as high up in the pelvis as is possible, after which the chin is to be turned from the os sacrum to either os ischium, and afterwards brought down to the inferior part of the last-mentioned bone. This done, the operator must pull the forceps with one hand, whilst two fingers of the other are fixed on the lower part of the chin or under-jaw to keep the face in the middle and prevent the chin from being detained at the os ischium as it comes along, and in this manner move the chin round with the forceps and the above fingers till brought under the pubes, which done, the head will easily be extracted."

3. Can the head be brought down by the forceps without turning the chin forwards? This is a practice against Nature. If the forceps grasp, and it will

generally slip, it will bring more of the base of the wedge into the brim. The head must be small, or the pelvis large, to admit of success by this mode.

4. Shall we extricate the head by perforating? The wedge may be lessened, but even after this, delivery is not always easy unless part of the cranial vault be removed, so as to allow of the flattening in of the head.

5. Can we turn simply? It is the best course, but if the head is low it may be difficult to accomplish.

6. The chin will sometimes turn forwards at the very last moment, when the face is quite on the floor of the pelvis. If not, it may be possible to hitch the chin over the perinæum by drawing the chin forwards by forceps, and pulling the perinæum backwards. The chin thus outside, the forceps or lever may be applied to draw the occiput down under the pubes and backwards, so as to make the head revolve on its transverse axis, thus restoring flexion. You are in fact decomposing the base of the wedge. You deliver by a process the reverse of that of ordinary occipito-anterior labour. In this, the occiput escapes by a process of extension. In the mento-sacral position you deliver by promoting flexion. Or, to take our illustration from the mechanism of face-labour, you obtain flexion by causing the chin to turn over the coccyx or sacro-sciatic ligament as a centre, instead of over the symphysis. The latter is the natural mode, but it may be that the first alone is possible. This is a case in which incision, bilateral, of the perinæum, here acting as an obstructing posterior valve, may be performed in order to facilitate the release of the chin.

LECTURE V.

THE FORCEPS IN DISPROPORTION OF THE PELVIS—DEGREES OF DISPROPORTION—INDICATIONS IN PRACTICE—THE MECHANISM OF LABOUR IN CONTRACTION FROM PROJECTING PROMONTORY—THE CURVE OF THE FALSE PROMONTORY—DEBATABLE TERRITORY ON THE CONFINES OF THE SEVERAL OPERATIONS—PENDULOUS ABDOMEN—THE CAUSE OF DIFFICULTY IN PENDULOUS ABDOMEN—SUSPENDED LABOUR—THE MODE OF MANAGEMENT—DYSTOCIA FROM FAULTY CONDITION OF THE SOFT PARTS—CONTRACTION OF THE CERVIX UTERI — RIGIDITY — SPASM — DEVIATION — HYPERTROPHY — CICATRIX — CLOSURE — ŒDEMA—THROMBUS — CANCER —FIBROID TUMOURS—THE NATURAL FORCES THAT DILATE THE CERVIX — THE ARTIFICIAL DILATING AGENTS — VAGINAL IRRIGATION — WATER-PRESSURE — INCISIONS — RESISTANCE OFFERED BY THE VAGINA, VULVA, AND PERINÆUM.

Now we have to consider what the forceps can do in cases of disproportion; for instance, where the brim is too small to allow the head to pass by the unaided powers of the uterus. This brings up the problem of the compressibility of the head under the forceps, and the comparison of the advantages of the forceps with those of turning. The degrees of contraction of the brim may be classified approximatively in the following manner:—

Scheme of Relation of Operations to Pelvic Contractions, Labour at Term.

	Conjugate diameter reduced to		
The first degree .	4 to $3\frac{1}{4}$ in.,	admits	the forceps, opposed to the bi-parietal diameter of $3\frac{1}{2}$ to 4 in.
The second degree .	$3\frac{3}{4}$ to 3 in.,	,,	of turning, opposed to the bi-mastoid diameter of 3 in.
The third degree .	$3\frac{1}{4}$ to $1\frac{1}{2}$ in.,	,,	of craniotomy and cephalotripsy.
The fourth degree .	below $1\frac{1}{2}$ in.,	,,	of Cæsarian section.

If you have the advantage of bringing on labour at seven months, then you may eliminate the Cæsarian section, and slide down the scale of operations, so that craniotomy shall correspond with the fourth degree, turning with the third, and the forceps with the second, whilst the first degree, being reduced to the conditions of natural labour, may require no operation at all.

Scheme of Relation of Operations to Degrees of Pelvic Contraction under Labour at Seven Months.

	Conjugate diameter reduced to		
First degree . .	4 to $3\frac{1}{4}$ in.,	admits	spontaneous labour
Second degree . .	$3\frac{3}{4}$ to 3 in.,	,,	of forceps.
Third degree . .	$3\frac{1}{4}$ to $1\frac{1}{2}$ in.,	,,	of turning.
Fourth degree . .	below $1\frac{1}{2}$ in.,	,,	of craniotomy.

Cæsarian section is eliminated.

The range of application of the forceps is, I believe, not great. The head cannot be compressed by it quickly. The proper use of it is to aid that natural process of moulding which always takes place in protracted labour. Now, this is a gradual, even a slow process. The head is seized by the long forceps in the way already described. The handles are firmly grasped with both hands, and especial care is required to extract well backwards in the axis of the brim, so as to make the head revolve round and under the projecting, overhanging promontory as a centre. Here I may pause to show that, in labour with conjugate contraction from rickets, the promontory possesses a like importance at the brim or entry of the pelvis to that

Fig. 29.

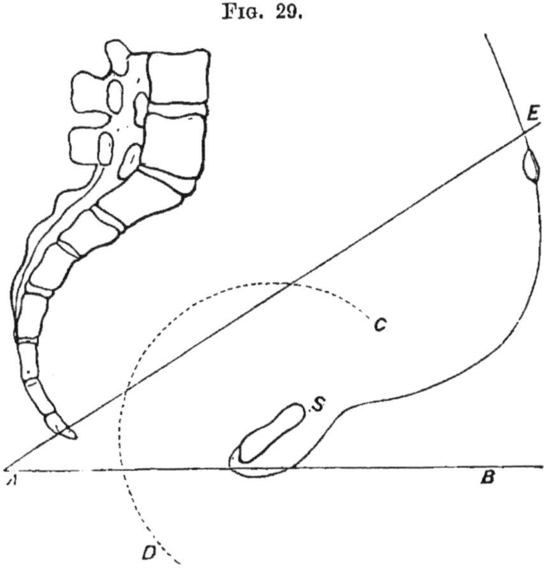

NORMAL PELVIS.

S, the symphysis pubis, the centre of Carus' curve C D; A E, the axis of the brim, forming an acute angle, not less than 30°, with the datum-line A B. The uterus and the child's body nearly corresponding with the axis of the pelvic brim, the head enters its natural orbit, represented by Carus' curve, at once.

which the symphysis pubis possesses at the outlet. The promontory is a turning-point—a centre of revolution of the head, just like the symphysis. The curve round the pubes, which Carus described, has its counterpart in a curve round the promontory. In ordinary labour, with a well-constructed pelvis, the head enters the pelvis, and reaches nearly to the floor, without deviating much from the straight line which represents the axis of the brim. Thus it enters its orbit, the circle of Carus, at once.

FIG. 30.

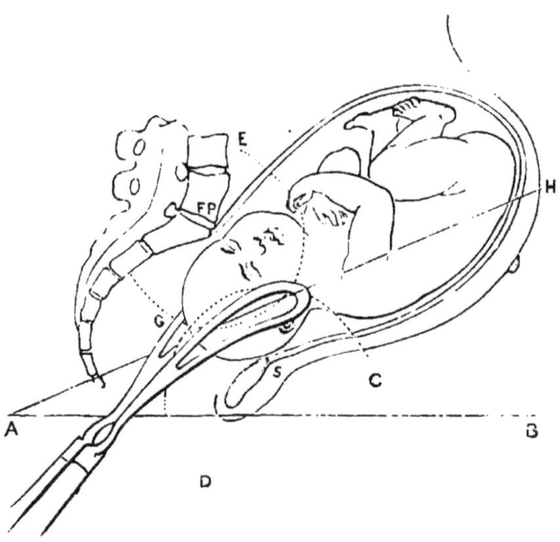

PELVIS CONTRACTED BY RICKETS TO SHOW THE CURVE OF THE FALSE PROMONTORY.

S, the symphysis, the centre of Carus' curve C D; F P, the false promontory, the centre of the false curve E G; G, the point of intersection of the two curves where the head passes from the false to the true orbit; A H, the axis of the brim, forming a very acute angle, varying from 30° to 20° or less, with the datum-line A B. The head is thrown over the pubic symphysis by the projecting promontory. The forceps draws backwards in line A H to bring the head under the promontory in the orbit of the false promontory.

But a projecting promontory, involving, as it commonly does, a scooped-out sacrum below, disturbs this course. The promontory must be doubled. The head must move round this before it can strike into its natural orbit. I propose to call this curve *the curve of the false promontory*.

This curve is the chart by which to steer in turning on account of contracted pelvis.

Bearing this in mind, and assuming that the head is seized nearly in its transverse diameter, which is very rarely effected, the blade corresponding to the anterior or pubic side of the head must describe a large circle,

FIG. 31.

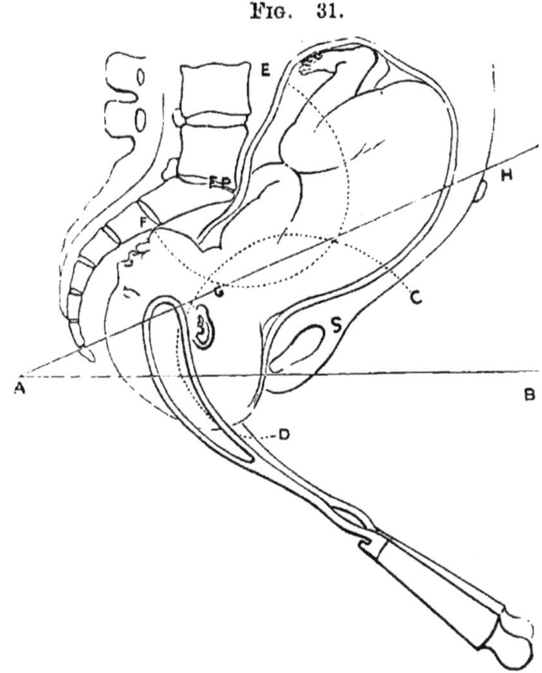

A B, datum-line; C D, Carus' curve; E F, curve of the false promontory; G, point of intersection of the two curves where the head passes from the false to the true orbit. The forceps now draws the head in the direction of the outlet in Carus' curve.

whilst the sacral side of the head, and the blade in relation with it, move but little until the promontory is rounded, and the head has entered the pelvis. When this point is reached, the direction of traction is that of Carus' curve. The head, which was compelled to traverse the brim nearly in the transverse diameter, will quickly rotate, face to sacrum. This turn, imparted to the handles of the forceps, and sudden transition from resistance to ease, a sort of jerk, mark the completion of the first circuit and the beginning of the second. The rest falls within the laws of natural labour.

It happens, however, in these cases of contracted conjugate diameter, that the head commonly presents at the brim, with its long diameter very nearly, if not quite, in correspondence with the transverse diameter of the pelvis. The blades of the forceps, also finding most room in this diameter, will grasp the head in its longitudinal diameter. In extraction, therefore, both blades will move equally around the false promontory.

What is the extreme degree of narrowing that will admit of the useful application of the forceps? I have stated it in the table at 3¼ inches, but it cannot be defined absolutely. A head slightly below the normal size, and less firmly ossified than usual, may be brought through a conjugate diameter of only 3 inches. And as we cannot know with sufficient precision what the properties of the head still above the brim are, we are justified in making tentative, experimental efforts with the forceps before resorting to turning, which is, perhaps, more hazardous to the child; or to craniotomy, which is certainly destructive to it. This uncertainty, or want of fixity, in the relations between the head and the pelvis, compels us to leave a range or borderland of debatable territory between the more clearly recognised or conventional limits assigned to the several operations. This debatable territory is further liable to invasion from either side, according to the relative skill with which the competing operations are carried out. And herein lies the source of the great controversies in obstetric practice. Thus one operator, possessing a good long forceps, and confident in his skill in handling it, will use this instrument with success where the contraction is $3\frac{1}{4}$ inches; whilst another, possessing only a single-curved forceps or a bad doubled-curved one, *must* either turn or perforate. So again, the region between the second and third degrees of contraction, the region assigned to turning, may be invaded on the one side by the forceps, on the other by the perforator, and become the subject of a partition treaty, which shall dispossess turning, the rightful power, altogether. It unfortunately happens that perforation, being an easy operation, is apt to carry its inroads further than the forceps; and thus the child falls under a wide, arbitrary, and fatal proscription.

There is a condition causing dystocia called the *pendulous abdomen.* It is most frequent in women who have borne many children, and in whom the abdominal walls are much relaxed. Where this exists, the uterus hanging down in front of the pubes, is out of the axis of the brim, and, if it contract, it will only direct the child over the brim, backwards against the promontory. This may sometimes be remedied by putting the patient on her back, and making up for the want of support from the abdominal muscles by applying a broad binder, so as to lift the fundus of the uterus upwards and backwards. This will restore the relation between the axis of the uterus and pelvic brim. But if contractile energy be still insufficient, the long forceps will come into use. And this is a case where the dorsal decubitus will much assist the delivery. If the patient continue on her side, the uterus not only hangs forwards, but swags downwards to the dependent side, constituting a further deviation, and increasing the obstacle to parturition.

How to determine the choice between forceps and turning? There are two cases. First, the liquor amnii has drained away, and the head is pressing into the brim: the forceps is strongly indicated here. Secondly, the head is mobile above the brim, and not easy to grasp in the blades: here turning may be preferable. I have several times rescued a living child by turning under these circumstances.

The second case may sometimes be reduced to the first, and thus brought within the more desirable dominion of the forceps. One result of the pendulous abdomen and uterus is to form a kind of reservoir, in which the liquor amnii is dammed up. Hence an added impediment to contraction of the uterus. Now, the waters can be drained off by lifting the fundus uteri up to its normal position against the spine, by laying the patient on her back, and making a channel past the head to the uterine reservoir, by introducing the lever or one blade of the forceps. Having accomplished this, the uterus, under the combined advantages of restoration to its natural axis, and of steady pressure by the hands or a belt, may recover its power, and expel the child. If not, the forceps supplies an easy remedy.

In the following diagrams (Figs. 32, 33), the mechanism of labour obstructed by this form of malposition of the uterus is illustrated.

Until the uterus is brought back to its normal

position, it is clear that two causes concur to render labour difficult. First, the uterus being thrown forwards, its fundus is carried away from the diaphragm and upper part of the abdominal walls. It loses, therefore, the aid which the expiratory muscles, acting powerfully when the glottis is closed and the chest is fixed, usually give. This expellent power of the expiratory muscles is so great, that it appears to be of itself sufficient in some cases to complete labour, the uterus remaining quite passive. When the uterus is thrown forwards across the pubes, any force propagated downwards from the diaphragm will strike the posterior wall of the uterus at a right angle with the body of the uterus and of the long axis of the fœtus. It will, in short, drive the uterus and its contents down upon the symphysis, or even more forwards still, since the body of the child, which lies in front of the symphysis, forms the longer arm of a lever, and the force is expended upon it.

Secondly, the uterus itself, if not paralyzed, acts in a wrong direction. It loses the stimulus to action which the normal pressure and support of the diaphragm and abdominal walls supplies, and therefore acts languidly. Its independent power is also weakened by another circumstance. It is a law, of which the patient observer will not fail to discover many proofs in the progress of difficult labour, that, whensoever a mechanical obstacle is encountered, before long, the uterus, conscious, as it were, of the futility of its efforts, intermits its action, takes a rest, lies dormant, until the time shall arrive when it can act with advantage. This provision protects, for a long time, against exhaustion from protracted labour. Indeed, what appears to be protracted labour is often simply suspended labour; and suspended labour may even pass into what Dr. Oldham has so aptly called "missed labour."

Fig. 32.

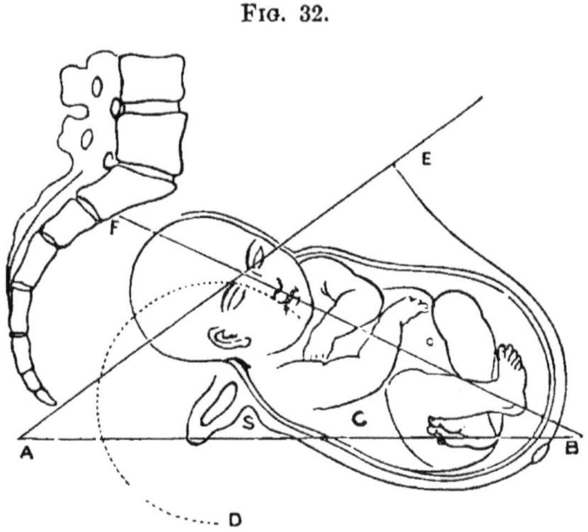

A B, datum-line; A E, axis of pelvic brim and normal axis of uterus; F D, axis of uterus pointing to sacrum; C D, Carus' curve.

The remedy is obviously to restore the uterus to its normal position. In Fig. 33, the uterus and child are represented lying across the symphysis pubis, like a sack across a saddle. H I is the line in which the proper uterine force would be exerted; F G is the line of force of the expiratory muscles striking the long axis of the uterus nearly at a right angle. These two forces, which ought to coincide, thus cross each other, and the error is but imperfectly compensated by the resultant force obtained between the two. But raise the uterus to its normal position, as indicated by the dotted outline, and, immediately, the expiratory force and the uterine force coincide with the axis of the child and of the pelvic brim, and both conspire to expel the contents of the uterus. Not even the forceps will act efficiently until this restoration is made.

FIG. 33.

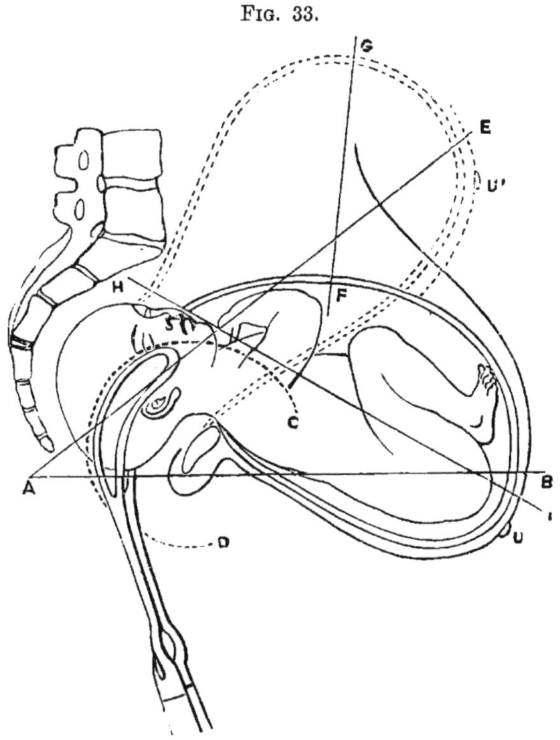

SHOWING THE MECHANISM OF LABOUR IN PENDULOUS BELLY.

A B, datum-line determining position of the pelvis; C D, Carus' curve; H I, axis of uterus and of child directed towards the promontory; F G, line of force of expiratory muscles cutting the axis of the uterus and of child; A E, normal axis of pelvis; U U′, the umbilicus.

Dystocia from faulty condition of the soft parts of the parturient canal is only incidentally and occasionally related to the history of the forceps. This incidental relation, however, makes it convenient to discuss the question in this place.

The cervix uteri, the vagina, or the vulva, including the perinæum, may refuse to yield a passage to the child, and to permit the application of the forceps. The conditions which lead to this result are various. First, as to the cervix uteri, including the os externum uteri, the following causes of obstruction may be observed:—1. Spastic annular contraction. 2. Thickening from œdema. 3. The cervix may have an abnormal direction and position. The cervix is not in the line of the axis of extrusion. Perhaps it is bent down at a more or less acute angle upon the body. A more frequent condition is the pointing of the os uteri backwards towards the promontory, and very high up, so that it is difficult to reach, perhaps impossible, without passing the hand into the vagina. In such a case, the head bears unduly upon the anterior segment of the lower part of the uterus. This is often the result of slight narrowing of the pelvic brim, which throws the head upon the anterior wall of the pelvis. This may persist so long that the tissues become worn and their texture softened, so that when the head is driven down into the pelvis the damaged cervix rends. 4. From contraction of the brim, or from the presentation of some part of the child—as the arm, face, or feet—not adapted to descend easily and fairly upon the cervix, there is insufficient dilatation. 5. The cervix may be organically diseased. The most marked causes of this kind are hypertrophy, occlusion from false membrane, fibroid tumour, or cancer. To this may be added abnormal formation. 6. The cervix may be closed by cicatricial atresia.

It is further customary to refer to cases in which no os uteri can be found.

There is another condition, which, although not in itself abnormal, will be properly considered in connection with the above. The os and cervix may be met with closed or only imperfectly dilated under circumstances which render speedy delivery eminently desirable. In such a case, the cervix must be treated as one that is rigid or otherwise diseased. It obstructs labour, and, just as in the cases where the closure of the os is the primary cause of obstruction, it must be opened.

The first of the conditions enumerated includes what is commonly understood as *rigidity*. It is really much more rare than is supposed. Most frequently when the os will not dilate, it is because the presenting part of the child cannot come down upon it. But if the membranes are ruptured prematurely, and the presenting part comes to press upon the os before this is at all dilated, then it often acts as a source of irritation, and produces this spastic annular contraction.

Before discussing the second cause of rigidity, it will be useful to examine *what are the forces that dilate the cervix.*

This study will throw light on the causes and pathology of rigidity, and furnish useful indications in treatment. By some it is held that the opening of the cervix is the direct result of the active contractions of the longitudinal uterine muscles, which, pulling the os towards the fundus, thus draw it open. I very much doubt if more than a very inconsiderable opening is effected in this manner.

It is a matter of observation that the os uteri does

not expand in any marked degree until either the bag of membranes or the child's head comes to bear upon it. These distend the cervix and os as a direct mechanical force; they are, in fact, wedges, themselves inert, but propelled by the contractions of the uterus and the abdominal muscles. Under this distending force, the circular fibres of the cervix yield, just as the sphincter ani or the sphincter vesicæ yields under the pressure from above. The yielding of the cervix uteri is indeed a question of the preponderance of the *vis à tergo* exercised by the body of the uterus and the expiratory muscles over the resistance offered by the cervix. Sometimes the normal harmony between this preponderance and resistance is disturbed; the active force or the passive resistance is in excess; or the resistance may become active, and the force may be reduced to inefficiency. There is, in fact, a translation or metastasis of the nervous energy from the body of the uterus to the neck. This disturbance most frequently arises from an inversion in time, in the order of succession of the parturient phenomena. Thus, if the liquor amnii escape prematurely, the presenting part of the child will bear too early upon the cervix, and excite it to irregular action. This, by diverting and disordering the nervous supply to the body of the uterus, disables this part of the organ; and concurrently the cervix itself, becoming congested and thickened by undue pressure, irritation, and action, loses its natural capacity for dilatation.

A very instructive illustration of the theory that the dilatation of the cervix uteri is essentially dependent upon the eccentric pressure exerted by the liquor amnii and fœtus driven into it is found in the equivalent action of my hydrostatic cervical dilator. This instrument is inserted within the cervix in a collapsed state, and then gradually distended with water, as seen in Fig 34. It very nearly represents the normal action of the liquor amnii distending the sac of the amnion. Under this pressure, the cervix yields smoothly and gradually, just as in natural labour; the speed, however, being very much within the discretion of the operator. In this instrument we possess a power in midwifery, at once safe and efficient, that brings the cervix, and therefore the course of labour, completely within the control of skill.

How shall we restore the due relation between the expulsive and the resisting forces? How, in other words, shall we overcome the rigidity of the cervix uteri? This may be done in one of two principal ways. We may increase the power of the body of the uterus, so as to restore its preponderance over the cervix, or we may apply direct means to the cervix to dilate it, doing ourselves the work that the uterus cannot do. Great judgment is necessary in selecting between these two courses. Before deciding in favour of the first, we must be satisfied that the resistance opposed by the cervix is of such kind and degree that it may be overcome by moderate force. We must also be satisfied that there is potential energy enough in the system and in the uterus to respond to the stimulus, to the whip we propose to administer. To give ergot, for example, when the frame and the uterus are exhausted, is to equal the folly of the heavy rider who drives his spurs into his jaded horse when he ought to dismount and lead him.

It will almost always be proper, as the first step—

that is, before seeking to rouse the uterus to increased action—to secure a more favourable condition of the cervix. How is this to be done? Let us take the case of spastic rigidity, the cause of which we have just glanced at. The first indication is to soothe, to subdue nervous irritability. Belladonna in the form of extract has been smeared upon the part. One sees the action of this drug in expanding the pupil. I have never felt it on the os uteri. The analogy is probably defective in theory. I believe it is not in the least degree to be relied upon in practice. It is at best an expedient for passing time.

Chloroform is often of signal service. It acts by annulling the sense of pain and the fear of pain, and by restoring the equilibrium of the nervous system by removing disturbing causes that divert the nerve-force from its appropriate distribution; the sphincteric spasm relaxes, the body of the uterus contracts as it ought to do, and the labour proceeds.

A remedy sometimes of equal value to chloroform is *opium*. Thirty drops of the tincture or twenty of the sedative liquor combined with thirty drops of compound sulphuric ether will assuage pain, procure rest, and restore the harmony of the distribution of nerve-force; and if not in itself sufficient, it will aid in the carrying out of other measures.

Tartar emetic in small doses to provoke nausea has been recommended. In some cases I have proved its use, but I am not now disposed to resort to it, at the cost of postponing means at once more prompt and less distressing in their action.

Bleeding has been much extolled. In certain cases, as of convulsions, apoplexy, or such states of system as threaten these catastrophes, this proceeding may be adopted. But, apart from a decided special indication of this kind, it is not wise to bleed a woman in labour. Nor can bleeding be depended upon, as may be frequently seen in cases of placenta prævia, where even flooding *ad deliquium* will sometimes fail to relax the rigid or spastic cervix.

Warm baths have been much praised, and no doubt have a certain degree of power in inducing relaxation of tissue. But a warm bath is rarely at hand, and, if it were, the inconvenience of putting a woman in labour into it must often be insurmountable.

The most valuable of all preparatory measures is the *irrigation of the cervix and vagina* with a stream of tepid water. We know that this is even efficacious in the induction of labour. And it is obviously useful to apply our knowledge of the agents that are effective in the solution of the major problem to the minor one, how to facilitate, to accelerate labour that has begun. In many cases this irrigation will be enough. Presently the cervix softens and yields, spasm is subdued, and abnormal action of the cervix is turned into normal activity in the body of the uterus. The mode of proceeding is simple. Introduce the vaginal tube connected with Higginson's syringe into the vagina, guided by the fingers of the left hand to the os uteri—*not into the os uteri*, there is danger in that—so that the stream of water shall play upon the cervix and fundus of the vagina. This may be continued for ten or fifteen minutes at a time, and repeated after an equal interval.

When the cervix has become disposed to yield, it may not yield. The dilating force has still to be found. You may now, perhaps, give *ergot.* But when you have given ergot you are likely to be in the position of Frankenstein. You have evoked a power which you cannot control. Ergotism, like strychnism, will run its course. If it acts too long or too intensely, you cannot help it. The ergotic contraction of the uterus, when characteristically developed, resembles tetanus. Then, woe to the mother if the cervix does not yield, if the pelvis is narrowed, if, in short, any obstacle should delay the passage of the child. And woe to the child itself if it be not quickly born. I very much prefer to use weapons that obey me, that will do as much, or even less, than I wish. I fear to use weapons that will do more.

The cervix may be dilated by the hand. Two or three fingers are insinuated within the os one after another, so as to form a conical wedge. This wedge is gently and gradually pushed forward into the cervix, and, widening as it goes, the cervix gives way. This wedge has the advantage of being a sentient force. It tells you what it is doing. But what it will tell you is sometimes this: it is that the fingers, with their hard joints, form a rather painful and irritating wedge. As it proceeds it is apt to renew the spasmodic contractions you have taken such pains to allay. If the head is pressing upon the cervix, you may, as has been already mentioned, help the dilatation by hooking down the anterior lip with one or two fingers, holding the os open, as it were, to allow the head to engage in it. But this application is limited; and, I think, what is called manual dilatation of the spasmodic cervix should be abandoned, except in the case of spasmodic contraction after the expulsion of the child —as, for example, when the placenta is retained, or clots are filling and irritating the uterus. In such a case, the steady onward pressure of the hand-wedge will in a few minutes wear out the spasm and effect a passage, enabling you to clear out the cavity.

FIG. 34.

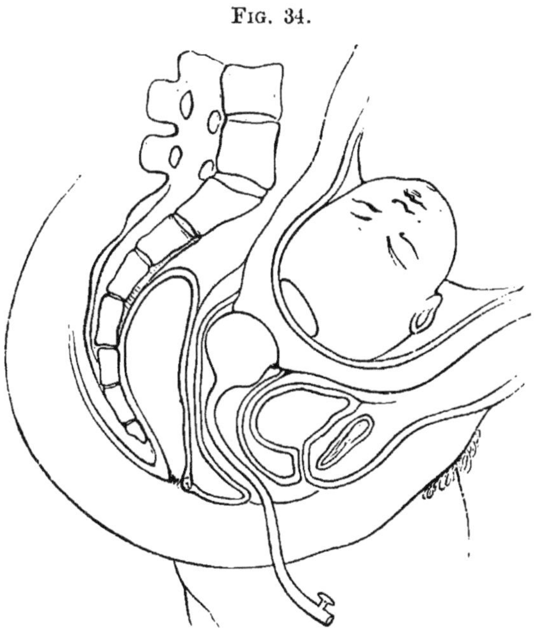

THIS FIGURE SHOWS THE HYDROSTATIC DILATOR DISTENDED *in situ* WITHIN THE CERVIX UTERI.

Water pressure is the most natural, the most safe, and the most effective. An os uteri that will admit one finger will admit No. 2 dilator in the collapsed state. The introduction is effected in this way: Insert the point of the uterine sound, of a male catheter, or any convenient stem, into the little pouch at the end of the bag; roll the bag round the stem, anoint it with lard or soap, then pass it into the cervix, guided by the forefinger of the left hand, which is kept on the os uteri. When the bag is passed so far that *the narrow or middle part is fairly embraced by the cervical ring*, withdraw the sound, keeping the guiding finger on the os to insure the preservation of the bag *in situ*. Then pump in water gradually. Continue distending the bag until you feel it is tightly nipped by the os. When this is done, wait a while; close the stop-cock, and give time for the distending eccentric force to wear out the resistance of the cervix. No muscle can long resist a continuous elastic force. From time to time inject a little more water, so as to maintain and improve the gain. But be careful not to distend the bag beyond its strength. There is of course a limit to the distensibility, even of india-rubber, and I have been told of cases where the bag has burst. I think this accident ought to be avoided. It has never happened to me, and I think it need not happen if the bags are well made. When you have got all the dilatation out of No. 2 that it is capable of giving, remove it, and introduce No. 3, which is larger and more powerful. The dilatation No. 3 will give is commonly enough to afford room for the forceps or the hand. The time required for this amount of dilatation will range from half an hour to two hours. But not to lose time it is desirable to keep your finger on the edge of the os, so as to be sure that the bag does not slip forward into the uterus altogether, or is not driven down into the vagina by uterine action. If it slips wholly into the uterus, it may displace the head. When you have gained your end, open the stop-cock, the water is ejected in a stream, and the bag is easily withdrawn. The cervical dilator serves yet another purpose. Taking the place of the liquor amnii, it does duty for the bag of membranes. It not only directly expands the cervix, but, setting up a quasi-normal reflex excitation, it evokes the regular action of the body of the uterus.

The proceeding I have described will succeed in the great majority of instances, especially where the closure of the cervix is due to spasmodic action, or where, the tissue of the cervix being normal, it cannot expand for want of an eccentric expanding force, as when the bag of membranes or the child does not bear upon it. But in certain cases where there is rigidity from alteration of tissue, as œdema, hypertrophy, cicatrix, something more is required; and that is found in the *knife*. There is nothing new in this use of the knife. It is an old resource too much neglected. Coutouly, Velpeau, Hohl, Scanzoni, indeed all the most eminent continental practitioners, advocate it. Judiciously employed, the knife can do no harm. It will save life when nothing else can.

You are sometimes in the presence of this alternative: exhaustion, sloughing, or rupture of the uterus, on the one hand, or the timely use of the bistoury on the other. It would be as absurd to hesitate as it would be to refuse to perform the Cæsarian section to give birth to a child which cannot be delivered by the natural passages. Indeed, it would be far more absurd, for the Cæsarian section is a most dangerous operation, whilst vaginal hysterotomy of the kind under discussion is free from danger.

There are various cases in which *vaginal hysterotomy*, or *dilatation of the cervix by incisions*, is necessary.

First, no os uteri is to be found. Of course, at the time of conception there was an os uteri; it may have been subsequently closed by a false membrane or by cicatricial contraction. You will rarely fail to feel a nipple or depression where the os ought to be. It is generally very high up and far backwards, near the promontory. Pressure with a sound or the finger will mostly break down a false membrane, and offer a sufficient opening to admit a hernia-knife or the special knife described in the first lecture. This is long and straight, probe-blunted at the end, having a cutting edge of about three-quarters of an inch near the end. The forefinger of the left hand is kept on or in the os uteri, as a guide. The knife is then slipped up, lying flat upon this finger, until its cutting edge is within the os. This edge is then turned up, the back supported by the guiding finger, which takes cognizance of what is to be done and of what is done; and an incision of about a quarter of an inch deep, a slight nick rather, is made in the sharp ridge of the os. The knife is then carried round to another part of the ring of the os, and another nick is made. In this way four or five nicks are effected. Each gives perhaps little; but the aggregate gain of these minute multiple incisions is considerable. I do not think it matters much at what particular points of the circumference of the os these incisions are made; perhaps the two sides are to be preferred.

Before extending or repeating these incisions, it is proper to observe the effect of uterine action in continuing the dilatation. And if nothing is gained in this way, introduce the hydrostatic dilator; distend this

FIG. 35.

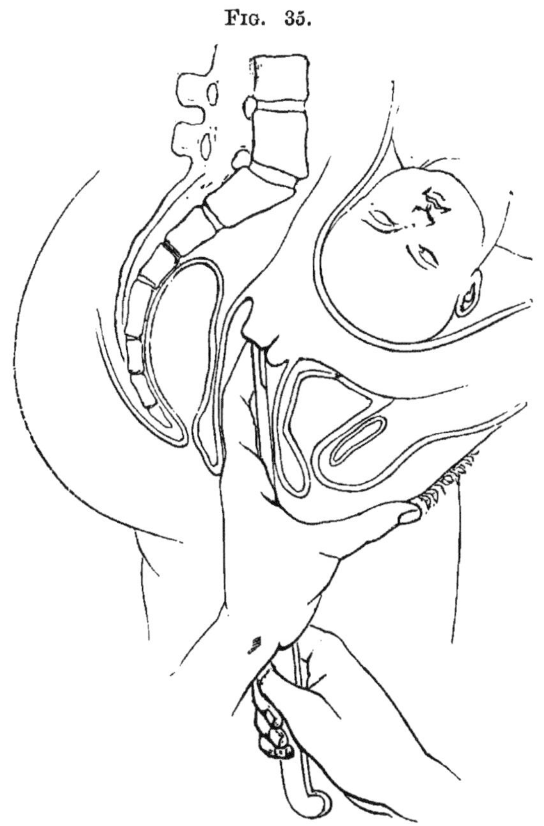

THIS FIGURE SHOWS THE OPERATION OF DILATING THE RIGID OR HYPERTROPHIED CERVIX UTERI BY INCISION.

gently, carefully testing by the finger its action. This plan of *combining the water-dilator with incisions* is especially valuable in cases of rigidity from hypertrophy of the cervix, or of atresia of the os or vagina from cicatrices.

When the forceps will pass—and it is quite possible to apply it when the os will allow the three fingers to pass as far as the knuckles—this instrument may serve to dilate further. But this must be done with great caution. The head being grasped, you may draw steadily down; and by keeping up gentle traction, the wedge formed by the blades and the head will gradually dilate the os, perhaps enough to allow the head to pass, and thus to save the child's life.

But it will occasionally happen that, neither by incision, water-pressure, the hand, nor the forceps, will you obtain an opening sufficient without danger of laceration or other mischief. In such a case, you are justified in reducing the head to the capacity of the cervix by perforation.

Narrowing and rigidity may exist in the vagina in consequence of similar conditions, and may be treated on the same principle. The small rigid vagina of a primipara is best dilated by irrigation and the hydrostatic dilator. This will often singularly shorten labour. Atresia from cicatrices presents a more formidable obstacle. I have found the passage constricted by dense cartilaginous tissues, so as to permit no more than a probe to pass. In such a case, a careful process of incisions, multiplied in all points of the circumference, alternating with water-pressure, is necessary; and it is, after all, probable that you will have to meet the difficulty half-way by perforating the head.

Lastly, obstruction may occur at the vulva and perinæum. In primiparæ, especially, the vulva may form a small rigid oblong ring, scarcely permitting the scalp of the presenting head to show through it. The expulsive pains cause the perinæum behind this ring to protrude; but the ring itself will not open; in fact, the perinæum will yield first. It bulges more and more, and may give way in the raphe, just behind the commissure, this part remaining, for a time at least, intact. A central rent is thus made, through which the child has occasionally been expelled, instead of through the vulva. Or if the perinæum does not yield, something else must. The uterus will cease to act, or, struggling in vain, may burst. Here, again, you may avoid mischief by incisions. The forefinger is passed between the head and the edge of the vulva, and two or three small nicks are made on either side, nearer to the posterior commissure than to the anterior. The relief sometimes gained in this way is surprising. Spasm, irritation, pain subside; the vulva dilates, and labour is soon happily at an end. The bleeding is insignificant; and the minute wounds left when the parts have contracted quickly heal.

Sometimes the vulva, including the labia majora, is so greatly swollen by serous infiltration, as to offer a serious obstacle to labour. This condition is commonly associated with albuminuria and convulsions. And out of this association a double difficulty arises. The convulsions urge to the acceleration of labour; the state of the soft parts forbids active interference. If the head comes down through tissues thus distended by fluid, not only laceration, but subsequent sloughing or gangrene may result. The obstacle to labour, and the local mischief, may be avoided by pricking the skin and mucous membrane in numerous points, so as to let the serum drain off. The operation is performed by a lancet held by the blade between finger and thumb, at a distance of a quarter of an inch from the point, so that the stabs made shall not exceed that depth.

Any point of the parturient canal may be swollen, so as to impede the descent of the child, by a submucous infiltration of blood—the so-called thrombus. I have seen a large tumour formed in this way on the os uteri; but the more common seat is the labia of the vulva. It is not desirable to open these collections of blood, if it can be avoided. But if the obstacle be so great that the head, in passing, threatens to burst and rend the tumour, it is better to open it with a lancet. As soon as the child is born, the part should be carefully examined to see if it bleeds; and pressure upon it by plugging will be necessary.

In cases of formidable obstacle from cancerous or fibrous growth, recourse to the *ultima ratio*—the Cæsarian section—may be indicated.

LECTURE VI.

TURNING—DEFINITION—THE CONDITIONS WHICH DETERMINE THE NORMAL POSITION OF FŒTUS — CAUSES OF MALPOSITION—FREQUENCY OF CHANGE OF POSITION OF FŒTUS IN UTERO.

Turning.

If we were restricted to one operation in midwifery as our sole resource, I think the choice must fall upon turning. Probably no other operation is capable of extricating patient and practitioner from so many and so various difficulties. In almost every kind of difficult labour with a pelvis whose conjugate diameter exceeds three inches, it would be possible to deliver by turning with a reasonable prospect of safety to the mother, and in many instances with probable safety to the child. We might very greatly restrict craniotomy. We might dispense with the forceps; but neither forceps nor craniotomy will serve as a substitute for turning in its special applications. It is difficult, therefore, to exaggerate the importance of carrying to the utmost limit the perfection of this operation. Yet the text-books exhibit a very inadequate appreciation of the subject. Turning by the feet was once said, not inaptly, to be the master-stroke of the obstetric practitioner. And still the operation was very imperfectly developed.

I propose to describe and illustrate with some fulness the conditions upon which mobility of the fœtus *in utero* depend, the various modes by which the fœtus may be made to change its position, and the applications of this knowledge to the practice of turning, embodying the teaching of Wigand, D'Outrepont, Radford, Simpson, D'Esterlé, Lazzati, Braxton Hicks, myself, and others.

Having regard to the various allied operations which it is convenient to class under a general description, I would define *Turning as including all those proceedings by which the position of the child is changed in order to produce one more favourable to delivery.*

There are three things which it is very desirable to know as much about as possible before proceeding to the study of turning as an obstetric operation—

1. What are the conditions which determine the normal position of the fœtus *in utero?*

2. What are the conditions which produce the frequent changes from the ordinary position?

3. What are the powers of Nature, or rather the methods employed by Nature, in dealing with unfavourable positions of the fœtus?

1.—*The Conditions that determine the Normal Position of the Fœtus in Utero.*

It would be idle to do more than glance at the fanciful ideas upon this subject that have obtained currency at various times, although most have an element of truth in them. Ambroise Paré believed that the head presented owing to the efforts made by the child to escape from the uterus. Even Harvey believed that the fœtus made its way into the world by its own independent exertions. Dubois endeavoured in a long argument to show that the fœtus has *instinctive power*, which determines it to take the head-position. Simpson, rightly concluding that the maintenance of normal position depends very much upon the life of the fœtus, observes that it has no power of motion except muscular motion, and infers that the fœtus adapts itself to the uterus by *reflex muscular movements* excited by impressions — as by contact with the uterus—upon its surface. Thus we come down by a curious scale of theories, in which the philosopher may trace the influence of contemporary physiological doctrines. First, the fœtus is endowed with the high faculty of *volition;* then it falls to the lower faculty of *instinct;* and lastly, it is degraded to the lowest nervous function — that of *reflex* motion. I should be disposed to estimate at a still lower point the influence of the fœtus as an active agent in maintaining its position during pregnancy or labour. It is incontrovertibly true that the normal position of the fœtus and the course of labour are intimately dependent upon the life of the fœtus. But I think I am enabled to affirm from very close observation that a fœtus, if full grown and *only recently dead*—that is, for a few hours—may be nearly as well able to maintain its position and to conduce to a healthy labour as one that is alive. How is this? It depends simply upon the preservation of sufficient tone and resiliency in the spinal column and limbs to maintain the form and posture of the fœtus. Whilst alive, or only recently dead, the spine is firmly supported in a slight curve, the limbs are flexed upon the trunk, the whole fœtus is packed into the shape of an egg, which is very nearly the shape of the cavity of the uterus. It has a long axis, represented by its spine. This long axis, being endowed with sufficient solidity, resembles a rod, rigid or only slightly elastic. It is a lever. Touched at either pole, the force is propagated to the opposite pole. If the head impinge upon one side of the uterus, the breech will be driven into contact with the opposite point of the uterus; head and breech will move simultaneously in opposite directions. In labour, when the uterus is open to admit of the passage of the fœtus, the propelling power applied to the breech is propagated throughout the entire length of the spine or long axis, so that the head, the end furthest from the direct force, is pushed along in the direction of least resistance, turning at those points where it receives the guiding impact of the walls of the canal.

When the fœtus has been some time dead, the elasticity and firmness of its spine are lost; flaccidity succeeds to tonicity. Force applied to one extremity is not propagated to the other extremity—or, at least, it is very imperfectly so; the long axis bends, doubles up like a rod of gutta-percha softened by heat. If, the fœtus *in utero* being in this state, pressure be applied to one side of the head, the head will simply move towards the opposite side of the uterus. And if labour be in progress, the propelling force applied to the breech will not be duly transmitted to the head, but will tend to double up the trunk, to make it settle down in a squash in the lower segment of the uterus or in the pelvis. The head—the cervical spine having lost its resiliency—will not take the rotation and extension turns. It will run into the pelvis like jelly into a mould. Or, at an earlier stage, the limbs, especially the arms, having lost their tonicity, drop or roll in any direction under the influence of gravity or of pressure; and hence may fall into the brim of the pelvis, constituting what are called transverse presentations. The influence of this law is clearly seen in the course of that process called "spontaneous expulsion," by which a dead child is expelled, a shoulder presenting.

Other factors besides the child have to be considered. Scanzoni correctly observes that the frequency of head-presentation is dependent on the operation of various causes. 1. There is the force of gravitation; 2. The form of the uterine cavity; 3. The form of the fœtus (to which must be added the properties I have described due to life or death); 4. The quantity of amniotic fluid; 5. The contractions of the uterus during pregnancy and the first stage of labour. In the early stages of pregnancy the embryo is so small relatively to the cavity containing it that it floats suspended in the liquor amnii. But about the middle of pregnancy the fœtus grows rapidly; it acquires form; and, at the same time, the uterus grows more in its longitudinal than in its transverse diameter. As soon, therefore, as the fœtus—an ovoid body—attains a size that approaches that of the capacity of the uterus, the walls of the uterus will impose upon the fœtus a vertical position. The fœtus has become too long to find room for its long diameter in the transverse diameter of the uterus. Mutual adaptation requires that the long diameters of fœtus and uterus shall coincide.

A condition not, to my knowledge, hitherto noticed, which has a powerful influence upon the determination of the child's position *in utero*, is the normal flattening of the uterus in the antero-posterior direction. In the non-pregnant uterus, the cavity of the body—the true and only gestation-cavity—is a flat triangular space, the angles of which are the orifices of the Fallopian tubes and the os internum uteri. A similar triangular superficies is marked out on each half of the uterus, anterior and posterior. The anterior superficies lies flat against the posterior superficies, touching it as if the two were squeezed together. When pregnancy supervenes, these surfaces are necessarily separated to form a cavity for the growth of the ovum. But the original form is never entirely lost. The cavity is always more contracted from before backwards than from side to side. This is proved by direct observation if the fingers are introduced after abortion, or the hand after labour at term. The uterine cavity is closed by the flattening of the anterior and posterior walls together. This takes place the moment the uterus contracts. If the finger or hand be in the uterus at the time, this is plainly felt. Now, this flattened form of the uterus is the reason why the fœtus takes a position with either its back or belly directed forwards. The fœtus is broader across the shoulders than from back to front, and therefore its transverse diameter is fitted to the transverse diameter of the uterus. There is a physiological design that dictates the downward position of the head. The fundus is the part designed for the implantation of the placenta, where it can grow undisturbed, and continue its function during the expulsion of the child. The lower part of the cavity is therefore left free for the development of the embryo. Why the back is commonly directed forwards to the mother's belly is this:—The child's back is firm and convex; its head is also firm and convex behind. The anterior aspect of the child's body is plastic and concave, and therefore fits itself better to the firm convexity of the mother's spine. It is clear that the two solid convex spines of mother and child would naturally repel each other; and the child being moveable, it is the child's back that recedes, turning forwards.

2.—*The Conditions which produce the frequent Changes in the Child's Position.*

Any considerable disturbance of the correlation of the factors which keep the fœtus in its due position of course favours malposition. The principal disturbing conditions may be stated as follows:—An *excess of liquor amnii* acts in two ways—first, it favours increased mobility of the fœtus; secondly, it tends to destroy the elliptical form of the uterus. The transverse diameter increasing in greater proportion than the longitudinal, the cavity becomes rounder. Hence the fœtus is no longer kept in a vertical position for want of the proper relation between its form and size and those of the uterus.

Obliquity of the uterus was considered by Deventer to be a main cause of malposition. It is now very much discredited, but I am disposed to believe that it has, not seldom, a real influence. Dubois and Pajot showed that in 100 women the uterus in 76 exhibited a marked lateral obliquity to the right, in four to the left, and in twenty an anterior obliquity. Wigand had shown that deviations of the uterus to the right and forwards were far the most frequent. The normal direction of the non-pregnant uterus is nearly that of the axis of the pelvic brim. As the uterus grows during pregnancy, rising above the brim, the projecting sacro-vertebral angle and the curve of the lumbar column deflect its fundus to one or other side; and, if the abdominal walls be very thin and flaccid, the fundus will fall forwards. The tendency of these obliquities, if carried beyond ordinary measure, is to throw the axis of the uterus out of the axis of the pelvic brim, and to bring some other part than the vertex of the fœtus to present. The probability of this will be increased by the irregular contractions of the uterus likely to be excited by parts of the fœtus pressing unequally upon its walls. For example, in extreme lateral obliquity the breech may press strongly upon one side of the fundus; contraction taking place here, will drive the head farther off the brim on to the edge, where, if it finds a *point d'appui*, it will rotate on its transverse axis, producing forehead or face presentation, and favouring the descent of the shoulder. Wigand explains how a too loose and shifting relation of the uterus to the pelvis disposes to cross-birth. In this condition it is observed that the head is now fixed in one place, now in another, and now not felt at all.

He further * says that any obliquity of the uterus

* "Die Geburt des Menschen." Berlin, 1820. Vol. ii. p. 137.

exceeding an angle of 25° is unfavourable; and that even a lesser obliquity, with excess of liquor amnii or a small child, is likely to cause the presenting head to be displaced, and to bring a shoulder into the brim, especially if strong pains or bearing-down efforts be made *early* in labour.

He explained that the os uteri might be brought over the centre of the brim by internal drawing upon the os, combined with external pressure upon the fundus in the opposite direction, thus putting in practice the principle of acting simultaneously upon the two poles of the uterus.

Deformity of the pelvis or lumbar vertebræ is often a powerful factor. The comparative frequency of transverse presentations in cases of deformed pelvis is certainly greater than where the pelvis is well formed. I think, however, that *slight* deformity has more influence in causing malposition than extreme degrees. In these latter, malpositions are rarely observed.

The attachment of the placenta to the lower segment of the uterus is, as Levret has clearly shown, a cause of malposition, by forming a cushion or inclined plane, which tends to throw the fœtal head out of the pelvic axis across the brim. Hence the frequency of cross-birth and of funis-presentation in cases of partial placenta prævia. But there are numerous cases in which the placenta dips into the cervical zone, growing downwards from the posterior and lateral walls of the uterus, without leading to hæmorrhage, and thus not suspected to be cases of placenta prævia, which, nevertheless, form an inclined plane behind or on one side, and produce malposition.

Then there is the *influence of external forces*, as of pressure applied to the uterus through the abdominal walls. The dress of a woman at the end of pregnancy is a matter of no small moment. The pressure of a rigid busk of wood or steel upon the fundus of the uterus, modified by the various movements and postures of the body, may flatten in the fundus, thus reducing the longitudinal diameter of the uterus, or it will push the fundus to one side, causing obliquity. It will, at the same time, press directly upon the breech, and thus tend to give the fœtus an oblique position, throwing the head out of the pelvic axis. Pluriparæ should do the reverse of this. They should wear an abdominal belt, which supports the fundus of the uterus from below upwards.

Want of tone in the uterus, which implies inability to preserve its elliptical form, and a tendency to fall into rotundity, a form which evidently favours malposition. Scanzoni says laxity of uterus is a chief cause. As soon as contraction begins, the uterus tends to resume its ovoid form.

Irregular or partial contractions of the uterus cause malposition. Naegele insisted upon this. He found that in some cases malposition was averted by allaying spasm.

The researches conducted by several German physicians, amongst whom I may cite Credé, Hecker, and Valenta,* establish the fact that the fœtus changes its position with remarkable frequency. Valenta examined 363 multiparæ and 325 primiparæ in the latter months of pregnancy. He found that a change of position took place in 42 per cent. Change was

* "Monatsschr. f. Geburtsk," 1866.

more frequent in multiparæ, and in these in proportion to the number of previous pregnancies. Narrow pelves very frequently cause change of position. Circumvolutions of the cord, so often observed, are produced by changes of position, and hence bear evidence to the correctness of the proposition. It is interesting to observe that the general tendency of changes of position is towards those which are most propitious. Thus, cranial positions are least liable to change. Oblique positions are especially liable to change. These mostly pass into the long axis by spontaneous evolution. *Self-evolution is a very frequent resort of nature.* In some cases several changes of position have been observed in the same patient. The presentations are made out by external manipulations. Valenta thus describes his method of ascertaining a breech-position during pregnancy :—He lays his right hand flat on the fundus uteri, and then strikes the tips of the fingers as suddenly as possible towards the cavity of the uterus, against the part of the child lying at the fundus. By this manœuvre he has always succeeded in recognising the head, if lying at the fundus, by its peculiar hardness and evenness. He detects the head in oblique and cross positions in the same manner. P. Muller* relates a case in which within five days a complete version of the fœtus was effected six times.

Yet the fact of the "spontaneous evolution" of a living child, as described by Denman from actual observation, has been doubted!

* "Monatsschr. f. Geburtsk," 1865.

LECTURE VII.

TURNING (*continued*): THE POWERS OF NATURE IN DEALING WITH UNFAVOURABLE POSITIONS OF THE CHILD—THE TRUTH OF DENMAN'S ACCOUNT OF SPONTANEOUS EVOLUTION—THE MECHANISM OF HEAD-LABOUR THE TYPE OF THAT OF LABOUR WITH SHOULDER-PRESENTATION—THE MODES IN WHICH NATURE DEALS WITH SHOULDER-PRESENTATIONS ANALOGOUS TO THOSE IN WHICH SHE DEALS WITH HEAD-PRESENTATIONS.

We have now to study—

3. *The powers of Nature, or rather the methods employed by Nature, in dealing with unfavourable positions of the fœtus.*

I will do no more at present than glance at those minor deviations from the natural position in which the long axis of the child's body still maintains its coincidence with the axis of the pelvic brim. With some additional difficulty, Nature is in most of these cases able to effect delivery without materially modifying the position. Forehead and face positions have, indeed, already been described in some detail. Difficult breech positions will be specially considered at a later period of the description of turning.

From the time of Hippocrates, who compared the child *in utero* to an olive in a narrow-mouthed bottle, it has been known that the child could hardly be born if its long axis lay across the pelvis. But before the time of Denman it was not clearly explained that a correction of the position, or a restitution of the child's long axis to coincidence with the axis of the pelvic brim, could be brought about by the spontaneous operations of Nature. And observations of this most deeply interesting of natural phenomena are so rare that many men, even at the present day, do not hesitate to deny the accuracy of Denman's description. I would, with all deference, suggest for the consideration of these sceptics, whether they do not carry too far their regard for the maxim, "*Nulla jurare in verba magistri.*" In matters of deduction, of theory, that maxim can hardly be too rigorously applied. But to reject as false or impossible matters of fact, observed and recorded by men of signal ability and conscientiousness like Denman, is to push scepticism to an irrational degree. There are subjects, and this is one, which are not questions of opinion, but of evidence. Shall we reject the testimony of Denman? Whose shall we, then, accept in contradiction? Shall it be the testimony of those who deny that Denman saw what he says he saw, because they themselves have never seen it? This is simply to give the preference to negative over positive evidence, to say nothing of the relative weight or authority of the witnesses. There is no man whose experience is so great that nothing is left for him to learn from the experience of others. Let us first call Denman into the box. He says:

"In some cases . . . . the shoulder is so far advanced into the pelvis, and the action of the uterus is at the same time so strong, that it is impossible to raise or move the child. . . . This impossibility of turning the child had, to the apprehension of writers and practitioners, left the woman without any hope of relief. But in a case of this kind which occurred to me about twenty years ago, I was so fortunate as to observe, though it was not in my power to pass my hand into the uterus . . . that, by the mere effect of the action of the uterus, an evolution took place, and the child was expelled by the breech. . . . The cases in which this has happened are now become so numerous, and supported not only by many examples in my own practice, but established by such unexceptionable authority in the practice of others, that there is no longer any room to doubt of the probability of its happening, more than there is of the most acknowledged fact in midwifery. As to the manner in which this evolution takes place, I presume that, after the long-continued action of the uterus, the body of the child is brought into such a compacted state as to receive the full force of every returning pain. The body, in its doubled state, being too large to pass through the pelvis, and the uterus pressing upon its inferior extremities, which are the only parts capable of being moved, they are gradually forced lower, making room as they are pressed down for the reception of some other part into the cavity of the uterus which they have evacuated, till the body turning, as it were, upon its own axis, the breech of the child is expelled, as in an original presentation of that part. I believe that a child of the common size, living, or but lately dead, in such a state as to possess some degree of resilition, is the best calculated for expulsion in this manner. Premature or very small children have often been expelled in a doubled state, whatever might be the original presentation; but this is a different case from that we are now describing."

Denman cited, in confirmation, the evidence of Dr. Garthshore, Consulting Physician of the British Lying-in Hospital, who related to him a case of the kind, in which the child was living, and the not less trustworthy evidence of Martineau of Norwich. But, before Denman's time, similar cases had been observed, although not understood. Thus, Perfect: "The arm presented; and after endeavours were ineffectually made to get at the feet to turn the child, the patient was thereupon left to herself, and delivered, in a few hours, of a live child, without any assistance whatever."

D'Outrepont* cites Sachtleben, Löffler, Christoph von Siebold, Wilhelm Schmitt, Wiedemann, Vogler, Saccombe, Ficker, Simons, Elias von Siebold, Hagen, Wigand, as all having witnessed self-turning, chiefly, indeed, by the head. He says he himself has frequently witnessed it.

Since Denman's time, evidence has accumulated. Professor Boer, of Vienna, a name in the first rank of the illustrious in Medicine, described, in 1801, a case of arm-presentation, the fingers having been seen at the vulva. He was preparing to turn, when he found the hand higher than when he had examined

* "Abhandlungen und Beiträge." Würzburg. 1822.

before. As the pains continued, Boer rested with his hand in the pelvis. The arm distinctly moved up. At this time the whole cavity of the pelvis was filled with the breech of the child. The body and head of a fresh living child were expelled. Velpeau, a man remarkable for the precision of his observations, is equally decided in corroboration.

What observations can be more positive, exact? Who can give evidence more carefully? Who is more worthy of belief? Upon what grounds is evidence so distinctly impeached? There are two grounds. In the first place, there is the observation of a fact, of a different method of spontaneous or unaided delivery under arm-presentation from that which Denman described. In the second place, there is the assumption that this different method is the only true one.

Now, let us admit the accuracy of the observation, which we may do unreservedly: does it follow that the assumption which excludes the possibility of the occurrence of any other mode of unaided delivery is to be received? Denman, more logical and more philosophical than his opponents, is not so ready to impose limits upon the resources of Nature. He not only observed the "spontaneous evolution" or version of living children, and described this as one resource, but he also observed the "spontaneous expulsion" of dead or premature children by doubling-up, and described this as a second and different resource. Not only, therefore, did Douglas fail to correct or to displace the explanation of Denman, but Denman had actually left nothing for Douglas to discover. In two papers published in the *London Medical Journal* in 1784, Denman relates several cases of spontaneous birth with arm-presentation—some observed by himself, some communicated to him. In these the child was born dead, and the shoulder remained fixed at the pubes. They are clearly described, and certainly anticipate the description given by Dr. Douglas in 1811. But it is fair to add that at this time Denman had not arrived at that sharp distinction which he afterwards drew (1805, see fifth edition "Introduction to Midwifery") between spontaneous version and expulsion.

These are the facts, the evidence. The assumption to which I have referred is further rebutted by abundant collateral testimony. The observation of Denman, so far from being incredible or improbable, is in entire harmony with the phenomena of gestation and labour.

We will now endeavour to trace, with more precision, the modes employed by Nature in dealing with shoulder-presentations. The mechanism of labour with shoulder-presentation is strictly analogous to that of ordinary labour. It is, therefore, desirable to set before our minds the picture of an ordinary head-labour.

In the first head-position, the occiput is directed to the left cotyloid foramen, the face looks to the right sacro-iliac joint, the vertex points downwards to the os uteri, whilst the long axis or trunk corresponds with the long axis of the uterus, which is coincident, or nearly so, with the axis of the pelvic brim. The head, in its progress to birth, undergoes five successive movements:—

1st. *A Movement of Flexion.*—The posterior fontanelle placed opposite the cotyloid cavity descends and approaches the centre of the brim, the chin is strongly pressed upon the chest, the back of the neck comes to bear upon the cotyloid wall, whilst the forehead rises on the right, and the anterior fontanelle is applied to the right sacro-iliac joint. By this movement the head fixes itself upon the trunk, and presents its smaller diameters to the greatest or oblique diameters of the pelvic brim.

2ndly. *A Movement of Descent or of Progression.*—This begins commonly with the escape of the head from the mouth of the uterus, the clearing of the brim, and ends with the total expulsion of the fœtus.

3rdly. *A Movement of Rotation.*—This takes place in the lower part of the pelvic cavity. The forehead and the anterior part of the region of the vertex resting on the right sacro-iliac ligament, or on the right posterior wall of the pelvic cavity, follow the incline backwards and downwards, turning towards the sacral cavity, whilst the back of the neck slides behind the left foramen ovale, or the left anterior wall of the pelvis, and, following the incline forwards and a little upwards, turns towards the upper part of the pubic arch.

4thly. *A Movement in a Circle.*—The back of the neck is arrested under the symphysis pubis; the posterior fontanelle is nearly in the centre of the pelvic outlet; the occiput, the vertex, the forehead, the face, and, lastly, the chin, roll successively over the posterior commissure of the vulva, traversing the concavity of the lower part of the sacrum and the distended perinæum.

5thly. *A Movement of Restitution.*—As soon as the head is freed from the pelvis, the occiput turns quickly to the left, and the face and forehead to the right. This last movement of the head is the effect of the first of a succession of movements similar to those described which is now pursued by the trunk.

The shoulders, entering the brim in the left oblique diameter, turn the head, now freed from all restraint, bringing the face forwards to the right.

The movements undergone by the trunk are three:—

1st. *A Movement of Descent or of Continuous Progression.*—The right shoulder is forward to the right, the left is behind to the left; the child's back is directed forwards to the left.

2nd. *A Movement of Rotation.*—The shoulders and the upper part of the trunk having descended into the excavation, the right shoulder turns towards the apex of the pubic arch, and the left rotates towards the concavity of the sacrum. The child's back, after the rotation, is turned to the left.

3rd. *A Movement in a Circle.*—The right shoulder remaining fixed beneath the pubic arch, the left shoulder, followed by the corresponding side of the trunk and the left hip, describe the arc of a circle; and gradually the right shoulder rises over the mons Veneris, whilst the parts placed behind traverse the sacro-perinæal concavity. These movements are governed by the form of the pelvis. Labour with shoulder-presentation must obey the same laws.

Shoulder-presentations may be *primitive* or *secondary*. The *primitive* exist before labour has set in, and are almost necessarily associated with obliquity of the uterus. The *secondary* are formed during the initiatory stage of labour, under conditions which lead to the deflection of the head from the pelvic brim when it is made to move under the influence of force applied to the breech or trunk.

In Nature we observe two chief shoulder-positions, and each of these has two varieties. In the *first position*, the head lies in the left sacro-iliac hollow. In the *second position* the head lies in the right sacro-iliac hollow. Now, in either position, either the right or the left shoulder may present. Thus, if the head is in the left ilium, the right shoulder will descend when the child's back is directed forwards; and the left shoulder will descend when the child's belly is directed forwards. In the case of the *second* or right cephalo-iliac position, the right shoulder will descend when the child's belly is turned forwards, and the left shoulder when the child's back is turned forwards.*

* Fig. 36.

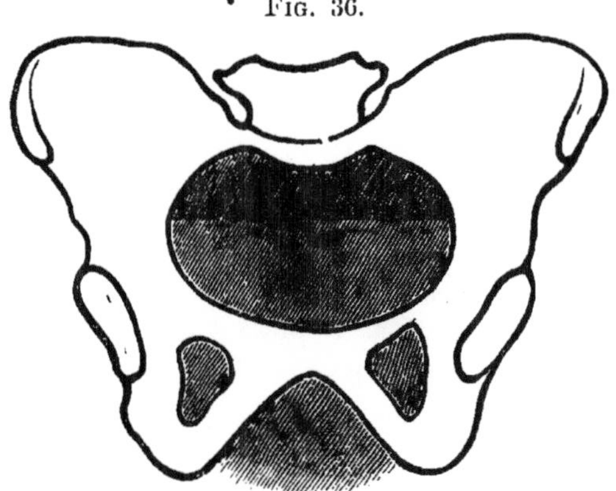

An easy method of realizing a description of some positions of the fœtus is to follow them with a pelvis and a small lay-figure, such as is used by artists. It is not even necessary to have a pelvis. A partial equivalent may be made by tracing a drawing of a pelvis on a piece of cardboard, and cutting out the oval which represents the brim and the space beneath the symphysis pubis. The above figure will serve as a model. Cut out the parts shaded dark. A small lay-figure of corresponding size must be procured.

LECTURE VIII.

DEFINITION OF SPONTANEOUS VERSION AND SPONTANEOUS EVOLUTION—VARIETIES OF SPONTANEOUS VERSION—MECHANISM OF SPONTANEOUS VERSION BY THE BREECH—SPONTANEOUS EVOLUTION—MECHANISM OF, IN THE FIRST SHOULDER-PRESENTATION, DORSO-ANTERIOR.

It is especially necessary, before we proceed, to define with precision the significance that attaches to the terms employed, the more especially that I find it desirable to use some terms in a different sense from that current in this country. Dr. Denman used the term "spontaneous evolution" to express the natural action by which the pelvis or head was substituted for the originally presenting shoulder. The term "spontaneous expulsion" has been applied to the process of unaided delivery described by Douglas, in which the child is driven through the pelvis doubled up. Neither of these terms is free from objection. The first especially is inaccurate, and has given rise to much misapprehension. The process described by Denman is a true *version* or *turning*. All German, French, Italian, and Dutch authors apply to this process the term "spontaneous version"—"*versio spontanea.*" It might be called *natural version*, to distinguish it from artificial version effected by the hand of the obstetrician. All continental authors likewise call Douglas's process by the name, "spontaneous evolution," the process being one of unfolding, as it were, of the doubled-up fœtus. It is of great consequence to bring our nomenclature into harmony with that of our brethren abroad, and it is of still greater consequence to bring our nomenclature into harmony with Nature. It is clear, therefore, that the change in terms should be made by us. I shall use the terms "version" and "evolution" in the correct sense.

There are *two varieties of spontaneous version*—one in which the head is substituted for the shoulder, the other in which the pelvis is substituted for the shoulder. These varieties of spontaneous version correspond with two similar varieties of artificial turning.

There are likewise *two varieties of spontaneous as well as of artificial evolution.* The head or the trunk may be evolved or extracted first.

These processes I will describe successively, beginning with the spontaneous or natural operation, since these are conducted in obedience to mechanical laws which must be respected in the execution of the artificial operations.

In Fig. 37 I have endeavoured to represent the very earliest stage or condition of things in shoulder-presentation. The long axis of the child, and of the uterus, stands obliquely to the plane of the pelvic brim. It is not, indeed, very distant from the perpendicular. It is a very serious error to regard these presentations as entirely cross or transverse. It is only in the advanced stages of labour with shoulder-presentation, when the liquor amnii has been long drained off, when the uterus has been contracting forcibly, driving the shoulder deeply into the pelvis, that the child can truly be said to lie across the pelvis. Diagrams copied from text-book into text-book seem to have fixed this false idea firmly in the obstetric mind. Wigand insists that transverse positions are rare. Esterlé and Lazzati say the same, and maintain that the oblique position is favourable to spontaneous version.* I venture to say that, except in cases of dead, monstrous, or small children, or with loss of form of the uterus through excess of liquor amnii, a true cross-birth, such as is commonly pictured and generally accepted, does not exist at the commencement of labour. It would be better, because certainly true as a fact, and because it does not commit us to any theory, to call these presentations *shoulder-presentations*, and to discard the terms "cross-birth" and "transverse presentation" altogether. In shoulder-presentation, an oblique position of the child *becomes* transverse in the course of labour; but the presentation is not transverse *ab initio*.

The neglect of this fact has been a main cause of the errors that prevail in the doctrine and practice of turning.

In the diagram (Fig. 37), the child and the uterus, E F, stand obliquely, at an angle of about 15° or 20° to

* It has been my habit, when making notes of cases coming under my observation, to record the position of the child by means of sketches. It is from these graphic memoranda that most of the illustrations in these lectures of the phenomena of shoulder-presentation and turning will be taken.

a perpendicular C D drawn upon the plane of the pelvic brim. The head is nearly in a straight line with the spine. It stands half over the brim, and half projecting beyond into the left iliac fossa. That is the *first act.* This act may pass into natural head-labour. Wigand, Jörg, and D'Outrepont say this position is common, and that the effect of the first uterine con-

Fig. 37.

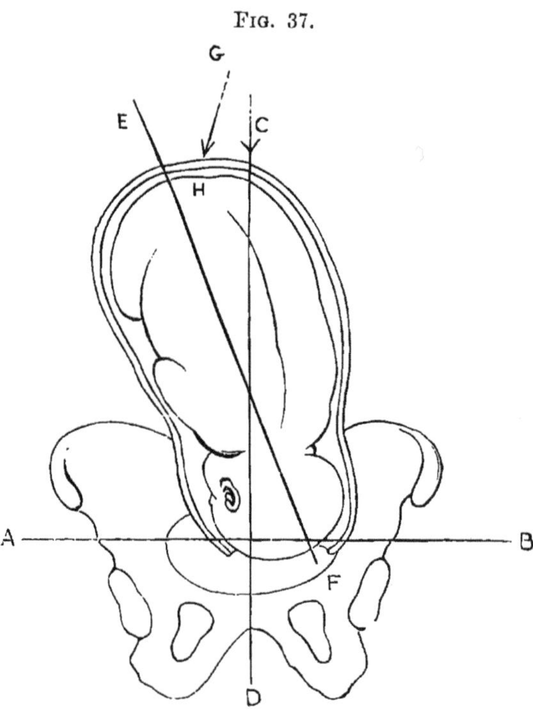

tractions is usually to bring the long axis of the uterus and of the child into due relation with the pelvic brim. This phenomenon is, in fact, a form of self-turning or natural rectification. If this attempt at rectification fail, then we have the transition into shoulder-presentation. The shoulder or arm cannot come down into the pelvis until the *second act*, a movement of flexion of the head upon the trunk, takes place.

This happens in the following manner:—The muscles of the fundus uteri contracting, aided or not by the downward pressure of the abdominal muscles

Fig. 38.

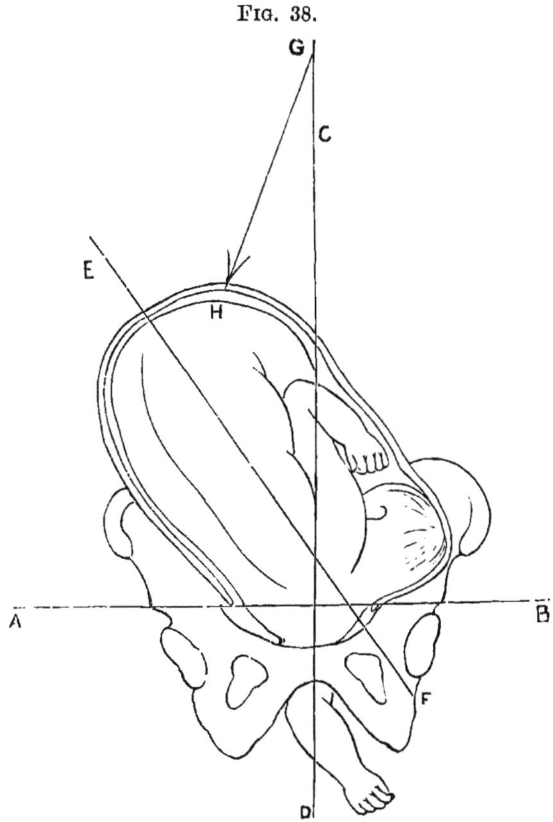

and diaphragm, bring a force acting primarily upon the breech which lies at the fundus. This force will strike with greatest effect upon the left or uppermost side of the breech, at an angle with the long axis of uterus and child. The line G H represents the direction of this force. The result is that the breech descends. And now mark what follows:—If the cavity of the uterus were as broad as long—that is, a flattened sphere or short cylinder like a tambourine—the child's long axis, formed by spine and head, might preserve its rectilinear character; and as the breech descended, the head would simply rise on the opposite side until it came round to the spot abandoned by the breech, performing, in fact, a complete version. But the uterus, we know, is narrower from side to side than from top to bottom. The head will find great difficulty in rising; it therefore bends upon the neck. The shoulder, pertaining to the trunk, is kept at the lowest point in a line with it. The head is thrown more into the iliac fossa, where it rests for a while. Fig. 38 represents this second position of the fœtus. A B is the plane of the brim; C D the perpendicular to the plane, representing the axis of entry to the pelvis; E F is the axis of the child, now a bent line; and G H shows the direction of the downward force, which now strikes the uterus and breech at a greater angle with the perpendicular.

Now, the arm will commonly be driven down, and the hand may appear externally. The observation of the hand will tell the position of the child. The back of the hand looks forwards, the palm backwards, the thumb to the left. All this tells plainly that the head is in the left iliac fossa, and that the child's back is turned forwards to the mother's abdomen. The right scapula will lie close behind the symphysis pubis; the acromion and right side of the neck will rest upon the left edge of the pelvic brim; and the right axilla and right side of the chest will rest upon the right edge of the pelvic brim; whilst the belly and legs of the child, turned towards the mother's spine, will occupy the posterior part of the uterus.

At this stage, even after the liquor amnii has been drained off, spontaneous or natural version may still be effected. The process described as the second act still continuing, the breech is driven lower down; the trunk bends upon its side; the curve thus assumed by the long axis carries on the propelling force in a direction across the pelvic brim; the head tends to rise still higher into the left iliac fossa; the presenting shoulder and prolapsed arm are drawn upwards a little out of the pelvis. The *third act*, one of increased lateral flexion of the child's body, and of movement across the pelvic brim, is represented in Fig. 39.

If spontaneous version is to be completed, the *fourth act* succeeds. The breech being the most moveable part, and the trunk being capable of bending upon itself, partly on its side, partly on its abdomen, is driven lower and lower, the right shoulder being forced well over to the left side of the pelvic brim, and the head being fairly lodged in the upper part of the iliac fossa, the brim is comparatively free for the reception of the trunk. This enters in the following manner:—The right hip comes first into the brim; it is forced lower, and is followed by the breech. As soon as the breech enters the pelvis—that is, as soon as it gets below the sacral promontory—a *movement of rotation* takes place analogous to the rotation which the head takes in head-labour.

There is most room in the sacral hollow, and there the breech will turn. This turn of the trunk brings the body from the transverse position it occupied above the brim to one approaching the antero-posterior; and commonly the head yields somewhat to the altered direction of the spine by coming more forward.*

Fig. 39.

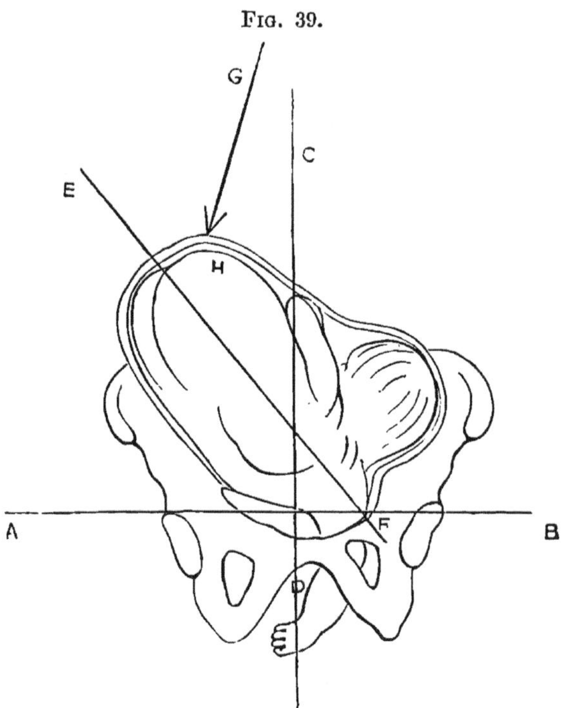

When this rotation-movement is effected, or rather simultaneously with it, a *movement of descent or progress* in an arc of a circle round the pubic centre

* This part of the mechanism of spontaneous version will be illustrated further on.

goes on. The flexion of the spine is now reversed. Above the brim the trunk was concave on its left side, as seen in Figs. 38 and 39. When the breech has dipped into the pelvis, the trunk becomes concave on its right side. The breech descends first. The right ischium presents at the vulva. Then the whole breech sweeps the sacral concavity and perinæum. The trunk follows. The right arm, which has not always completely risen out of the way, comes next; then the left arm; and lastly the head, taking its rotation-movement, and its movement in a circle.

Fig. 40.

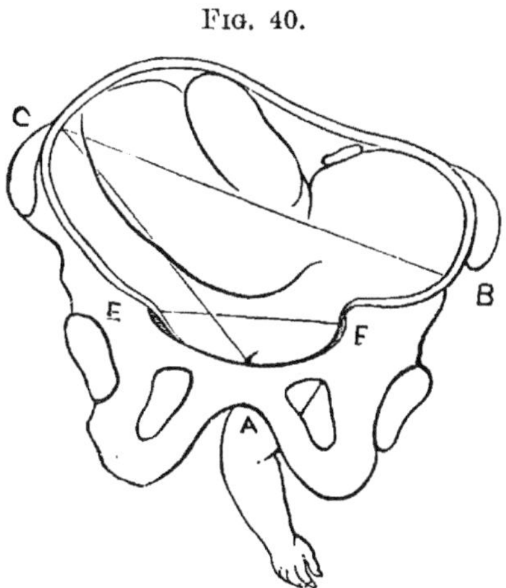

The cause of the difficulty that opposes delivery in shoulder-presentation is obvious. The pelvic canal is too narrow to permit the child to pass freely when its long axis lies across the entry. On looking at the diagram (Fig. 40), we see the shoulder driven into the pelvis, forming the apex A of a triangle whose base B C is considerably longer than E F, the transverse diameter of the pelvic brim. To overcome this difficulty, Nature struggles to shorten the base B C. To a certain extent she generally succeeds, and occasionally she succeeds completely.

The uterus contracts concentrically, tending to shorten all its diameters, especially its transverse diameter. The axis formed by the trunk and head of the child, which go to make up the resisting base of the triangle, is flexible; therefore B and C admit of being brought nearer to each other. But, when the utmost approximation has been obtained in this manner, we still have the entire thickness of the head, equal to four inches, and only very slightly compressible, plus the thickness of the body, which, after all possible gain by compression is effected, is equal to at least two inches more, being an inch or more in excess of the available space in the pelvic brim. As a general rule, it may be stated that no part of the child, except a leg or an arm, can traverse the pelvis along with the head, the head alone being quite large enough to fill the pelvis.

One result of the great compression exerted by the concentric contraction of the uterus is to cause such pressure upon the chest and abdomen of the child, and so to compress the placenta and cord, that the child is asphyxiated and killed. The death of the child, leading to the loss of resiliency, will, after sufficient time, admit of a much further degree of compression, and then, possibly, the child may be so doubled up and moulded that it may enter the pelvis.

The condition, therefore, of spontaneous evolution is the death of the child. If not already dead at the commencement, the child will almost certainly, if of medium size or larger, be killed in the course of the process. Herein lies a great distinction between spontaneous evolution and spontaneous version. A living child is favourable to version, a dead one to evolution.

FIG. 41.

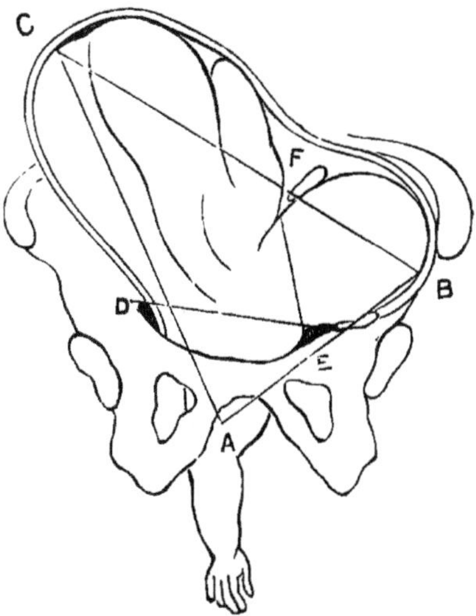

SHOWS THE POSITION OF THE CHILD AFTER THE ESCAPE OF LIQUOR AMNII.

The head is strongly flexed upon the trunk, forming together the base of a wedge too large to enter the brim. The line E F represents the line of decapitation, by which proceeding the base of the opposing wedge is decomposed. The head thus being put aside, the axis of the trunk will easily be brought into coincidence with the axis of the brim, permitting delivery.

Spontaneous evolution from the first position proceeds as follows:—At *first* we have the oblique position of fœtus and uterus represented in Figs. 37 and 38. *Secondly*, strong flexion of the head upon the trunk, and descent of the shoulder into the pelvis (see Figs. 39 and 40); the head is in one iliac fossa, the trunk and breech in the other. At this stage, commonly, the membranes burst, and the arm falls into the vagina, the hand appearing externally. *Thirdly*, increased descent of the shoulder and protrusion of the forearm, doubling with compression of the body, so that the breech is driven into the pelvis; as soon as this takes place a movement of rotation succeeds (see Fig. 42). The inclined planes of the ischia direct the breech backwards into the sacral hollow; this backward movement of the trunk throws the head forwards over the symphysis pubis; from

FIG. 42.

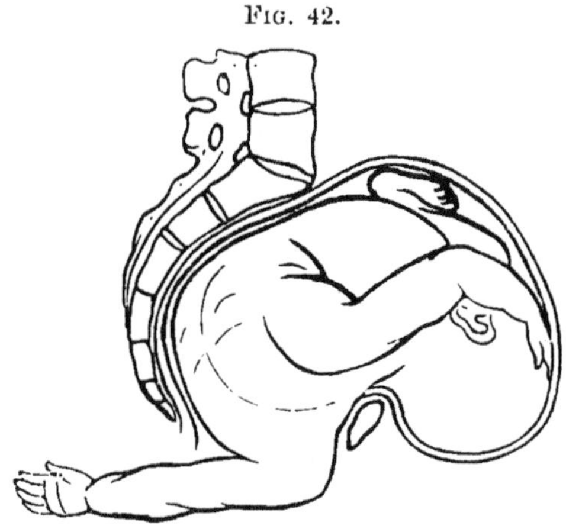

RIGHT SHOULDER; FIRST POSITION AFTER ROTATION.

transverse, as the child was above the brim, it now approaches the fore-and-aft direction; the right side of the head, near its base, is forcibly jammed against the symphysis; the side of the neck corresponding to the presenting shoulder is fixed behind the symphysis pubis, and the shoulder itself is situated under the pubic arch. *Fourthly*, the expulsive force continuing, can only act upon the breech and trunk, the shoulder being absolutely fixed; the trunk bends more and more upon its side, the presenting chest-wall bulges out, and makes its appearance under the pubic arch. Then, *lastly*, the movement in a circle of the body round the fixed shoulder is executed. The side of the trunk and of the breech sweep the perinæum and concavity of the sacrum; the legs follow. When the whole trunk is born, the movement of restitution is effected, the back turning forwards, the belly backwards. The head escapes from its forced position

FIG. 43.

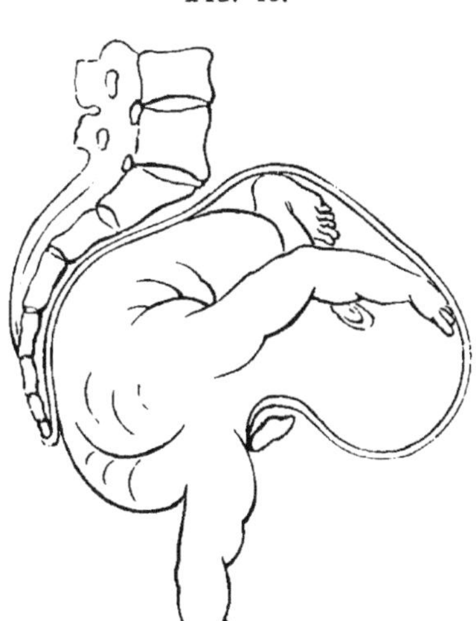

RIGHT SHOULDER, FIRST POSITION; DURING MOVEMENT IN CIRCLE AROUND SYMPHYSIS.

above the symphysis, the chin turns downwards, the occiput looks upwards to the fundus uteri, the nucha is turned to the right foramen ovale. It enters in the left oblique diameter, it takes the rotation movement in the pelvis, the occiput coming under the pubic arch; then the movement in a circle is executed. The chin first appears, followed by the mouth, nose, and forehead, which successively sweep the perinæum. The occiput, which had been applied to the symphysis, comes last. So strict is the subjection throughout this process to the laws which govern the mechanism of ordinary labour, that Lazzati* does not hesitate to describe spontaneous evolution as the natural delivery by the shoulder.

* "Del parto per la Spalla," 1867.

LECTURE IX.

TURNING (*continued*): SPONTANEOUS EVOLUTION—MECHANISM OF, IN FIRST SHOULDER-PRESENTATION, ABDOMINO-ANTERIOR; IN SECOND SHOULDER PRESENTATION, DORSO-ANTERIOR AND ABDOMINO-ANTERIOR —SPONTANEOUS EVOLUTION BY THE HEAD—THE MECHANISM OF SPONTANEOUS VERSION AND OF SPONTANEOUS EVOLUTION FURTHER ILLUSTRATED—THE CONDITIONS REQUISITE FOR SPONTANEOUS VERSION—EXAMPLES OF SPONTANEOUS VERSION BY THE HEAD AND BY THE BREECH.

THE case we have just described is the most common form of spontaneous evolution. It is the type of the rest. Keeping its mechanism well in mind, there will be little difficulty in tracing the course of spontaneous evolution when the child presents in any other position.

If the head lies in the right iliac fossa, constituting the second shoulder-position, as in the first position, the child's back may be directed forwards or backwards. In the first case we have exactly the counterpart of the process described. It would be superfluous to repeat the description, when all is told by simply substituting the words "right" for "left" and "left" for "right." It is, however, useful to trace the course of a labour in which the child's belly is directed forwards. Let us take the second position—head in right iliac fossa. This will involve the presentation of the *right* shoulder. (*See* Fig. 44.)

FIG. 44.

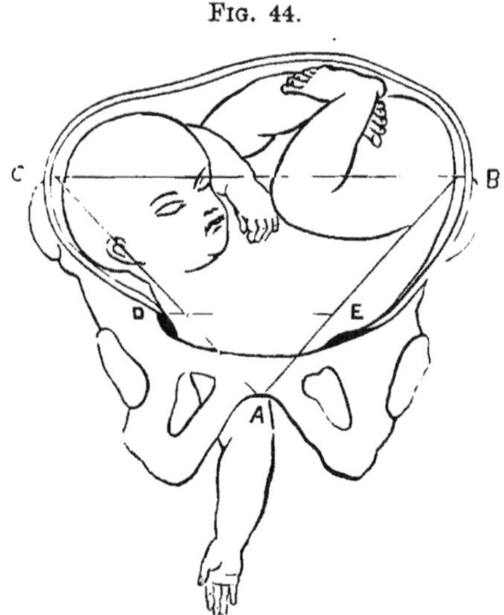

SECOND POSITION OF SECOND SHOULDER-PRESENTATION ABOVE THE BRIM; STAGE OF FLEXION. A, apex of triangle wedged into pelvis; B C, base of triangle opposing entry into D E, brim of pelvis.

A represents the presenting shoulder forming the apex of the triangle, whose base B C is formed by the long axis of the child's body. The expulsive force and the concentric contraction of the uterus draw the head towards the breech, shortening the opposing base by bending the head upon the chest and the trunk upon itself. This is the movement of flexion. This movement continuing, is combined with movement of descent. The right side of the chest is driven more deeply into the pelvis, and is followed by the breech. Then rotation takes place. (*See* Fig. 45.) The head comes forward over the symphysis; the breech rolls into the sacral hollow. The right side of the chest emerges through the

FIG. 45.

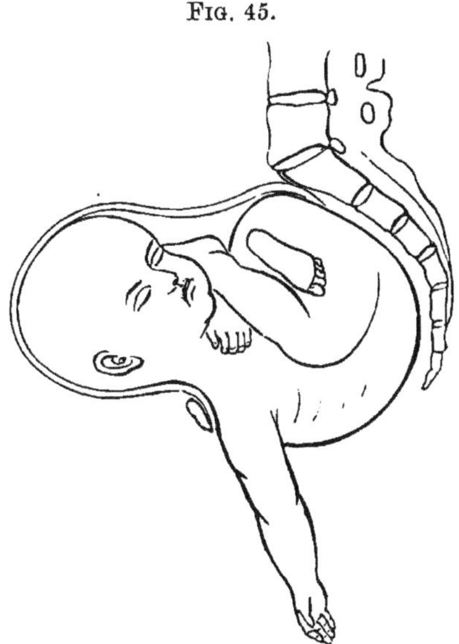

RIGHT SHOULDER—SECOND POSITION AFTER ROTATION.

vulva; the trunk and breech sweep the perinæum; the left arm follows, and lastly the head, the occiput taking up its fixed point at the pubic arch forming the centre of rotation.

The presentation of the left shoulder in the *first position* offers no essential difference in its course from that pursued in the case of right shoulder with dorso-anterior position. The fœtal head is in the left iliac fossa; its sternum is directed forwards;

the thumb of the prolapsed arm is turned to the left, the back of the hand looks backwards, the palm towards the pubes. The lateral flexion of head upon trunk and of trunk upon itself taking place, the left shoulder with the corresponding side of the chest descending into the pelvic cavity, the rotation-move-

FIG. 46.

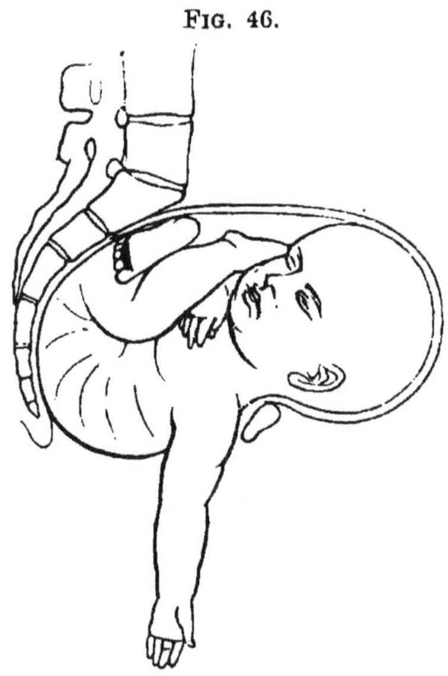

LEFT SHOULDER—FIRST POSITION AFTER ROTATION.

ment takes place, and carries the head over the symphysis pubis. (*See* Fig. 46.) The basilar part of the left temporal region will be applied to the anterior part of the brim; the sternum will turn to the right, the dorsum to the left and backwards. Then the movements of descent and in a circle follow. The side of the chest, trunk, and breech sweep the sacrum and perinæum. The body having escaped, the movement of restitution is performed—

the back will be directed to the left and forwards. The head will be above the brim, with the nucha turned to the left and forwards behind the left foramen ovale, the face looking to the right sacro-iliac joint. Thus it will be born according to the mechanism observed in breech-labour.

Fig. 47.

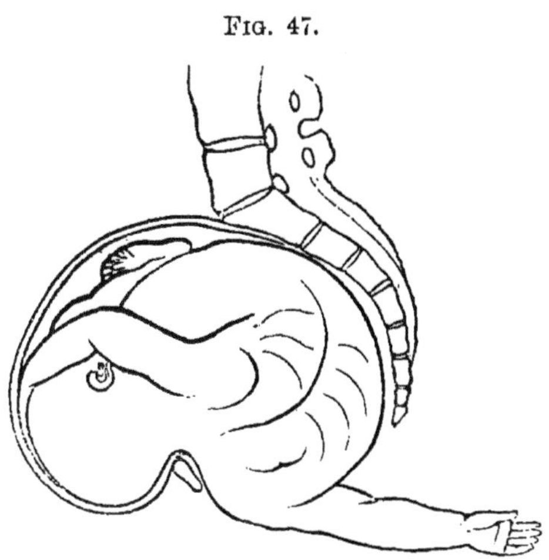

LEFT SHOULDER—SECOND POSITION AFTER ROTATION.

In the case of the dorso-anterior position, with the head in the right ilium, we have, as has been stated, simply the reverse of the dorso-anterior position with the head in the left ilium—the left shoulder becomes wedged in the brim, the left side of the head gets fixed upon the symphysis, the left side of the chest bulges out of the vulva. (*See* Fig. 47.)

Such, in brief, is the description of spontaneous evolution. The process is the normal type of labour in shoulder-presentation. Were it more often justifiable to wait and watch the efforts of Nature, we should probably not seldom enjoy opportunities of observing it; but the well-grounded fear lest Nature should break down disastrously impels us to bear assistance. To be useful in the highest degree, that assistance must be applied in faithful obedience to the plans of Nature. In seeking to help, we must take care not to defeat her objects by crossing the manœuvres by which she attains them. Whenever we lose sight of this duty, whenever we try to overcome a difficulty by arbitrary operations, greater force, running into violence, is required, and the risk of failure and of danger is increased.

It has been already said that spontaneous evolution may be effected by the head traversing the pelvis first. The case is indeed rare, but the process and the conditions under which it occurs deserve attention. The essential idea of spontaneous evolution is that the presenting shoulder remain fixed, or, at least, shall not rise up out of the pelvis into the uterus. Therefore, if the head comes down, it must do so along with the prolapsed arm. This simultaneous passage of the head, arm, and chest can hardly take place unless the child is small. If the child is very small, the difficulty is not great. If the child be moderately large, it will be far more likely to be born according to the mechanism already described and figured, in which the side of the chest corresponding to the presenting shoulder emerges first, and the head last. But some cases of head-first deliveries have been observed. Pezerat* relates a case that seems free from ambiguity. The child

* "Journ. Complémentaire," tome xxix.

was large, the shoulder presenting. Pezerat tried to push it up, but could not. A violent pain drove the head down. Fichet de Flichy * gives two cases. In both the midwife had pulled upon the arm. Balocchi relates a case.† It was an eight-months' child. He says the case is unique rather than rare, but still regards it as a natural mode of delivery in shoulder-presentation. Lazzati thinks the descent of the head in these cases is always the result of traction upon the presenting arm. As the expelling power is exerted mainly upon the breech, tending to drive the head away from the brim, it is indeed not easy to understand how spontaneous action can restore the head, if the shoulder is forced low down in the pelvis. Monteggia ‡ held the same opinion. He relates two cases, in both of which tractions had been made. I myself have seen an instance of the kind.

Fielding Ould (1742) relates the following: He was called to assist a midwife, who had been pulling at the child's arm, which came along with the head, yet she could not bring it forth. The head was so far advanced that it could not be put back, in order to come at the feet. However, after an hour of excessive toil, he brought forth a living child, with a depression of the parietal and temporal bones proportional to the thickness of the arm. Next morning the bones had recovered. Child and mother did well.

It is so important, as a guide to the artificial

* "Observ. Méd. Chir."

† "Manuale Completo di Ostetricia." Milano, 1859.

‡ "Traduzione de l' Arte Ostetricia di Stein," 1796.

means of extricating a patient from the dangers of shoulder-presentation, to possess accurate ideas of the mechanism of spontaneous version and evolution, that I am led to present a further illustration of these processes.

To make the mechanism of spontaneous version clearer, let us represent the child's body by a rod, flexible and elastic, as the spine really is. In Diagram

FIG. 48.

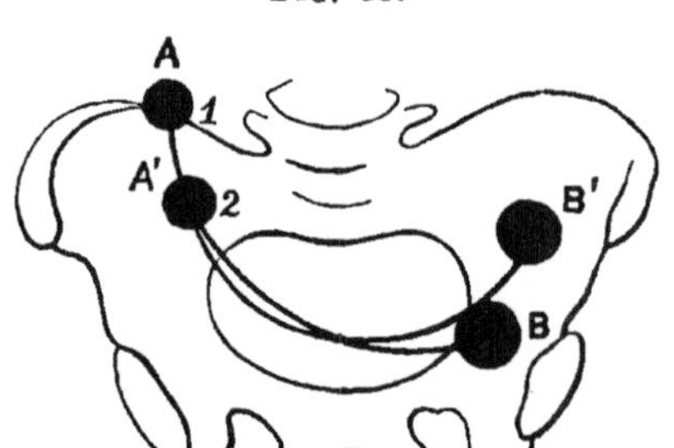

48, A B 1 is the rod fixed at B by a sort of crutch, formed by the head and neck against an edge of the pelvis. A, the breech, being moveable, receives the impulse of the force, and is driven downwards. This rod, or spine, therefore bends. But the rod, being elastic, constantly tends to straighten itself. This effort will, if the head is not immovably fixed, lift the head off the edge of the pelvis, and carry it higher into the iliac fossa. The force continuing to press upon A, as in 2, will drive it still lower, and the rod still bending, and tending to recover its straightness, the head will rise further from the edge of the pelvis. At last (*see* Fig. 49) there will be room for the end A to enter the pelvis, and the rod springing into straightness by the escape of A from the pelvis, the whole may emerge, B coming last. For this process to take place, it is obvious that the rod must be endowed with elasticity or spring; and, therefore, as Denman said, a live child is best adapted to undergo spontaneous version.

Fig. 49.

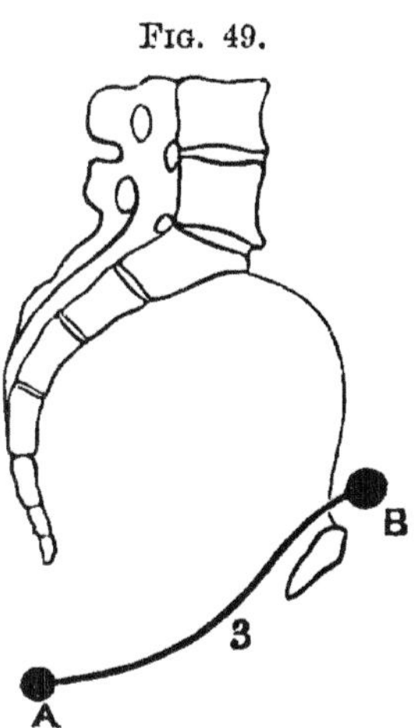

The mechanism of spontaneous evolution may also be illustrated in like manner. Let us represent the child's body by a rod, flexible, but almost without elasticity. In Diagram 50, one end of the rod, B, is fixed against the edge of the pelvis; the other end, A, being moveable, receives the impulse of the downward force, and is driven first to 2; the rod continuing to bend, A falls to 3, and, as B is fixed, the rod forms a strong curve, with its convexity downwards, in the cavity of the pelvis. This convexity will be the first part of the rod to emerge.

The force urging on the end A, more and more of the convex rod will emerge, until A itself escapes. Then, and not till then, can the rod recover its straightness, and the end B will follow. (*See* Diagram 51).

In the case of spontaneous version, as well as in that of spontaneous evolution, it is necessary to exhibit first a pelvis seen from the front, then a section as seen from the side; because in the earlier stages the movement is across the pelvis, and in the

FIG. 50.

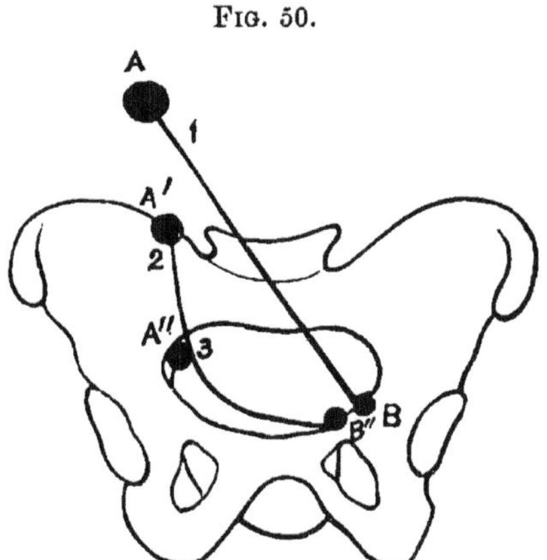

later stages the head comes forward above the symphysis, and the movement in a circle around this centre is from behind forwards.

Now, we may ask, What are the conditions required for the execution of spontaneous version, or natural turning? Some of them, probably, are not understood. Certain it is that we are hardly yet in a position to predicate in any given case of shoulder-presentation, seen at an early stage, that spontaneous version will take place, as we might be, if all those conditions were known and recognisable. They would be more familiar if the law to turn were not laid down in such imperative terms—if the dread of evil as the consequence of neglect of that law were not so overwhelming. But if Nature be always superseded, if the physician always resort to artificial turning as soon as he detects a shoulder presenting, how can we obtain sufficient opportunities for discovering the resources of Nature, and how she acts

Fig. 51.

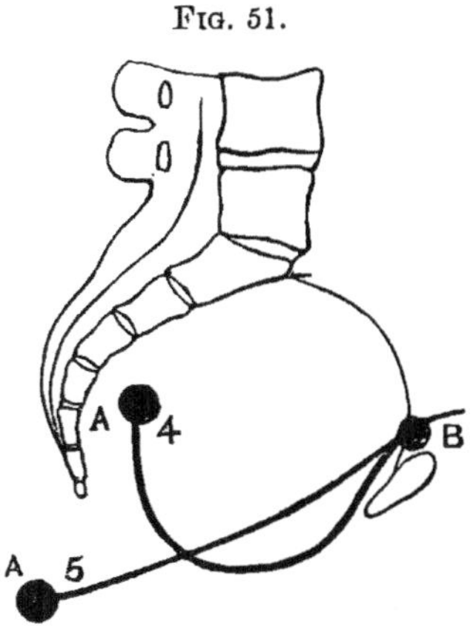

in turning them to account? The principal conditions seem, however, to be these: 1. *A live child*, or one so recently dead that the tone or resiliency of its spine is still perfect. 2. A certain degree of *mobility of the child* in utero. 3. Strong action of the uterus and auxiliary muscles. A roomy pelvis does not appear to be always necessary.

Spontaneous version is not likely to take place when the shoulder has been driven down in a point with a part of the chest-wall low in the pelvis, and the uterus is strongly grasping the fœtus in every part, bending its long axis by approximating the head and breech. It is not likely to take place when the head has advanced towards a position above the symphysis pubis, that is, when the movement of rotation has commenced.

But the practical question will arise, Is spontaneous version ever so likely to occur, that we shall be justified in trusting to Nature? Ample experience justifies an answer in the affirmative. But it appears to me that the great lesson taught by the observation of the phenomena of spontaneous version is this: If Nature can by her unaided powers accomplish this most desirable end, we may by careful study and appropriate manipulation assist her in the task. We shall be the better ministers to Nature in her difficulties as we are the better and humbler interpreters of her ways. *Natura enim non nisi parendo vincitur.*

It has been already stated that spontaneous version may take place either by the head or by the pelvis. It may be interesting to cite further examples of either process occurring under the observation of competent practitioners. I will first give an example of spontaneous turning by the head.

Velpeau* relates the following case of cephalic version. A woman was in labour at the Ecole de Médecine (1825). The os was little dilated. The left shoulder was recognised. The waters escaped five hours after this examination. Four students recognised the shoulder. The pains were neither strong nor frequent; and "*being not without confidence*

* "Traité complet de l'Art des Accouchements," 1835.

in Denman," Velpeau did not search for the feet. In five hours later the shoulder was sensibly thrown to the left iliac fossa. The pains increased, and the head occupied the pelvic brim. The vertex came down, and the labour ended naturally.

Dr. E. Copeman, of Norwich, records the following case:* Some time after the waters had escaped in great quantity, the child was found lying across the pelvis, with the back presenting; neither shoulders nor hips could be felt. At a later period, preparing to turn, Dr. C. was surprised to find the pelvis filled. He endeavoured to pass his hand over the right side of the child towards the pubes, but in so doing he felt the child recede, and therefore confined himself to raising the child's pelvis with his flat hand and fingers; whilst the pains forced down the occiput, the head descended, and delivery was quickly completed. Dr. C. thinks, if he had waited a little longer, spontaneous evolution would have occurred, and the child would have been born even without manual interference. The child was a full-grown male, lively and vigorous.

Here is a remarkably clear case of spontaneous version by the pelvis. Dr. H. Scholefield Johnson, of Congleton, communicated to Dr. Murphy † the following history:—Attending a patient in her first labour, he diagnosed a head-presentation in the third position. At this time the os uteri was somewhat larger than a crown-piece, and the membranes were unbroken. No further examination was made until the liquor amnii had escaped, when the os uteri was

* J. G. Crosse's "Cases in Midwifery," 1851.

† "Dublin Quarterly Journal of Medical Science," 1863.

found three parts dilated and the breech presenting. The funis also descended. The child was nearly still-born, but was restored; it had a swelling on the upper part of the left parietal bone extending towards the occipital, thus confirming the first diagnosis of head-presentation. In reply to questions addressed to Dr. Johnson, he informs me that the child was of full size and healthy; that the liquor amnii was not remarkably above the usual quantity, and that he does not think any external pressure was concerned in the production of the version. Here is another case, equally instructive, communicated to me by the same gentleman:—"In December last (1862), I had a labour. I found a compound presentation. I felt both feet with the backs of the legs towards the left sacro-iliac synchondrosis, the right hand with the palm lying in front of the right ankle, a small loop of funis anterior to the wrist. Above the os I could feel the head, but could not make out the part. I waited for the membranes to burst. The feet descended lower, and the head passed out of my reach. I brought down the feet and delivered. When the child was born, I carefully examined it. The right hand was slightly swelled, and there was a distinct swelling (gone the next day) on the right side of the head and up the right margin of the anterior fontanelle. Now, I do not doubt, if I had come to this case later, I should have only found a footling case."

LECTURE X.

APPLICATIONS OF THE KNOWLEDGE OF THE MECHANISM OF SPONTANEOUS VERSION AND SPONTANEOUS EVOLUTION TO THE PRACTICE OF ARTIFICIAL VERSION AND ARTIFICIAL EVOLUTION—THE BI-POLAR METHOD OF TURNING, HISTORY OF—ARTIFICIAL VERSION BY THE HEAD—REASONS WHY VERSION BY THE BREECH IS COMMONLY PREFERRED—ILLUSTRATIONS OF HEAD-TURNING, OR CORRECTION OF THE PRESENTATION, BEFORE AND DURING LABOUR IN OBLIQUITY OF THE UTERUS AND FŒTUS—SHOULDER PRESENTATION—FOREHEAD AND FACE PRESENTATIONS—DESCENT OF HAND OR UMBILICAL CORD BY THE SIDE OF THE HEAD.

From the observation of the spontaneous or accidental changes of position of the fœtus *in utero*, the transition is natural to the account of those changes which can be effected by art. The observations already referred to prove that the fœtus *in utero* may, under certain conditions, change its position with remarkable facility. It follows that the judicious application of very moderate forces may, under favourable circumstances, effect similar changes.

We have seen that spontaneous version may be effected by the substitution of the head for the shoulder, and of the pelvic extremity for the shoulder; also that spontaneous evolution may be effected by the descent of the head with the presenting shoulder and arm, or by the descent of the chest and trunk with the presenting shoulder and arm. Now, each of these natural or spontaneous operations for liberating the child may be successfully imitated by art. Let us study the conditions which guide us in the selection of the natural operation we should imitate, and the methods of carrying out our imitations.

A successful imitation of natural version by the head or by the inferior extremity demands the concerted use of both hands. You must act simultaneously upon both poles of the fœtal ovoid. This combined action may be exerted altogether externally—*i.e.*, through the walls of the abdomen—or one hand may work externally, whilst the other works internally through the os uteri. The first method—that practised by Wigand, D'Outrepont, D'Esterlé, and others — has been called the bi-manual proper. The second, which has been most clearly taught by Dr. Braxton Hicks, has been called by him combined internal and external version. But the same principle governs both. As I have already said, you must act at the same time upon both poles of the long axis of the fœtus. It would be more correct to describe them both as forms of *the bi-polar method of turning*. It is an accident, not a fundamental difference, if, in one case, it is more convenient to employ the two hands outside; and in another, to employ one hand outside, and the other inside. Each form has its own field of application. We should be greatly crippled, deprived of most useful power, if we were restricted to either form. At the same time, I am of opinion that the combined internal and external bi-polar method has the more extensive applications to practice.

I have found the bi-polar method serviceable, adjuvant in every kind of labour in which it is necessary to change the position of the child. It is true that a rather free mobility of the fœtus *in utero* is most favourable to success; it is true that the external bi-polar method can hardly avail unless at least a moderate quantity of liquor amnii be still present; it is true that the internal and external bi-polar method requires, in its special uses, if not the presence of liquor amnii, at any rate a uterus not yet closely contracted upon the fœtus. But I am in a position to state that amongst nearly 200 cases of turning of which I have notes, there was scarcely one in which I did not turn the bi-polar principle to more or less advantage; and in not a few cases of extreme difficulty from spasmodic concentric contraction of the uterus upon the fœtus, with jamming of the shoulder into the pelvis, where other practitioners had been foiled, I have, by the judicious application of this principle, turned and delivered safely.

The history of the bi-polar method of version, the steps by which this the greatest improvement in the operation has been brought to its actual perfection, deserve to be carefully recorded.

From what has been already said it is clear that Wigand, D'Outrepont, and others who took up Wigand's views, had acquired an accurate perception of the theory of bi-polar turning, and had, moreover, successfully applied that theory in practice. They had applied it to the purpose of altering the position of the child before labour, chiefly by bringing the head over the centre of the pelvis, restoring at the same time the uterus and fœtus from an oblique to a right inclination. This they did generally by external manipulation; but not exclusively, for sometimes one or two fingers introduced into the os uteri served to drag the lower segment or pole of the uterus to a central position, whilst the hand outside acted in the opposite direction upon the upper pole. Here the application seems to have stopped short. At least, I am not aware of any distinct description of the application of the bi-polar principle to produce version.

In one form, indeed, the bi-polar principle of turning by the feet has been in use for a long time. It not uncommonly happens, when turning is attempted after the waters have escaped, and when the uterus has contracted rather closely upon the child, that, even when one or both legs have been seized and brought down, the head will not recede or rise from the pelvis—that is, version does not follow. It then becomes obvious that by some means you must push up the head out of the way. The operation by which this is effected—an exceedingly important one—will be fully explained hereafter. It is enough to say in the present place that it consists in holding down the leg that has been seized, whilst a hand or a crutch introduced into the pelvis pushes up the head and chest. In this operation it will be observed that both hands work below the pubes, whilst in the true bi-polar method one hand works below and inside, and the other above and outside.

In several obstetric works (Moreau, Caseaux, Churchill, &c.), diagrams illustrating the operation of turning are given, representing one hand applied to the fundus uteri outside, and the other seizing the feet inside. But it would be an error to infer that these indicate an appreciation of the principle of bi-polar turning. They simply indicate the principle of *supporting the uterus*, so as to prevent the risk of laceration of the cervix whilst pushing the hand through the uterus and up towards the fundus. The true bi-polar method does not involve passing the hand through the cervix at all.

The following passage from the late Dr. Edward Rigby ("Library of Medicine," Midwifery, 1844), may be taken as a description of the diagrams referred to:—"In passing the os uteri . . . we must at the same time fix the uterus itself with the other hand, and rather press the fundus downwards against the hand which is now advancing through the os uteri. In every case of turning we should bear in mind the necessity of duly supporting the uterus with the other hand, for we thus not only enable the hand to pass the os uteri with greater ease, but we prevent in great measure the liability there must be to laceration of the vagina from the uterus in all cases where the turning is at all difficult."

The same precept is even more earnestly enforced by Professor Simpson*:—"Use both your hands," he says, "for the operation of turning. In making this observation, I mean that whilst we have one hand *internally* in the uterus, we derive the greatest possible aid in most cases from manipulating the uterus and infant with the other hand placed externally on the surface of the abdomen. Each hand assists the other

* "Lond. and Edin. Monthly Journal of Med. Science." Feb., 1845.

to a degree which it would not be easy to appreciate except you yourselves were actually performing the operation. It would be extremely difficult, if not impossible, in some cases to effect the operation with the single introduced hand; and in all cases it greatly facilitates the operation. The external hand fixes the uterus and fœtus during the introduction of the internal one; it holds the fœtus *in situ* while we attempt to seize the necessary limbs, *or it assists in moving those parts where required towards the introduced hand*, and it often aids us vastly in promoting the version after we have seized the part which we search for. Indeed, this power of assisting one hand with the other in different steps of the operation of turning forms the principal reason for introducing the left as the operating hand."

Here the consentaneous use of the two hands is well described. But the bi-polar principle is at best but dimly foreshadowed.

Dr. Robert Lee, in his "Clinical Midwifery," relates several cases in which he succeeded in converting a head or shoulder presentation into a pelvic one by introducing one or two fingers only through the os uteri, when, indeed, this part was so little expanded that to introduce the *hand* would have been impossible. These cases were mostly cases of placenta prævia, the fœtus being premature and small. He managed this by gradually pushing the presenting part towards one side of the pelvis until the feet came over the os uteri. Then he seized the feet and delivered. But there is no mention of the simultaneous or concerted use of the other hand outside, so as to aid the version by pressing the lower extremity of the child over the os, or to carry it within reach of the hand inside. It is a manœuvre of limited application. It differs in principle from the bi-polar method, which requires the consentaneous use of both hands, and which enjoys a far wider application.

A process of synthetical reasoning, especially if informed by the light of experience in practice, might construct out of the elements thus contributed by Wigand and his followers, by Rigby, Simpson, and Robert Lee, a complete theory and practice of bi-polar turning in all its applications to podalic as well as to cephalic version. I am conscious myself of having in this manner evolved that theory, and applied it in practice. Dr. Rigby's was the work I had adopted as my guide from the commencement of my career; and my attention was especially directed by Dr. Tyler Smith to the admirable lecture of Professor Simpson, from which the passage above quoted is drawn, at the time of its appearance. Since then I must have turned at least two hundred times. In no case have I failed to observe the precept of using both hands; and gradually I found out that the external hand often did more than the internal one—so much so, indeed, that the introduction of one or two fingers through the os uteri to seize the knee pressed down upon the os by the outside hand was all that was necessary. I feel that I am entitled to say this much; and not a few of my professional brethren who have honoured me by seeking my assistance can bear witness to the fact that it was by the application of the bi-polar method that I have been enabled to complete deliveries where others had failed.

But, in saying this, I should be sorry indeed if it

were interpreted as a desire on my part to detract in any degree from the merit of my colleague, Dr. Braxton Hicks. His claim to originality in working out and expounding the application of the external and internal bi-polar method of podalic version is indisputable. I know of few recent contributions to the practice of obstetrics that possess greater interest or value than his memoirs on "Combined External and Internal Version," published in the *Lancet* in 1860, in the *Obstetrical Transactions*, 1863, and in a special work in 1864.

If the proposition which I have already urged with reference to the forceps be true—namely, that the carrying to the greatest possible perfection of an instrument that saves both mother and child is an object of the highest interest—it is scarcely less true of turning, also a saving operation. I cherish a fervent hope that the exposition of the principles and methods of turning which will be made in the following lectures will, in conjunction with those on the forceps, be the means of materially enlarging the field of application of the two great saving operations, and, as a necessary result, of supplanting, in a corresponding degree, the resort to the revolting operation of craniotomy.

As head-presentation is the type of natural labour, it follows that to obtain a head-presentation is the great end to be contemplated by art. It seems enough to state this proposition to command immediate assent. But in practice it is all but universally contemned. No one will dispute that the chance of a child's life is far better if birth takes place by the head than if by the breech or feet. Yet delivery by the feet is almost invariably practised when turning, or the substitution of a favourable for an unfavourable presentation, has to be accomplished. Why is this?

The answer is not entirely satisfactory. It rests chiefly upon the undoubted fact that in the great majority of instances, at the time when a malpresentation comes before us, demanding skilled assistance, turning by the feet is the only mode of turning which is practicable. Frequent experience of one order of events is apt so to fill the mind as to exclude the reception of events that are observed but rarely. Many truths in Medicine escape recognition because the mind is preoccupied by dogmas and narrowed by an arbitrary and enslaving empiricism. Many things are not observed because they are not sought for with an intelligent and instructed eye. And then, reasoning in a vicious circle, some men will boldly deny the existence of that which their untrained faculties cannot perceive. They go further: by doggedly and consistently following a practice which arrests Nature in her course, substituting a violent proceeding of their own, they never give Nature a chance of vindicating her own powers, and they consequently never give themselves a chance of learning what those powers are, or of realizing the imperfection of their own knowledge. They close the shutters at noon-day, and say the sun does not shine.

In the seventeenth and in the beginning of the eighteenth centuries, Velpeau remarks, cephalic turning was hardly ever mentioned unless to be condemned. But if the practice of podalic turning was then so general, it was justified because the forceps was not known. In many cases it is not enough to correct the position—it is also necessary to extract. Without the forceps our predecessors could only extract by the legs. But now, if the head is brought to the brim, the forceps affords a ready means of extraction.

Flamant appears to have been amongst the first to revive the practice of turning by the head; he did it by external manipulation. Osiander and Wigand (1807) investigated the subject with remarkable sagacity and skill. D'Outrepont pursued it; and many other names might be cited. The researches of Wigand, however, contain the germ of all the subsequent inquiries.

Flamant strenuously contended that head-turning was best. In two cases of arm-presentation, he raised the breech towards the fundus uteri. The head thus made to descend was seized by the hand. The liquor amnii had long escaped. He worked in these cases entirely by *internal* manipulation. Wigand accomplished the same object by *external* manipulation, saving the children. D'Outrepont had a case of a woman who had lost three children by foot-turning. In her fourth labour she had a shoulder-presentation. There was slight conjugate contraction. The head lay to the right, the feet to the left; the back of the chest was above the brim. He seized the child by the back, placed his right thumb and the right side of four fingers on its left side; then he pushed it to the left and upwards; then he released the back, and seized the neck, whilst he pressed upon the shoulder with his thumb, and the palm and four fingers on the back. The head came over the brim, and the child was safely delivered. In a second case, the breast was on the brim, the head to the left; he pushed up the chest and brought down the head, which entered by the face, and was so delivered. Strong pains prevented his reducing the face to a cranial position. In a third case he was equally successful. D'Outrepont afterwards practised with success Wigand's method of head-turning by external manipulation.

Here is a case of bi-manual and bi-polar head-turning by D'Outrepont. The head lay in the right side. He placed the patient on her left side raised. During each pain he imparted gentle pressure on the side where the head lay, directing it towards the brim; and *at the same time* he pressed with his other hand in the opposite direction upon the fundus where the breech lay. In the intervals of pain, he planted a pillow in the side where the head lay. The head was brought into the pelvis, and a large living child was born.

Professor E. Martin * has carefully described the operation, and practised it with great success.

Hohl † says turning by the head is much less esteemed than it ought to be, and it would be more esteemed if more pains were taken to instruct pupils how to do it on the phantom.

Head-turning, or simple rectification of the presentation, may be indicated under the following circumstances:—

A. *Before the Accession of Labour.*—When the uterus and fœtus are placed obliquely in relation to the pelvic brim; and in some cases where the shoulder is actually presenting.

* "Froriep's Notizen," 1850.

† "Lehrbuch der Geburtshülfe," 1862.

B. *When Labour has Begun.*—1. When the uterus and fœtus are placed obliquely in relation to the pelvic brim, which obliquity may be preparatory to the complete substitution of the shoulder for the head.

2. In some cases of shoulder-presentation, the membranes being still intact.

3. In some cases of shoulder-presentation, the membranes having burst, but considerable mobility of the child being still preserved.

4. The forehead or face presenting.

5. Descent of the hand by the side of the head.

6. Prolapse of the umbilical cord by the side of the head.

A. *Head-turning, or Rectification before Labour.*—This has been often practised by Wigand, D'Outrepont, and others. I will describe the operation after Esterlé. It was the observation of the frequent occurrence of spontaneous version in the eighth and ninth months of gestation that led this eminent obstetrician to practise external bi-polar version.* He observed that a large number of shoulder-presentations in the last two months, if left to themselves, were converted into natural presentations, either on the approach of labour or after the beginning of labour. He had further remarked that spontaneous version had occurred after the escape of the liquor amnii, and the shoulder was sensibly down. The most efficient cause of spontaneous version, he says, is the combined action of the movements of the fœtus and of its gravity, the centre of gravity not being far from the head. The extension of the feet must drive the breech away from the uterine wall as the feet strike it, and so the head is

* "Sul Rivolgimonto Estorno." "Annali Universali di Medicina," 1859.

brought nearer to the brim. His method was as follows :—

The patient must be placed in such a posture as to produce the greatest possible muscular relaxation. Bearing in mind the conditions which take part in spontaneous version, it is necessary to imitate them as much as possible. Amongst these is the lateral and partial contraction of the uterus, which diminishes the transverse diameter, and which exerts a convenient pressure upon the ovoid extremities of the fœtus; and the movements of the fœtus, the re-percussion of its head, and its descent when the centre of gravity of the fœtal body favours its fall. To imitate this, the lateral contractions must be replaced by lateral pressure. This is applied towards the fundus or the neck, according to the situation of the part which it is sought to raise or to depress. This pressure is assisted greatly by gentle strokes or succussions made by the palm of the hand alternately towards either ovoid extremity. These strokes are then made, in rapid succession, simultaneously upon the two extremities, one giving a movement of ascent, the other a movement of descent; or we may act upon the head alone, whilst the other hand makes a steady pressure on the contrary side, the more to diminish the transverse diameter. The desired position being effected, it is necessary to maintain it. This may be done by the adaptation of a suitable bandage.

Lazzati operates in a similar manner. He maintains the uterus and fœtus in due position by the adaptation of cushions or pads to the sides of the opposite poles of the fœtal ovoid.

B. 1, 2, and 3.—Head-turning or correction of the presentation may be attempted in cases of moderate obliquity, where the liquor amnii is still present or has only recently escaped. It is also necessary that the action of the uterus be moderate. Correction, as we have seen, consists in restoring the head, which has passed across the brim of the pelvis into the ilium, back to its due relation to the brim. This operation involves the rectification of the uterus, as well as of the child. It may in certain cases be effected entirely by external manipulations. Supposing the case be one in which the head is deviated to the left ilium, and the fundus, with the breech, are directed to the right of the mother's spine, the first step is to place the patient in a favourable position. Now, by laying her on her left side, the fundus of the uterus, loaded with the breech, and being moveable, will tend to fall towards the depending side. This will act as a lever upon the uterine ovoid, and raise the lower or head end of the uterus, so as to facilitate its return to the brim. In such cases Wigand recommends that the posture should be repeatedly changed, so as to ascertain which is the best to maintain the head in the central line of the pelvis. When this is found, the sooner the membranes are ruptured the better. The patient must thenceforward be kept carefully in the same posture, the uterus being supported in due relation by the hands externally. But I believe that in many cases the dorsal position will lend the greatest facility.

We must apply pressure to the uterus towards the median line of the mother, both at its fundus and at the lower part which contains the head. The head will thus be pushed by one hand to the right, whilst the fundus uteri is pushed by the other hand to the left. When the head has been thus brought over the brim, the difficulty is to secure it there. If the correcting pressure be removed, the uterus tends to resume its obliquity.

If labour has begun, we may combine internal with external manipulation. We may press upon the fundus with one hand, whilst with a finger in the os uteri we pull this over the centre of the brim (Wigand). External pressure by a cushion or pillow laid in the hollow of the ilium in which the head lay will aid this manœuvre. Then, having got the head into proper position, and whilst it is kept so by aid of an assistant, rupture the membranes. *The contraction of the uterus tends to restore its natural ovoid shape.* And this will tend to keep the child's long axis in relation. If by this contraction the head should happily become fixed in the brim, the manœuvre has succeeded; the labour has become natural. Velpeau and Meigs relate instances of the successful application of this practice. But if the head still shows a disposition to recede, grasp it at once with the long double-curved forceps, and hold it in the brim until it is sufficiently engaged to be safe.

4. The mode in which *forehead- and face-presentations* arise out of excess of friction or resistance encountered by the occiput has been described in Lecture IV.

Sometimes correction of these presentations may be effected by restoring the equilibrium of resistance to the anterior part of the head. Sometimes this is effected by simply keeping the tips of the fingers upon the forehead, trusting to the expulsive efforts propagated through the child's spine to cause the head to rotate upon its transverse axis, and bring down the occiput. Sometimes further aid is necessary. The tips of two fingers of the left hand are applied internally upon the forehead, and at the same time the occiput must be pressed down by the fingers of the right hand applied externally in the iliac fossa.

In some cases a rougher method has been pursued. The hand introduced into the uterus has seized the head by the occiput, and brought it down. This manœuvre is by no means easy, and, if the child is mature, will rarely succeed.

Wigand, when the head was not too low in the pelvis, first pushed the face upwards, so as to convert the face into a forehead-presentation, if not into a cranial; then he applied the forceps.

Smellie had already deliberately put in practice the restoration of a lost head-position.* In one case, feeling the face presenting through the membranes, he raised the forehead; then letting the waters escape, the head was fixed in its proper position, and the labour terminated successfully. In the second case, a hand presented. Smellie grasped the head and brought it into the brim, having pushed up the shoulder. In this position the head was fixed by the escape of the liquor amnii and bearing-down pains. The child was delivered naturally. In a third case, in which the breast presented, he was equally successful in bringing down the head.

5. *Descent of the hand by the side of the head.* When this accident occurs, it is apt to proceed to shoulder-presentation, the hand and arm slipping

* "Cases and Observations," vol. ii. 1754.

down and wedging the head off the brim to one or other iliac fossa. Hence the importance of correcting this presentation as early as possible. Whilst the parts are still moveable, it is commonly possible to push up the presenting hand by means of your left fingers in the vagina; and at the same time, by pressing down the head by the external hand towards the brim, you make the head fill the space until the double-curved forceps is applied. Then, drawing the head into the brim, the hand cannot again descend.

6. *Prolapse of the umbilical cord by the side of the head* may sometimes be managed successfully in a similar manner. The first thing to be done is to replace the cord above the presenting head. The postural or knee-elbow position will much facilitate this operation; but it is not always available or necessary. Braun's repositorium, or, better still, Roberton's, may be used to carry up the loop of prolapsed cord. As soon as this is done, press the head down upon the brim, and whilst it is supported by the two hands of an assistant, seize it with the double-curved forceps.

LECTURE XI.

TURNING (*continued*): THE MANAGEMENT OF CERTAIN DIFFICULT BREECH-PRESENTATIONS—LOCKED TWINS —DORSAL DISPLACEMENT OF THE ARM—DOUBLE MONSTERS.

BEFORE proceeding to the discussion of podalic turning, strictly so called, it will be both convenient and useful to deal with certain cases of difficult breech-presentation. It will be remembered that I defined "*turning as including all those proceedings by which the position of the child is changed in order to produce one more favourable to delivery.*" (See Lecture VI.) Now, the cases of breech-presentation to which I refer cannot be brought to a satisfactory conclusion unless the position of the child, or at least of some of its parts, be changed. They, therefore, fall within our definition. But since the breech or podalic extremity of the child is already presenting, a great part of the end contemplated in podalic turning is already accomplished. The problem, so far, then, is simpler than that of effecting complete version, and may therefore logically precede the latter in the order of discussion. The simplicity is, indeed, more apparent than real, more theoretical than practical. The task of delivering a breech case such as I shall presently describe, vies in difficulty with that which has to be encountered in the most severe forms of shoulder-presentation.

In a considerable proportion of breech cases, the labour is premature. In these generally there is no difficulty. Indeed, I have commonly observed in these premature breech-labours a remarkably active, even stormy character in the uterine contractions, driving the child through with unexpected rapidity. But when the child is mature and well developed, a breech-labour is by no means easy.

Fig. 52.

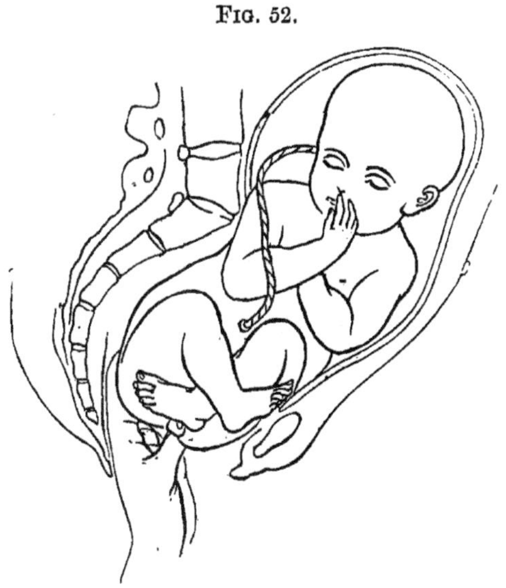

There are two principal conditions of breech-presentation under which labour may become arrested or difficult. Whether the position of the fœtus be dorso-anterior or abdomino-anterior, the legs may be disposed in one of two ways. First, and it is the most common case, the legs may be placed upon the thighs so that the heels are near the nates, and, what is very important to recollect, therefore not far from the os uteri. Secondly, the legs may be extended so that the toes are pointed close to the face.

Several causes concur in obstructing delivery. The breech is not nearly so well adapted as the head to traverse the pelvis. Instead of taking a movement analogous to the extension of the head forwards under the pubic arch, the breech tends to bend backwards in the hollow of the sacrum. The spine, tending to curve in a sigmoid form, is not so well fitted to transmit the expulsive force applied to the head by the fundus uteri. Then there is the wedge formed by the legs doubled up on the abdomen, which does not easily allow of more than the apex, represented by the breech, descending into, or traversing, the pelvis.

Now the apex of this wedge, represented by the breech and the thighs bent on the abdomen, can enter the pelvis very well. But then comes the widest part or base of the wedge, formed by the chest, shoulders, arms, head, and legs. This often exceeds the capacity of the brim in mere bulk. But, in addition, there is an impediment to rotation of the child on its long axis, which rotation is necessary to easy descent.

There is yet another obstacle. It arises out of the condition of the uterus. The cervix opens just in proportion to the dimensions of the body which traverses it. The breech, being of less bulk than the head and other parts constituting the base of the wedge, does not open the cervix widely enough to allow this base to descend. The uterus is apt to contract firmly upon the parts still retained in its cavity; and, the cervix encircling the wedge about its middle, a state of spastic rigidity ensues, which tends to lock up the head and chest and to impede descent and rotation. In Fig. 53 I have endeavoured to depict some of the conditions described.

FIG. 53.

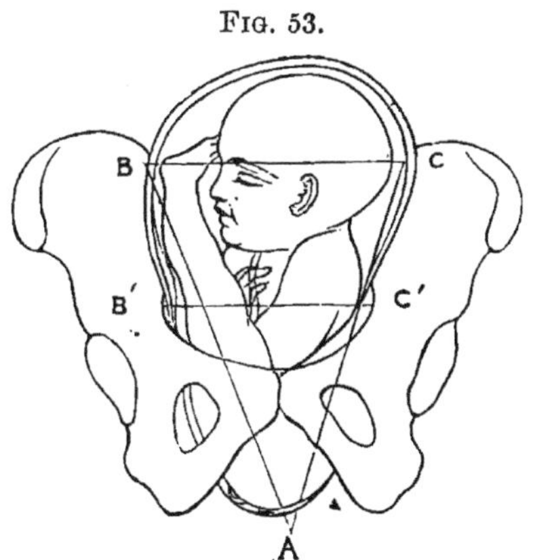

THE POSITION OF THE FŒTUS WITH THE LEGS EXTENDED AS SEEN FROM A FRONT VIEW.

The breech has descended into the pelvis. The fœtus forms a wedge, of which the apex A is turning forwards under the pubic arch. The base B C formed by the head and legs is wider than B′ C′, the transverse diameter of the pelvis.

Sometimes the cause of arrest is simple inertia: a little *vis à fronte* to compensate for defective *vis à tergo* may be all that is necessary. It is in the hope of extricating the child by this means that traction in various forms is resorted to. If this is unsuccessful, the case is rather worse than it would have been if left alone. The apex is dragged down a little more, the mother's pelvis is more tightly filled, and the uterus has become more irritable. I have on this account arrived at the conclusion that it is better not to resort to direct traction upon the breech in any case where there is arrest. The proper course is, I believe, to bring down a foot in the first instance. Then traction, if still indicated, can be exerted by aid of the leg with safety and with increased power, and under the most favourable conditions for the descent and rotation of the child.

I have seen fruitless and injurious attempts made to extract by fingers, hooks, and forceps. I believe that all the best authors — that is, of those who have encountered and have had to overcome this difficulty, for it is little considered in our text-books—condemn the use of hooks and forceps. Chiari, Braun, and Spaeth,* Ramsbotham, H. F. Naegele, are decided in their reprobation; Hohl† says the forceps is neither necessary nor effectual. The breech is already in the pelvis. To apply the blades safely, the hand must be passed into the vagina, and, having done this, it may as well do the right thing at once—that is, bring down a foot. Special forceps made to seize the breech are also superfluous.

I have always succeeded in delivering these cases by the simple use of the unarmed hand; and since the cases in which I did so succeed were the most difficult that can be encountered, it follows that the unarmed hand is sufficient to overcome the cases of minor difficulty of the same kind. To determine us to reject hooks and forceps, it should be enough to remember that the child is probably alive, and that, under proper skill, it may be born alive. Now hooks and forceps will, in all likelihood, either destroy the child

* "Klinik der Geburtshülfe," 1855.

† "Lehrbuch der Geburtshülfe, 1862.

or involve its death through the delay arising out of their inefficiency, or they may seriously injure the child. The blunt hook may fracture the femur, contuse the femoral vessels, or at least inflict severe bruises on the soft parts. The forceps may injuriously press upon the abdominal viscera.

The difficulty is seldom manifest until the breech has entered the pelvis, and this is the great cause of the obstacles opposing operative measures. To traverse the pelvis, the child's body must take a sinuous course, represented in Fig. 54.

FIG. 54.

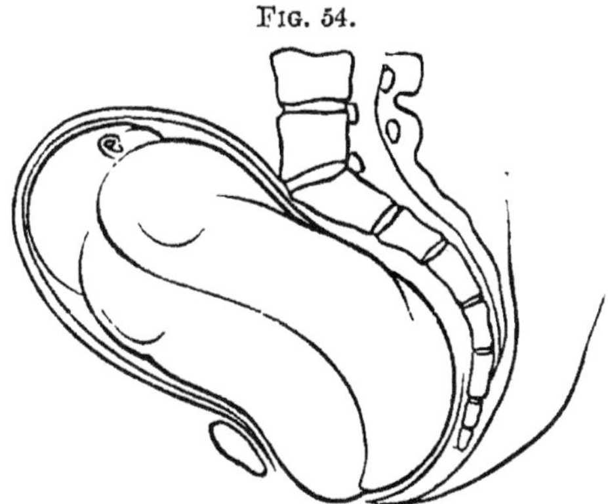

REPRESENTS A SIDE VIEW OF A BREECH-PRESENTATION, IN WHICH THE BREECH HAS ENTERED THE PELVIS.

It shows the sigmoid form imparted to the trunk in its effort to traverse the pelvis.

The clear indication is to break up or decompose the obstructing wedge. This is done by bringing down one foot and leg. For this purpose, pass your hand through the os uteri in front of the breech where the feet lie; seize one by the ankle with two fingers; draw it down, and generally the breech will soon descend. It is better to leave the other leg on the abdomen as long as possible, as it preserves greater rotundity of the breech, and helps to protect the cord from pressure. It will escape readily enough when the breech comes through the outlet.

The first thing to do is to determine the position of the breech in its relation to the pelvis, in order that you may know where to direct your hand to the feet. The breech simulates the face more than any other part, and so it is from the face that the breech has chiefly to be distinguished. There are four principal diagnostic points in the breech: the sacrum and anus behind, the genitals in front, an ischiatic protuberance on either side. The sacrum is distinguished by its uneven spinous processes from anything felt in a face presentation; and this is the most trustworthy characteristic, for the malar bones may pass for the ischia, and the mouth for the anus. In all cases of doubtful diagnosis it is well to pass the fingers, or hand, if necessary, well into the pelvis, so as to reach the higher presenting parts. In a breech case you will thus reach the trochanters, and above them the groins, where a finger will pass between the child's body and the thigh flexed upon it. Then in front will be the fissure between the thighs themselves; and here, if the legs are flexed upon the thighs, will be the feet to remove all doubt. These are what you are in search of. But you only want one. It is much more easy to bring down one foot than both; and it is, moreover, more scientific. The question now comes, Which foot to bring down? I believe the one nearest to the pubic arch is the proper one to take. To seize it, pass the index finger over the instep; then grasp the ankle with the thumb, and draw down backwards to clear the symphysis pubis. When the leg is extended outside the vulva, it will be found that traction upon it will cause the half-breech to descend, and the child's sacrum to rotate forwards. The further progress of the case falls within the ordinary laws of breech-labour.

FIG. 55.

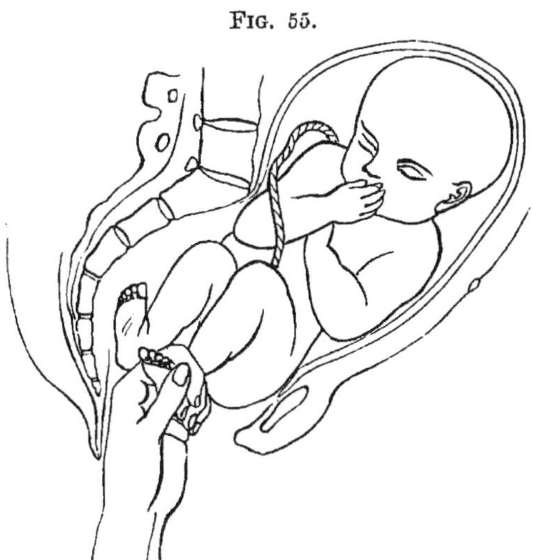

REPRESENTS A BREECH-PRESENTATION WITH THE LEGS FLEXED UPON THE THIGHS, AND THE MODE OF SEIZING A FOOT.

The second case—that in which the feet lie at the fundus of the uterus close to the face—is far more difficult. The wedge formed by the extended legs and the upper part of the trunk must, in some instances at least, be decomposed before delivery can be effected. The cause of the difficulty will be understood on looking at the diagram, Fig. 53; and on reflecting that the breech or wedge may in great part be driven low down into the pelvis, leaving but little space for the operator's hand to pass; and further, that the hand must pass to the very fundus of the uterus to reach a foot. No ordinary case of turning involves passing the arm so far.

The mode of proceeding is as follows (*See* Figs. 56, 57):—Give chloroform to the surgical degree; support the fundus of the uterus with your right hand on the abdomen; pass your left hand into the uterus, insinuating it gently past the breech at the brim, the palm being directed towards the child's abdomen, until you reach a foot—the anterior foot is still the best to take—a finger is then hooked over the instep, and drawn down so as to flex the leg upon the thigh. Maintaining your hold upon the foot, you then draw it down out of the uterus, and thus break up the wedge. The main obstacle is thus removed, and you have the leg to exert traction upon, if more assistance is necessary. One caution is necessary in performing this operation. It is this: the finger *must* be applied to the instep. It is of no use to attempt to bend the leg by acting upon the thigh or knee. You must therefore carry your finger nearly to the fundus of the uterus. This, and the filling up of the brim —and even of a part of the pelvic cavity sometimes—by the breech, render the operation one of considerable difficulty, demanding great steadiness and gentleness. I have brought a live child into the world by this proceeding on several occasions, where forceps, hooks, and various other means had been tried in vain for many hours. The *reason* of the operation, you will see, is analogous to that which indicates turning in arm-presentation. The further management of podalic or feet-first labours will be described under "Turning."

Fielding Ould (1742) seems to have clearly understood these cases. When the feet, he says, are near the outlet, seize them; and at the same time that they are drawn forwards, the buttocks must be proportionately thrust into the womb by the fingers of the left hand; for want of this precaution the thigh-bone of many an infant has been broken. "Both legs and thighs may be extended along the child's body, so as to have a foot over each shoulder, which much increases the difficulty. In this case each leg must be taken separately, and the knee bent." I must observe, that it is both superfluous and injurious to take each leg; one is enough, and better.

Fig. 56.

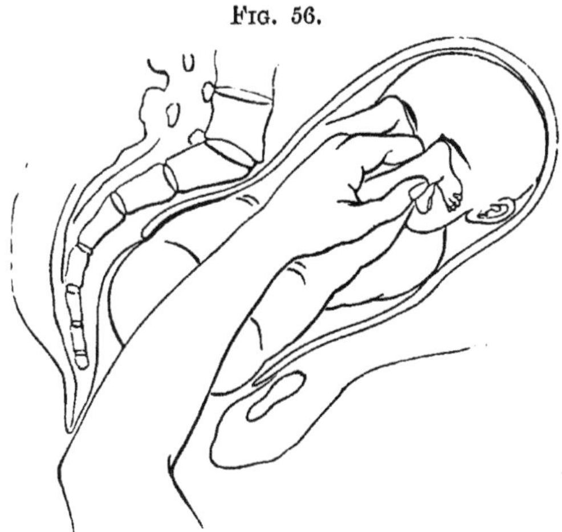

It is quite excusable, before proceeding to so difficult an encounter, to try some other method. The child may be small and the pelvis large, and so a moderate degree of tractile force may be enough to bring the wedge through without decomposing it. Various manœuvres have been adopted. You may hook one finger in a groin and draw down: or, what I have found better, you may with the forefinger *hook down each groin alternately*. (*See* Fig. 58.) In this way the breech will sometimes move. Or you may pass a piece of tape or other soft cord over the groins, as Giffard did in a case quoted by Perfect. The left

Fig. 57.

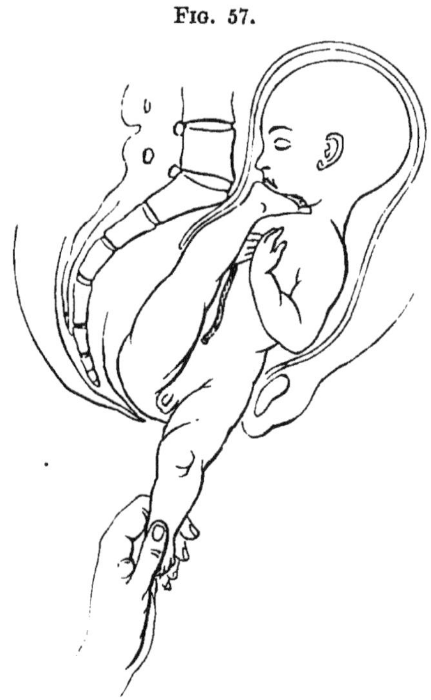

buttock presented. Giffard, not being able with the forefinger of each hand, placed on each side of the thigh near the groin, to draw out the feet, succeeded by putting a soft string over the end of his finger; and getting that up on one side over the thigh and a finger on the other side, he drew the string out, and fixing it close up to the hips, he took hold of the ends that hung out, and thus extracted, being aided by the pains. An apparatus, having a curved flexible spring, might be used to carry the string over the hips, or the object might be accomplished by a catheter first carried across, and then, having tied the string to the

FIG. 58.

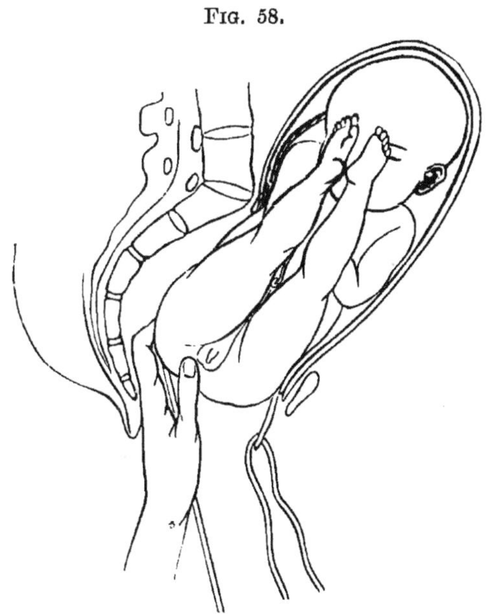

end, it could be drawn through—the proceeding resembling that adopted to plug the posterior nares for epistaxis. (*See* Fig. 58.)

Dr. Ramsbotham recommends the slipping a silk handkerchief over the groins. But it is possible that these and like measures may fail, and that you have nothing left but to break up the wedge by separating its component parts; and this, I repeat, is the proper thing to do in the first instance.

Amongst the most puzzling and difficult cases requiring operative interference are *certain cases of twins.* Commonly, when twins are found, the embryos are lodged each in its own bag of amnion and chorion, apart from each other, and so packed in the uterus that when labour occurs, one presents at a time in the brim, and traverses the pelvis before the bag of liquor amnii of the other is ruptured. Under such circumstances, labour is indeed apt to be tedious, because the uterus, being over-distended, acts at a great disadvantage.

There is another cause for this lingering, imperfect action, which I have not seen noticed: it is, that uterine force is lost by being transmitted to the first child through the fluid contained in the membranes of the second. But in this case there is no mechanical interference of the children with each other. All that may be necessary is to supply a little *vis à fronte* by means of the forceps, to make up for the defective or wasted *vis à tergo.*

But matters may be very different when the embryos are both lodged in the same sac, that is, in one common chorion and amnion. Before or during labour, the limbs and heads may become so entangled or locked as to form virtually one body, which is too large to pass through the pelvis. The embryos may perform the most remarkable evolutions. Cases have been observed where one embryo has dived through a loop in the other's umbilical cord; and knots have been formed involving the two cords.

In the more ordinary and favourable course of twin-labour, the first child presents by the head, and is entirely expelled before any part of the second becomes engaged in the pelvis. Indeed, the membranes of the second do not burst until after the

Fig. 59.

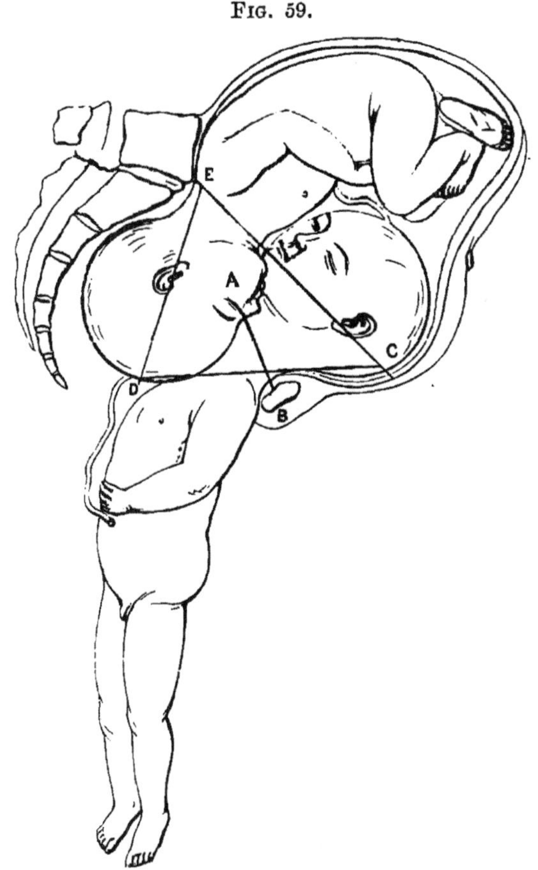

SHOWS HEAD-LOCKING, FIRST CHILD COMING FEET FIRST; IMPACTION OF HEADS FROM WEDGING IN BRIM.

D apex of wedge, E C base of wedge which cannot enter brim, A B line of decapitation to decompose wedge, and enable trunk of first child and head of second to pass.

first is wholly born. This second child presents either by the feet, or breech, or by the head.

The most common form of locking occurs through the hitching of one head under the chin of the other; and this may happen whether both children present head first, or one by the breech, the other by the head. The latter case appears to be the more frequent. A child presents by the feet or breech; and when born as far as the trunk or arms, it is found that the labour does not proceed, and on making traction to accelerate the delivery, unexpected resistance is encountered. You pull, but the child sticks fast in the pelvis. The first surmise is, probably, that the head is too large, from hydrocephalus, or that the arms have run up by the sides of the head, wedging it in the brim. You liberate the arms, and pull again, and still the head refuses to move. And now you must explore fully. You may get information in two ways. First, give chloroform, and pass your left hand into the cavity of the pelvis, so as to reach above the child's breast, feeling for its chin or mouth. Instead of feeling this first, you may be surprised at meeting a hard, rounded mass, jammed in the neck and chest of the presenting child, which can be nothing else than the head of another child, which has got in the way of the first. Secondly, by external palpation, you may succeed in making out through the abdominal walls the head of a child above the symphysis pubis, inclined to one or other side, in a position which its relation to the trunk partly born, and to the head you have felt whilst exploring the interior of the pelvis, will satisfy you is the head of the first child.

If the children are small, they may, with more or

less difficulty, come through the pelvis together in this fashion.

Sometimes it has been possible to seize the second head by the forceps, and to extract it and the embryo to which it belongs without disturbing the second child. But if the children be at all large, this proceeding is not likely to save them. The pressure to which both must be subjected is too hazardous. Even with children of the average size, the head of the second child, resting on the neck and chest of the first, form a wedge too large to clear the brim. You get the state of things represented in Fig. 59. D is the apex of the wedge driven into the pelvis; E C is the base, too large to enter; A B is the point at which the wedge may be decomposed, by detruncating the first child at the neck.

The apex of the wedge formed by the trunk of one child has traversed the pelvis; the base, formed by the head of one child pressed against the neck of the other, is too large to enter the pelvis; and if traction is exerted on the apex, the only effect is to jam the head tighter against the neck, hooking the two heads more firmly together. The problem is, how to extricate one head from the other, so as to allow one child to pass at a time. There are several methods of accomplishing this. But, before deciding upon one, it is well to study how the children are affected by the complication. Is one child in greater jeopardy than the other? If so, which? If we find that the situation involves extreme peril or death to one child, we shall of course not hesitate to mutilate this one, if, by so doing, we can promote the safety of the other.

The first thing you will try to accomplish is, to disentangle the heads without mutilating either child. It is still possible that both may be born alive. The patient being under chloroform, you press back the trunk into the pelvis as much as possible, so as to lift the heads off the brim, and so to loosen the lock. Then by external manipulation, aided by a hand in the pelvis, you try to push the heads apart in opposite directions. If you succeed in unlocking them, support the head of the second child out of the way, whilst you or an assistant draw down the body of the first child and engage its head in the pelvis. If you can manage this, the difficulty is over.

Now, experience and reflection concur in showing that the first child whose trunk is partly born encounters by far the greatest danger. Its umbilical cord is likely to be compressed; its neck and chest are forcibly squeezed. On the other hand, the umbilical cord of the second child is comparatively safe, and the pressure upon its neck is less severe. You may, moreover, find, by feeling the cord of the first child, that it is pulseless and flaccid, and that tickling its feet excites no reflex action. Having thus determined that there is no hope for the first child, you turn to the best means of rescuing the second. You may decompose the wedge formed by the two heads by detaching the head of the first child. This is done by drawing the body of the child well backwards, so as to bring its neck within reach. Being held in this position by an assistant, you pass the fingers of the left hand into the pelvis, so as to hook them over the neck, and serve as a guide to Ramsbotham's or Braun's decapitator, or the wire-écraseur. If these are not at hand, the task can be accomplished by strong scissors, or even by a penknife.* (*See* Fig. 59; A B represents the line of decapitation.)

As soon as the neck is severed, the trunk will be extracted easily enough. The head of the first child will then slip up or on one side, or you may make it do so by passing your hand inside the uterus. If the head of the second child do not descend by the spontaneous action of the uterus, you may either seize it by the long double-curved forceps, or seize a leg and turn. The first head will follow last of all. If it offer any difficulty, it may be dealt with as described in Lecture XIII.

If there be reason to conclude that the second child is dead, it would be justifiable to perforate its head, and lessen its bulk by help of the crotchet. This is another mode of breaking up the base of the wedge. The head will then flatten in, and permit the trunk and head of the second child to be delivered.

In the other case, where the head of the first child presents, and gets locked by the head of the second, as in Fig. 60, a similar rule of action will apply. You may disentangle the heads by external and internal manipulation. Failing this, you may seize the foremost head by the forceps; and whilst an assistant pushes away the second head, you can extract the first child. A good case, in which this plan succeeded, is related by Dr. Graham Weir.†

Dr. Elliot (Obstetric Clinic) relates an interesting case of twins with contracted pelvis, in which imminent locking of the heads was prevented by

* See a case by H. Raynes, Esq., "Obst. Trans." 1863.

† "Edin. Journ. of Med." 1860.

placing his hand on the abdominal wall over that head which was superior and to the left, and forcing it into the pelvis in advance of the other.

One rule of practice in all cases of twin-labour there is, which I think it important to insist upon. It is, as soon as the first child is born, to apply a binder firmly on the abdomen to support the uterus, to aid it in recovering energy to complete the delivery, and to lessen the risk of hæmorrhage. That this latter risk is

Fig. 60.

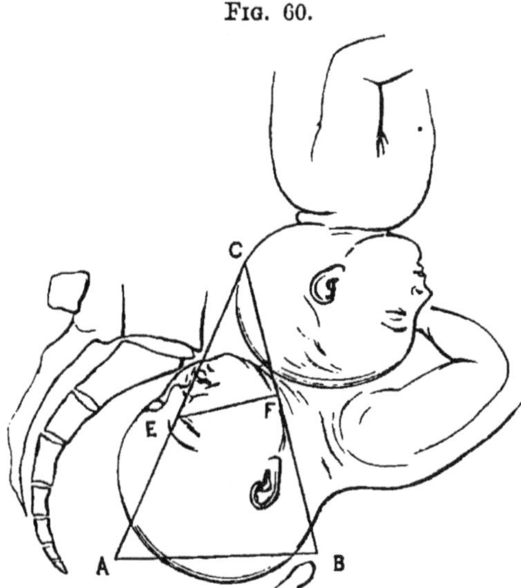

SHOWS HEAD-LOCKING, BOTH PRESENTING HEAD FIRST.

far greater after twin-labour than after single labour is well known. Not only is the uterus weakened by excessive distension, but there is a greatly-increased area whence blood may flow. I have ascertained that the superficies of an ordinary single placenta is about sixty square inches. If the uterine walls are relaxed, an equivalent surface may pour out blood. In twin-pregnancy, this bleeding area may equal 100 square inches or more. Thus, there is a vastly larger bleeding surface, and less contractile energy to close it.

Another rule may be stated. When the first child is born, do not pull upon the cord, or you may do mischief in two ways. It is possible that the cord may be entangled round the neck or a limb of the child still *in utero*, or the cord of the child *in utero* may be entangled in it, so that by pulling on the first cord you may strangle the second child, or arrest the circulation in its cord; or you may detach the placenta prematurely, thus giving rise to hæmorrhage, and, in the probable event of the placenta being united or common, again imperilling the child *in utero*.

Dorsal Displacement of the Arm.

There is a curious cause of dystocia resulting from the locking of an arm behind the neck, to which special attention has been drawn by Sir James Simpson* and Caseaux. As the difficulty is to be met by altering the position of the displaced limb, it comes under our definition of Turning.

Fig. 61 represents this position of the arm. This displacement more commonly occurs in cases of podalic or breech-labour, as after turning; and I am very much disposed to think it is then most frequently produced by unskilful manipulation. I have already insisted upon the importance of avoiding the error of rotating the child upon its axis during extraction. If you commit this fault, this is what is likely to happen: the trunk revolving under your manipulation, the arm is

* "Obstetric Works," vol. i.

caught against the wall of the uterus, and does not move round with the trunk, but comes to be applied to the nape of the neck. Dugès and Caseaux explain that this may happen in two different ways. First, the arm may cross behind the nucha after having been raised above the head; the crossing then takes place from

FIG. 61.

SHOWING NUCHAL HITCHING OF THE ARM.

above downwards and from before backwards relatively to the fœtus. Secondly, it may take place from below upwards, the arm rising on the back of the fœtus, and stopping below the occiput. This last method requires a little explanation. The arms are habitually placed by the sides of the chest. In rotating, the attempt is made to carry the abdominal aspect of the child towards the loins of the mother; the trunk only moves; the arm, therefore, remains behind; the operator, in performing extraction, draws the trunk down; the arm is caught by the symphysis pubis, where it is detained until the nucha comes down to clench it. It is obvious, therefore, that this dorsal displacement of the arm is manufactured by too great diligence on the part of the obstetrician. Those who anticipate Nature, thwarting her operations, must be prepared for the penalty. I am entitled to say, that by observing the rules I have laid down for version, this accident will not occur.

But to extract light from our errors is true wisdom. By reflecting on the mode in which the displacement is produced, we shall see how to remedy it. You must retrace your steps. By rotating the child back in the contrary direction, so as to restore the original position, you may possibly liberate the arm. At any rate, you will render the further proceeding that may be necessary more easy. You carry the trunk well backwards, so as to give room to pass your forefinger in between the symphysis pubis and the child's shoulder; then hooking on the elbow, draw this downwards and then forwards. It may be useful, as a preliminary step, to gain room by first liberating the other arm.

If the arm cannot be liberated, you may be driven to perform craniotomy.

In Professor Simpson's case, the head presented. It is not easy to imagine how the arm of a living child can get behind the neck when the head presents. Professor Simpson throws out the suggestion that this occurs more frequently than is suspected; and that it accounts for many cases of arrest of the head where there is no disproportion, and which oppose even traction by the forceps. Dr. Simpson recommended to bring down the hand and arm forward over the side of the head, converting the case into one of simple presentation of the head and arm. Or recourse might be had to turning, as was done successfully in a case related by J. Jardine Murray.*

The Delivery of Monsters.

We may here consider the mode of delivery of monsters. Double monsters may give rise to the same kinds of difficulty as sometimes occur in the case of twins. In many cases, Nature is able to deal with them. They are most frequently born dead — a circumstance not, perhaps, much to be regretted. The death is often the result of the mode of birth, one part of the monster pressing injuriously upon another. If, in any present case of obstructed delivery, it could be with certainty diagnosed that the cause of obstruction was a monstrous embryo, the indication would be clear to do our best to spare the mother, even at the cost of mutilating the embryo. But this cannot always be known in time to influence our choice of proceeding. As in the case of locked twins, we are therefore led to postpone mutilation, in the hope that the delivery may be accomplished without. Dr. W. Playfair has discussed this subject in an excellent memoir.† He divides monsters according to their obstetric relations into four classes.

* "Med. Times and Gaz." 1861.

† "Obstetrical Transactions," vol. viii.

A. Two nearly separate bodies are united in front to a varying extent by the thorax or abdomen.

In this case, the feet or heads may present. The most favourable presentation appears to be the feet. The trunks come down nearly parallel. The arms can be liberated without much trouble. It is when the heads come to the brim that the difficulty arises. The object then is to get one head at a time to engage in the brim. This has been done successfully, after turning, by Drs. Brie* and Molas,† in the following manner. When the shoulders were born, the bodies were carried strongly forwards over the mother's abdomen. This manœuvre has the effect of placing the two heads on a different level, bringing the posterior head lower than the anterior one, which for the time is fixed above the symphysis. When the posterior head is in the pelvis, traction then will bring it through, and the second head will follow. If not, either the first or second head can be detruncated by Ramsbotham's hook, or by scissors or knife.

The command thus obtained over the course of labour in podalic presentations renders it desirable to turn if the heads present.

But sometimes, when the heads present, a mutual adaptation takes place, which permits them to pass without mutilation. As in a case reported by Mr. Hanks,‡ one head got packed between the shoulder and head of the other body, so that both passed without great difficulty. One head is born first, either by

* "Bulletin de la Faculté de Méd.," vol. iv.
† "Mémoires de l'Acad.," vol. i.
‡ "Obstetrical Transactions," vol. iii.

aid of forceps or spontaneously, and the corresponding body may be expelled by a process of doubling up, or spontaneous evolution. If this does not proceed with sufficient readiness, decapitation of either the first or second child must be practised.

B. Two nearly separate bodies are united nearly back to back by the sacrum, or lower part of the spinal column.

In this case the mode of delivery is essentially the same as in A.

C. Dicephalous monsters, the bodies being fused together.

One head will come down first. The body follows, by doubling or spontaneous evolution. If this does not take place, decapitate the first head, and bring down the feet.

D. The bodies are separate below, but the heads are partially united.

Whether the head or feet present, if there be obstruction, perforate the head.

In the case of *dystocia from excessive size of the abdomen of the fœtus*, as happens sometimes from dropsy, perforation and evisceration may be indicated.

LECTURE XII.

PODALIC BI-POLAR TURNING—THE CONDITIONS INDICATING ARTIFICIAL TURNING IN IMITATION OF SPONTANEOUS PODALIC VERSION, AND ARTIFICIAL EVOLUTION IN IMITATION OF SPONTANEOUS PODALIC EVOLUTION—THE SEVERAL ACTS IN TURNING AND IN EXTRACTION—THE USE OF ANÆSTHESIA IN TURNING—PREPARATIONS FOR TURNING—THE STATE OF CERVIX UTERI NECESSARY—THE POSITION OF THE PATIENT—THE USES OF THE TWO HANDS—THE THREE ACTS OF BI-POLAR PODALIC VERSION.

THE conditions indicating podalic turning are:—

1. Generally, those which are not suited for head-turning, or for the imitation of spontaneous evolution.

2. And more especially, shoulder-presentations of living children, in which the knees or feet are nearer to the os uteri than is the head.

3. Cases in which the shoulder has entered the brim of the pelvis, and especially those in which the arm is prolapsed.

4. Most cases in which the cord has prolapsed with the arm or hand, and some cases where the cord alone is prolapsed, and cannot be returned or maintained above the presenting part of the child.

5. Cases of shoulder-presentation in which the liquor amnii has drained off, and in which the uterus has contracted so much as to impede the mobility of the fœtus.

6. Certain cases in which it is desirable to expedite labour on account of dangerous complications, present or threatening—as hæmorrhage, accidental or from placenta prævia; convulsions. In these cases it is indifferent what the presentation may be.

7. Some cases of inertia, the head presenting, as in pendulous belly and uterus, where the head cannot well be grasped by the forceps.

8. Certain cases of face-presentation. (*See* Lecture IV.)

9. Certain cases of minor contraction of the pelvis, or of the second degree (*see* Lecture V.), which are beyond the power of the forceps, and which ought not to be given over to craniotomy.

10. Certain cases of morbid contraction of the soft parts.

11. As a part of the operation for the induction of premature labour in certain cases in which the pelvis is contracted, or other circumstances do not permit the spontaneous transit of the fœtus with sufficient ease and quickness to secure a live birth.

12. Some cases after craniotomy, as the readiest mode of extracting the fœtus.

13. Certain cases of rupture of the uterus, the child being still in the uterine cavity.

14. Certain cases of death of the mother during labour, in the hope of rescuing the child, when Cæsarian section cannot be performed.

Here, then, is a wide range for the exercise of skill in podalic turning.

We will now discuss *what are the conditions necessary or favourable to turning in imitation of the spontaneous podalic version?*

These are: 1st. The pelvis must be capacious enough to permit the passage of the fœtus without mutilation. 2nd. The vulvo-uterine canal must be dilated, or sufficiently dilatable to permit of the necessary manipulations and of the passage of the fœtus. 3rd. The presenting part must not be deeply engaged in the pelvic cavity. 4th. The uterus must not be contracted to such an extent that the fœtus has been in great part expelled from its cavity, which is hence so diminished that the presenting shoulder or head cannot be safely pushed on one side into the iliac fossa. If the *shoulder is free* above the brim, the hand not descended, it will be easy to push it across to the nearest iliac fossa. If the *shoulder is moveable*, even if the hand has fallen into the vagina, the operation is practicable, often not even difficult. If the *shoulder has been driven low down* into the pelvis, near the perinæum, the body being firmly compressed into a ball by the spasmodic contraction of the uterus, the child is almost certainly dead, and turning may be difficult or impossible without extreme danger to the mother. This is the indication for imitation of the natural spontaneous evolution.

Let us, first of all, take the more simple order of cases where turning is resorted to on account of symptoms indicating danger to the mother, as hæmorrhage from placenta prævia, the head presenting, and the cervix uteri sufficiently dilated. I take such a case first because it requires *complete turning*, and therefore best illustrates the mechanism of the bi-polar method.

It is important, at the outset, to bear in mind that turning—that is, the changing the position of the child in order to produce one more favourable to delivery—is one thing, and that extraction, or forced delivery, is another thing. Sometimes turning alone is enough, Nature then taking up the case and completing delivery. Sometimes extraction, or artificial delivery, must follow, and complement turning.

It will, however, give a more complete exposition of the subject to describe the two operations of turning and extraction continuously, assuming a case in which both are necessary.

Each operation again admits of useful division into stages or acts. *The several acts in turning are these:—*

1st Act.—The removal of the presenting part of the child from the os uteri, and the simultaneous placement there of the knees.

2nd Act.—The seizure of a knee.

3rd Act.—The completion of version by the simultaneous drawing down of the knee, and the elevation of the head and trunk.

These three acts complete turning.

The several acts in extraction are:—

1st Act.—The drawing the legs and trunk through the pelvis and vulva. An incidental part of this act is the care of the umbilical cord.

2nd Act.—The liberation of the arms.

3rd Act.—The extraction of the head.

Before commencing the operation, there are certain preparatory measures useful or necessary to adopt.

The question of *inducing anæsthesia* arises. It would partake too much of the nature of a digression to discuss at length the indications for chloroform as an aid in turning. I will do no more than glance at the principal points.

Chloroform is resorted to in the hope of accomplishing two objects:—The first is to save the patient pain; the second is to render the operation easier to the operator. The attainment of both objects is sometimes possible; sometimes not. It is not difficult to render the patient insensible; but you will not always at the same time make the operation more easy. It will commonly be necessary to push anæsthesia to the surgical extent. If you stop short of this degree, the introduction of the hand will often set up reflex action, and you will be met by spasmodic contraction of the vaginal and uterine muscles, and perhaps by hysterical restlessness of the patient. You will have lost the aid of that courageous self-control which Englishwomen pre-eminently possess. You must then carry the anæsthetic further, to subdue all voluntary and involuntary movements, and to lessen the reflex irritability of the uterus. Then, but not always, you will secure passiveness, moral and physical, on the part of the patient; the uterine muscles will relax; they will no longer resent the intrusion of the hand. These advantages are not, indeed, always obtained without drawbacks. A perfectly flaccid uterus indicates considerable general prostration, and predisposes to flooding.

The *state of the cervix uteri* has to be considered. It is one of the natural consequences of a shoulder-presentation that the cervix uteri is but rarely found dilated sufficiently for turning and delivery until after —perhaps long after—the indication for turning has been clearly present. A shoulder will not dilate the cervix properly. The same may be said of many cases where turning is indicated by danger to the mother from convulsions, hæmorrhage, &c. To wait for a well-dilated cervix would be to wait till the child or mother is dead. It follows, therefore, that we must be prepared to undertake the operation at a stage when the cervix uteri is only imperfectly dilated.

What is the degree of dilatation necessary? If the question be simply one of turning, it is enough to have a cervix dilated so as to admit the passage of one or two fingers only, since it is not necessary in the class of cases we are now discussing to pass the hand into the uterus.

But since the ulterior object contemplated is delivery —with the birth of a live child, if possible—we must have a cervix dilated or dilatable enough to allow the trunk and head of the fœtus to pass without excessive delay. The modes of dilating the cervix artificially have been described in Lecture V. It is sufficient here to call to mind the two principal modes—viz., by the hydrostatic dilators and by the hand. The water-bag properly adjusted inside the cervix, if labour has begun at term, will commonly produce an adequate opening within an hour. Sometimes the fingers alone will succeed as quickly. Quite recently, in a case where the head presenting could not bear upon the cervix to dilate it because of slight conjugate contraction, I expanded it by the fingers sufficiently to admit the narrow blades of Beatty's forceps within a few minutes. The instrument, however, was not powerful enough to bring the head through. I therefore turned and made the breech and trunk complete the dilatation. The head required considerable traction to bring it through the narrow conjugate; but the child was saved. At the beginning the os barely admitted one finger; yet the patient was delivered within an hour. But we cannot always proceed so quickly. Nor is it commonly possible to effect by artificial means that complete dilatation which is required to permit the head to pass freely.

The average obstetric hand will easily traverse a cervix that is too small to allow the head to pass; so that after all, even in head-last labours, as in head-first labours, the head must generally open up the passages for itself. What we have to do is to take care that the parts shall be so far prepared when the head comes to be engaged in the cervix that the further necessary dilatation may take place quickly, for this is the stage of danger to the child from compression of the funis between the os uteri and the child's neck. The management of this stage will be described further on. It is enough now to point out that a cervix uteri expanded so widely as to admit of three fingers manœuvring without inconvenience is *enough for turning* under the circumstances of the case we have assumed.

The general rule of *emptying the bladder and rectum* applies even more strongly to turning than to forceps or craniotomy operations.

What shall be *the position of the patient?* I have generally performed the operation of turning, under whatever circumstances, the patient lying on her left side. It is of importance, I think, not to raise alarm in the patient or her attendants by adopting any great departure from the usual rules of the lying-in chamber. To place the patient on her back involves very considerable, even formidable preparations. The patient must be brought with her nates to rest on the very edge of the bed; she must be supported at her head; and two assistants must hold the legs. Still, there are cases in which this position may be preferable or unavoidable. There is another position, also in some cases useful—the knee-elbow position. But this precludes the use of chloroform. We may obtain all the necessary facilities by keeping the patient on her left side. The nates must be brought near the edge of the bed; the pillows are removed so as to allow the head and shoulders to fall to the same level as the nates. The head is directed towards the middle of the bed, so that the operator's arm may not be twisted during manipulation; the knees are drawn up; and the right leg is held up by an assistant, so as not to obstruct or fatigue the operator's right hand, which has to pass between the thighs to work on the surface of the abdomen.

The *presence or absence of liquor amnii in the uterus* is a matter of accident. If it be still present, so much the better; but you must be prepared to act all the same if it be not there. It is needless to state that the child will revolve more easily if it be floating in water; but it must not only be made to revolve; you have to seize a limb. At some time or other, therefore, the membranes must be ruptured. What is the best time to do this? If you are proceeding to turn in the old way—that is, by passing the whole hand into the uterus before seizing a foot—it is an advantage to follow the plan recommended by Peu, of slipping the hand up between the uterine wall and the membranes until you feel the feet, and then to break through and seize the limbs. During this operation the arm, plugging the os uteri, retains the liquor amnii, and on drawing down the legs, the body revolves with perfect facility.

But if you are proceeding to turn by the bi-polar method, with a cervix perhaps imperfectly opened, the membranes must be pierced at the os. In this case you may perhaps accomplish the first act in version—that is, of removing the head or shoulder from the brim, and of bringing the knees over the os, whilst the membranes are intact. This you can try first, only rupturing the membranes when you are ready to seize the knee. But sometimes an excess of liquor amnii imparts too great mobility to the child; you are unable to fix it sufficiently to keep the pelvic extremity steady upon the os; it will bound away as in *ballottement*, the moment you touch it through the os. In such a case it is better to tap the membranes first, and *allow a part of the liquor amnii to run off.* While doing this you should keep your fingers on the presenting part to ascertain how its position and mobility are being influenced by the escape of the waters and the contraction of the uterus, so as to seize the right moment for proceeding.

Now, if you assent to what I have stated, you will find that you are committed to *the use of your left hand as the more active agent* in the operation. You want the right hand to work outside on the abdomen; therefore, the left hand must be introduced into the vagina. It is a case where ambi-dexterity is eminently required. The left hand in most people is smaller than the right. The patient lying on her left side, the left hand follows the curve of the sacrum far more naturally than the right. It meets the right hand outside, the two working consentaneously with comfort, involving no awkward or fatiguing twisting of the body. Moreover, in the great majority of cases the anterior surface of the fœtus, and consequently its legs, are directed towards the right sacro-iliac joint—that is, inclining backwards and to the right, so that the left hand passed up in the hollow of the sacrum will reach the legs with the utmost convenience.

I strenuously advise every young man who is preparing for Obstetric Surgery to put his left hand into training, so as to cultivate its powers to the utmost. There are a thousand ways of doing this, and I hope it will not be considered idle to mention some. In all athletic exercises or games requiring manual skill, use the left arm as well as the right. It is an excellent practice to dissect with the left hand. Shave the right side of the face with the razor in the left hand; use your toothbrush with the left hand; and, if you now and then come to grief through left-handedness, think how much less is this evil than injuring a woman or breaking down in an operation.

Well, all things being ready, we will proceed to the operation. In the case we have assumed, the head is over the os uteri; the os uteri is open enough to admit the play of two or three fingers; the liquor amnii is still present, or has been only recently and partially discharged. Turning is indicated by symptoms threatening danger to the mother. The preparations necessary have been made: One thing more I have to insist upon: it is to avoid all parade or fuss in your conduct. Make your preparations as quietly and unostentatiously as possible. Do all that is essential, and no more. Tell the patient and her attendants that you find it necessary to help the labour. But let your help be so given as to involve the least possible changes from the usual proceedings in ordinary labour.

When the patient is in position, and under chloroform, if it be determined to give help, slip off your coat, turn up the shirt-sleeves above the elbows, anoint with oil, lard, or pomade the back of the left hand and all round the wrist; insinuate a piece of lard into the vulva.

The Introduction of the Hand.—Bring the fingers together in the form of a cone; pass in the apex of this cone, gently pressing backwards upon the perinæum, and pointing to the hollow of the sacrum. If you find any difficulty—as you probably will if the case be a first labour—you must watch for the most fitting opportunity. Wait till a pain comes on. There is good reason for deferring to the popular idea of "taking a pain." The pain caused by expulsive action will partly mask that caused by the manœuvre; and expulsive action tends to produce sphincteric relaxation, so that the passage of the hand will be actually facilitated. A source of difficulty is the tendency of the labia and hair to turn inwards before the fingers. This is counteracted by drawing the labia open by the thumb of the right hand, by an action similar to that you would use to lift up the closed upper eyelid. The passage of the vulva is often the most difficult part of the operation. It is commonly necessary to pass the entire hand into the vagina; and great gentleness and patience are required. I have, indeed, turned and extracted a mature child without passing in more than two fingers, without even turning back or soiling the cuff of my coat; but the circumstances must be favourable to admit of this.

We have now got as far as the orifice of the uterus, and it is an immense improvement in obstetric art that we are able to complete the operation without passing the whole hand through this part. The *first act* begins by passing the tips of the first two fingers through the os to the presenting part, which we assume to be the head. We ascertain to which side of the pelvis the occiput is directed, for it is to that side that we must send the head. At the same time, an assistant holding up the right leg at the knee, so as to give you freedom of action, you apply your right hand spread out over the fundus uteri where the breech is. And now begins the simultaneous action upon the two poles of the fœtal ovoid; the fingers of the hand inside pressing the head-globe across the pelvic brim towards the left ilium, the hand outside pressing the breech across to the right side and downwards towards the right ilium. The movements by which this is effected are a combination of continuous pressure and gentle impulses or taps with the finger-tips on the head; and a series of half sliding, half pushing impulses with the palm of the hand outside. Commonly, you may feel the firm breech through the abdominal walls under the palm, and this supplies a point to press against. A minute sometimes, seldom much more, will be enough to turn the child over to an oblique or nearly transverse position; the head quitting the os uteri, and the shoulder or chest taking its place.

This act may be divided, for the sake of illustration, into two stages.

FIG. 62.

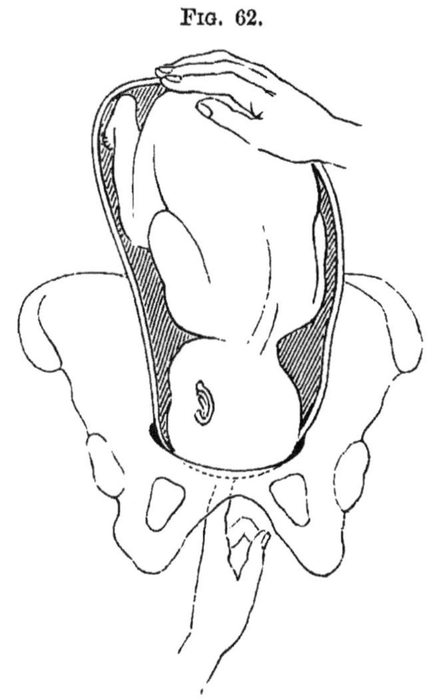

REPRESENTS THE FIRST STAGE OF BI-POLAR PODALIC VERSION.

The right hand on the fundus uteri pushes the breech to the right and backwards, bending the trunk on itself. The left-hand fingers on the vertex push the head to the left ilium, away from the brim.

At this stage it is important to keep the breech well pressed down, so as to have it steady whilst you attempt to seize a knee. This is the time to puncture the membranes, if not already broken. The fingers in the os uteri are pressed through the membranes during the tension caused by a pain, and you enter upon the

Second Act, the seizure of a knee. Which knee will you take? In the particular case we have to deal with, it is not of much importance which you seize,

FIG. 63.

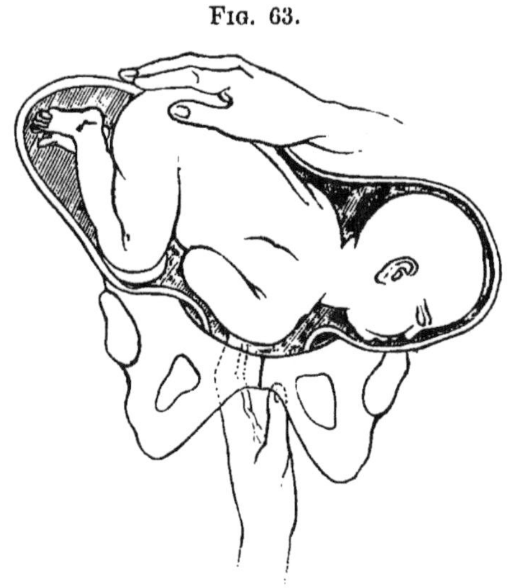

REPRESENTS THE SECOND STAGE OF THE FIRST ACT.

The right hand, still at the fundus uteri, depresses the breech, so as to bring the knees over the brim, whilst the left hand pushes the shoulder across the brim towards the left iliac fossa.

but the further one is, on the whole, to be preferred. You will observe in the diagrams, Figs. 62, 63, that the legs, doubled up on the abdomen, bring the knees near the chest, so that, as soon as the head and shoulder are pushed on one side, the knees come near the os uteri.

The knee being seized, the further progress of the case is under your command. By simply drawing down upon the part seized, you may often complete version. But it will greatly facilitate the operation to continue to apply force to the two poles. You will observe in Fig. 64 that the hands have changed places in relation to the two poles of the fœtal ovoid. Although the left hand has never shifted from its post in the vagina, the ovoid has shifted; and the

Fig. 64.

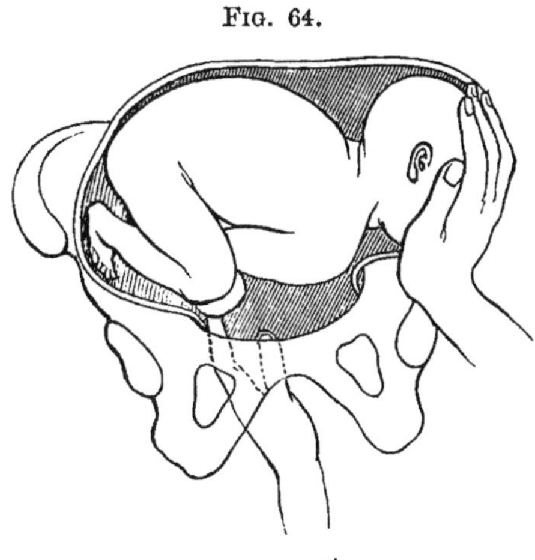

REPRESENTS THE SECOND ACT.

The trunk being well flexed upon itself, the knees are brought over the brim; the forefinger of the left hand hooks the ham of the further knee, and draws it down, at the same time that the right hand, shifted from the fundus and breech, is applied, palm to the head-globe, in the ilium, and pushes it upwards.

forefinger, drawing down the left knee, virtually acts upon the pelvic end of the ovoid. The right hand, therefore, is at liberty to quit this end; it is transposed to the head-end of the ovoid, which has been carried over to the left iliac fossa. The palm is applied under the head, and pushes it upwards in response to, and in aid of, the downward traction exerted on the child's leg. This outside manœuvre singularly facilitates the completion of version. It may be usefully brought into play in almost every operation of podalic turning. If it is neglected, as I shall show on a special occasion, you will sometimes fail in effecting complete version; for the head will not always quit the iliac fossa by simply pulling upon the legs.

FIG. 65.

REPRESENTS THE THIRD ACT IN PROGRESS.

The right hand continues to push up the head out of the iliac fossa; the left hand has seized the further leg, and draws it down in the axis of the brim. Version is now nearly complete.

Continuing to draw upon the leg, as soon as the breech nears the brim a movement of rotation of the child on its long axis begins, the design of which is to bring the back to the front of the mother's pelvis. This rotation depends upon a natural law of adaptation of the two parts. You are not to trouble yourselves in "giving the turns," as some authors imagine they can. I cordially agree with Wigand when he says, "Nature knows better than we do how to impart the proper turns."

FIG. 66.

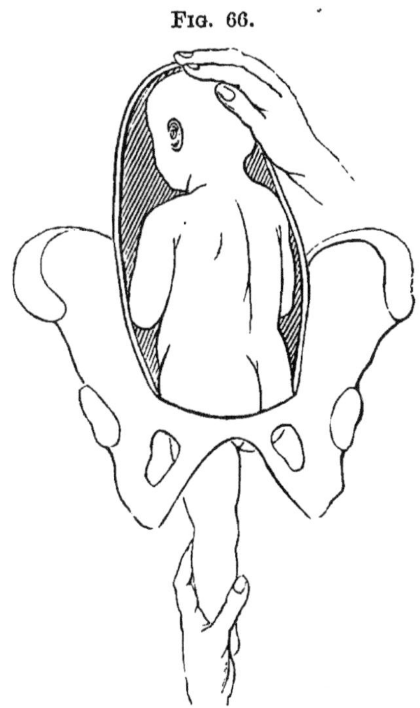

REPRESENTS THE COMPLETION OF THE THIRD ACT.

The right hand still supports the head, now brought round to the fundus uteri. The left hand draws down on the left leg in the direction of the pelvic axis. Version is complete. Rotation of the child on its long axis has taken place, the back coming forward as the breech enters the pelvis.

What you have to do is simply this—*to supply onward movement.* If the uterus be doing its own work, propelling the child breech first, we know we may rely upon Nature so to dispose the child in relation to the pelvis as to enable it to pass with the greatest facility. So it is when we supply the moving force from below. If this force is wanted, supply it; but do not attempt to do more. Avoid that fatal folly of encumbering Nature with superfluous help. Keep the body gently moving in the direction of the pelvic axis by drawing upon the leg, and Nature will do the rest. You will feel the leg rotate in your grasp, and the back will gradually come forward.

I have said that, upon the whole, the further knee is the better one to seize; but if you compare Figs. 64 and 66 you will see that, by drawing the nearer or anterior knee, you would directly secure the rotation of the child's back forwards; so that, as I have before said, it is not worth while to lose time in trying to seize the further knee if you find the anterior one the more easy to seize.

This completes version. The breech is substituted for the head. Nature may effect expulsion; but, if she fail, we have it in our power to effect delivery by extraction. We have assumed that extraction is necessary, and will proceed to this operation.

LECTURE XIII.

THE OPERATION OF EXTRACTION AFTER PODALIC VERSION, OR OTHER BREECH-FIRST LABOURS—THE THREE ACTS IN EXTRACTION—THE BIRTH OF THE TRUNK, INCLUDING THE CARE OF THE UMBILICAL CORD—THE LIBERATION OF THE ARMS—THE EXTRACTION OF THE HEAD.

THE operation of turning being completed by engaging the pelvic extremity of the child in the brim, we have next to consider the question of delivery. This, as I have already pointed out, is a distinct operation. Nature unaided may accomplish it. It is only in her default that we are called upon to undertake it. It is very desirable that as much of this operation be trusted to Nature as possible. Our duty is to watch the progress of the labour closely, interposing aid when that progress is too slow, or when the interest of the child demands it. As a general rule, the natural forces will carry the child through with more safety than the forces of art. But, even in the most favourable breech-first labours, whether the breech or feet have originally presented or have been brought to present by art, care on the part of the practitioner is necessary to avert certain dangers incurred by the child in its transit; and in some cases serious difficulty to the transit arises to demand the exercise of active skill.

The description I now propose to give of the operation of podalic extraction will embrace, and apply to, all the cases in which this operation is called for. We will begin with the most simple case—that in which there is no serious complication, in the shape of pelvic contraction, excessive size of the child, or resistance by the soft parts. It is either a case of inertia, or one in which prompt delivery or the acceleration of labour is indicated in the interest of the mother or child.

We possess, in our hold upon a leg, a security for the further progress of delivery, of which we can avail ourselves at pleasure. In this security consists one of the main arguments in favour of podalic version. We have divided the operation of extraction into three acts: drawing down the trunk through the vulva; liberation of the arms; extraction of the head.

The *first act* is effected by simply drawing down upon the extended leg in the axis of the brim. Two rules have to be observed. The first is to draw down simply, avoiding all attempts to rotate the child upon its long axis. You must not only not make such attempts; you must even be careful not to oppose the natural efforts at rotation. This is secured by holding the limb so loosely in the hand that the limb may either rotate within your grasp under the rotation imparted to it by the rotation of the trunk, or that the limb in its rotation will carry your hand round with it. The other rule is to draw well in the direction of the axis of the brim, and especially to avoid all premature attempts to direct the extracting force forwards in the axis of the outlet.

When the breech has come to the outlet, the extracting force is directed a little forwards, so as to enable the hip which is nearest the sacrum to clear the perinæum. This stage should not be hurried.

FIG. 67.

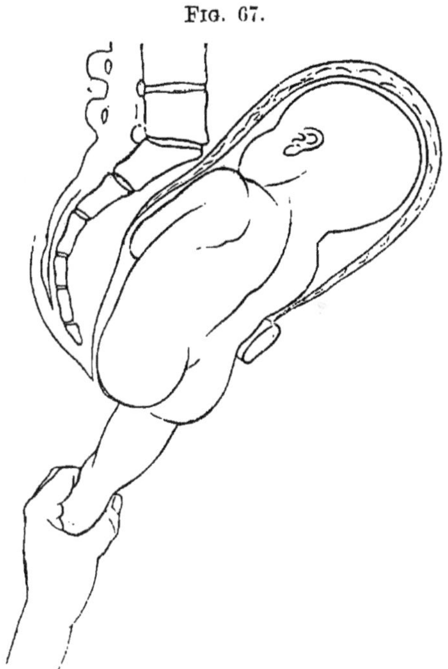

REPRESENTS THE FIRST ACT OF EXTRACTION.

The gradual passage of the breech has been doing good service in securing free dilatation of the vagina and vulva, an essential preparation for the easy passage of the shoulders and head. When the hips have cleared the outlet, you may pass the forefinger of your left hand into the groin, and gently aid extraction by this additional hold; and, at the same time, by pressing the knee forwards across the child's abdomen, you may facilitate the liberation of the leg.

When both legs and breech are outside the vulva, you have acquired a considerable increase of extracting power. You must, however, use it with discretion. You may now draw upon both legs, holding them at the ankles between the fingers and thumb of one hand.

And if you still want more power, you can grasp the child's body just above the hips with the other hand. It is generally desirable to interpose a thin soft napkin between and round the ankles. It gives a better hold, and lessens the risk of contusion.

Traction must now again be directed in the axis of the brim, in order to bring the shoulders through that aperture. The shoulders will enter in the same oblique diameter, back forwards, as that in which the breech traversed.

As soon as the belly comes to the vulva, your attention will be turned to the umbilical cord. This is apt to be put upon the stretch, by slipping up under the influence of friction as the body is drawn down; and, besides being stretched, it is liable to direct compression. The way to lessen these risks is to seize the cord near the umbilicus and draw down very gently a good loop; this loop should be laid where it is least exposed to pressure, that is, generally, on one side of the promontory of the sacrum; and you must further take care to keep off the pressure of the vulvar sphincter upon it by guarding it with your fingers. From time to time feel the cord, to ascertain if it continues to pulsate. If you find the pulsations getting feeble or intermittent, you have an indication to accelerate extraction.

The observations of May and Wigand upon this point are worthy of attention. Reasoning that the pressure suffered by the cord affects the vein more than the arteries, and hence that the access of blood to the fœtus is hindered, whilst the removal of the blood from the fœtus is little obstructed, so that a fatal anæmia results, they advise to tie the cord, as soon as the body is born, as far as the navel, and then to complete extraction. The apparent asphyxia so produced is easily remedied by the usual means. Von Ritgen says he has often done this, and affirms that when done there is little need to hurry extraction.

I refer to Lecture IV. for some observations upon the length of time a child is likely to survive after arrest of the circulation in the cord. I have there stated my belief that the prospect of a live child is very small if three or five minutes elapse before the head is born. This may be thought too narrow a limit, but certainly there is not a moment to be lost in starting aërial respiration.

The *second act* comprises *the liberation of the arms.* In the normal position of the fœtus the arms are folded upon the breast, and if the trunk and shoulders are expelled through a normal pelvis by the natural efforts, they will commonly be born in this position. But if ever so little traction-force be put upon the trunk, the arms, being freely moveable, encountering friction against the parturient canal as the body descends, are detained, and run up by the sides of the head. Hence often arises a serious delay in the descent of the head, for this, the most bulky and least compressible part of the fœtus, increased by the thickness of the arms, forms a wedge which is very apt to stick in the brim. This is one great reason for not putting on extraction-force if it can be avoided. If, however, we find the arms in this unfortunate position, we must be prepared to liberate them promptly, and, at the same time, without injury. It is very easy to dislocate or fracture the arms or clavicles if the proper rules are not observed. What are these rules ?

The cases vary in difficulty, and therefore in the means to be adopted. In some cases the arms do not run up in full stretch along the sides of the head. The humeri are directed a little downwards, so that the elbows are within reach. In such cases it is an easy matter to slip a forefinger on the inner side of the humerus, to run it down to the bend of the elbow, and to draw the forearm downwards across the chest and abdomen, and then to bring the arm down by the side of the trunk. But many cases require far more skill.

The cardinal rule to follow is to observe the natural flexions of the limbs, always to bend them in the direction of their natural movements. The arms, therefore, must always be brought forwards across the breast. The way to do it is as follows:—Slip one or two fingers up along the back of the child's thorax, and bend the first joints over the shoulder between the acromion and the neck; then slide the fingers forwards, catching the humerus in their course, and carrying this with them across the breast or face. This movement will restore the humerus to its natural flexion in front of the body. Of course, as the humerus comes forwards the forearm follows. Your fingers continuing to glide down will reach the bend of the elbow, and, still continuing the same downward and forward movement across the child's breast and abdomen, the arm is extended and laid by the side of the trunk.

That is what has to be done. But is it indifferent *which arm you shall bring down first?* The most simple rule is to take that first which is the easiest, for when

Fig. 68.

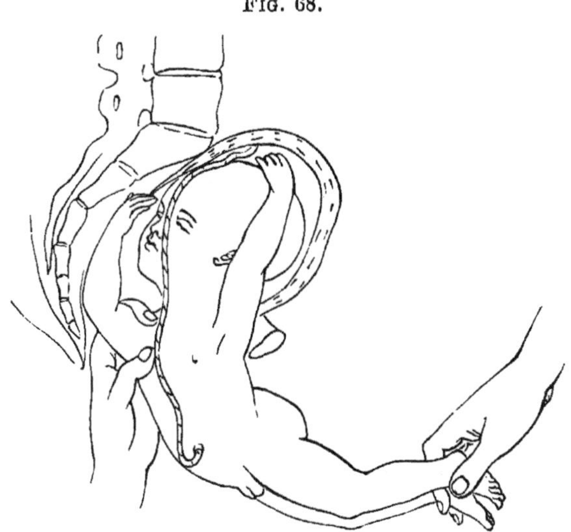

REPRESENTS THE MODE OF LIBERATING THE POSTERIOR OR SACRAL ARM.

one is released the room gained renders the liberation of the second arm easy enough. Generally there is most room in the sacrum; therefore it is best to take the posterior arm first.

Now I have to describe manœuvres for overcoming the difficulties which not seldom oppose your efforts to release the arms. There are two principal ones. The first is this: You want to bring the posterior or sacral arm within reach of your finger. Carry the child's body well forwards, bending it over the symphysis pubis. The effect of this is a twofold advantage. Space is gained between the child's body and the sacrum for manipulation; and as the child's body

FIG. 69.

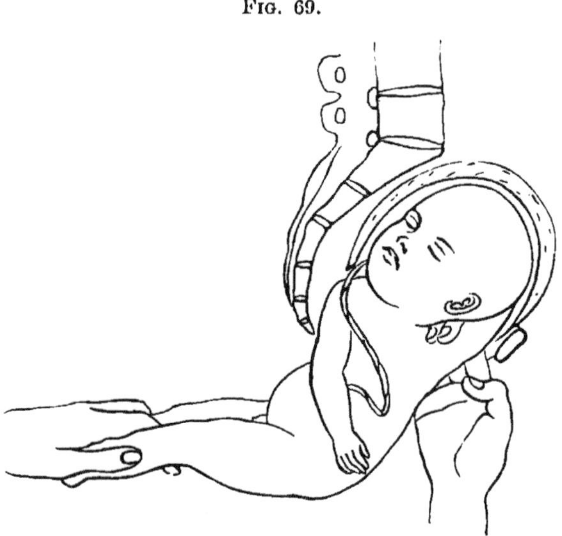

REPRESENTS THE MODE OF LIBERATING THE ANTERIOR OR PUBIC ARM.

revolves round the pubic centre, the further or sacral arm is necessarily drawn lower down, commonly within reach. When the sacral arm is freed, you reverse the manœuvre, and carry the child's trunk backwards over the coccyx as a centre. This brings down the pubic arm.

The second manœuvre may be held in reserve

should the first fail. To execute it you must bear in mind the natural flexions of the arms. You grasp the child's trunk in the two hands above the hips, and give the body a movement of rotation on its long axis, so as to bring its back a little to the left. The effect of this is to throw the pubic arm, which is prevented

FIG. 70.

REPRESENTS A MODE OF LIBERATING THE ARMS.
The trunk is rotated an eighth of a circle from right to left, so as to throw the left arm across the face.

by friction against the canal from following the trunk in its rotation, across the breast. Then, your object being accomplished so far, you call to your aid the first manœuvre, and bring this arm completely down. This done, you reverse the action and rotate the trunk in the opposite direction. The sacral arm is thus brought to the front of the chest, and, by carrying the trunk back, your fingers will easily complete the process.

It is desirable, for reasons we shall presently explain, to avoid this rotation if possible; but under

FIG. 71.

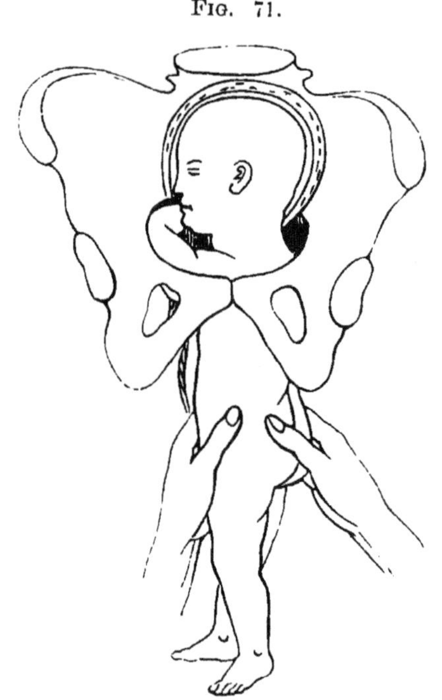

REPRESENTS THE RESULT OF THE MANŒUVRE BEGUN IN FIG. 70.
By rotating the trunk from right to left, the left arm is thrown across the face.

certain circumstances of difficulty it is exceedingly valuable. The rotation need not be considerable; an eighth of a circle is commonly enough, and as it is neutralized by reversal, an objection that might otherwise be urged against the manœuvre is removed.

A paramount reason why you should be careful in imparting rotation to the trunk, or "giving the turns," is this: the union of the atlas with the occipital condyles is a very close articulation; it permits flexion and extension only. The atlas forms with the axis a rotatory joint, so constructed that if the movement of rotation of the head be carried beyond a quarter of a circle, the articulating surfaces part immediately, and the spinal cord is compressed or torn. Thus, if the chin of the fœtus pass the shoulder in turning backwards, instant death results. I have no doubt that many children have been lost through oblivion of this fact. Sometimes the arm will hitch on the edge of the pelvic brim, or just above the imperfectly expanded os uteri. Never attempt, by direct hooking on the middle of the humerus, to drag it through. You would almost certainly break it. Press it steadily against the child's face, and under its chin, running your finger down as near the elbow as possible, so as to lift this part, as it were, over the obstruction.

The arms liberated, now begins the *third act, the extraction of the head*, often a task of considerable difficulty, and always demanding the strictest observance of the laws which govern the mechanism of labour. This act differs from the two first in that, whilst these are sometimes effected by Nature, the liberation of the head must almost always be conducted by art. When the head is last, and has entered the brim, it is very much removed from the influence of expulsive action. The uterus can with difficulty follow it into the pelvis, and the trunk, unless supported by the hands, would, by its mere *vis inertiæ* and friction against the bed, retard the advance of the head. Moreover, this is the stage of chief danger from compression of the cord. The round head fills the brim and the cervix uteri, so that the cord can hardly escape. It would be folly, therefore, to sit by and trust to Nature in this predicament, at the risk of losing that for which the whole operation of version and extraction has been performed—namely, the child's life. Let us suppose for a moment that the head is in the pelvis, and that you cannot extract it at once. If you can get air into the chest, which, being outside the vulva, is free to expand, there is no need to hurry the extrication of the head. You may sometimes get the tip of a finger in the mouth, and drawing this down, whilst you lift up and hold back the perineum, you may enable air to enter the chest. In this way I have kept a child breathing for ten minutes before the head was born. Another plan is to pass a catheter or other tube up into the mouth, so as to give, by means of a kind of artificial trachea, communication with the external air. But I must warn you not to trust to these or similar plans, lest the golden opportunity be irretrievably lost. The real problem is to get the head out of the pelvis.

There are two principal modes of doing this. One is to apply the forceps. This operation I have described. (*See* Lecture IV.) It has been advocated by Busch, Meigs, Rigby, and others. I have practised it successfully, but think it is inferior in celerity and convenience to the second mode, by manual extraction. Remember that the head has to perform a double rotation in its progress. It must revolve round the symphysis pubis as a centre; it must rotate in the cavity on its vertico-spinal axis, so as to bring the face into the hollow of the sacrum. You must then, in extracting, respect these natural movements. You will better follow or guide these movements if you

FIG. 72.

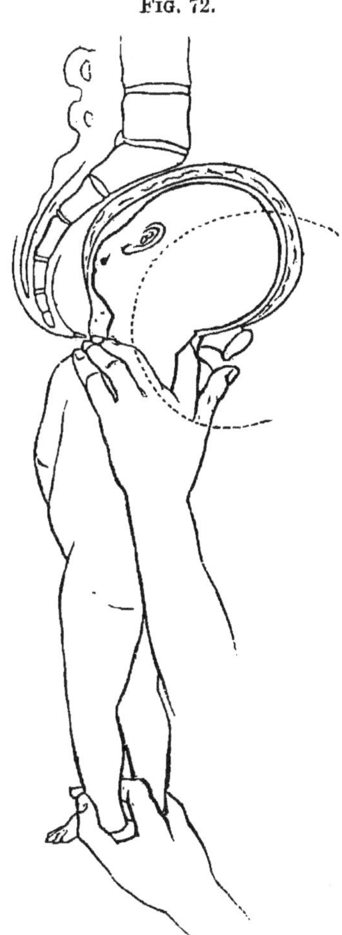

REPRESENTS THE EXTRACTION OF THE HEAD.
The dotted line is the curve of Carus, which indicates the direction to be observed in extraction.

fork the fingers of one hand over the neck behind, and at the same time, holding the legs with the other hand, draw down with careful attention to the curve of Carus. If you carry the body forward too soon, you simply convert the child's head and neck into a hook or crossbar, which, holding on the anterior pelvic wall, will effectually resist all efforts at extraction.

When there is little or no resistance to the escape of the head, it is enough to support the trunk with one hand by holding it at the chest, whilst the other hand on the nucha regulates the exit of the head.

Sometimes it requires considerable force to bring the head through the brim; but whilst force will never compensate for want of skill, it is astonishing how far skill will carry a very moderate force. The modes of extricating the head under circumstances of unusual difficulty will be discussed hereafter. But before passing on I must refer to one practice commonly taught, which is, I believe, based on erroneous observation. You are told to pass a finger into the mouth, or to apply two fingers on the upper jaw, to depress the chin, in order to keep the long axis of the child's head in correspondence with the axis of the pelvis. Now this is a piece of truly "meddlesome midwifery," because it is perfectly unnecessary. The chin is not likely to be caught on the edge of the pelvis or elsewhere, unless, by a previous piece of "meddlesome midwifery," you have been busy in "giving the turns." The truth is, Nature has taken care to arrange the convenient adaptation of means to end in head-last labour as in head-first. It is true that the occipito-spinal joint is seated behind the centre. It might, *prima facie*, appear that the occiput, forming the shorter arm of the head-lever, would tend to roll back upon the nucha. But this is not so in practice. The broad, firm expanse of the occiput, forming a natural inclined plane directed upwards, is surely caught by the walls of the parturient canal as the head descends. The greater friction thus experienced by a larger superficies favourably disposed virtually converts the shorter arm of the lever into the more powerful one; it is more retarded in its course; and therefore the chin is kept down near the breast, and therefore, again, there is no need for the obstetrist to meddle in the matter.

LECTURE XIV.

TURNING WHEN LIQUOR AMNII HAS RUN OFF, THE UTERUS BEING CONTRACTED UPON THE CHILD—THE PRINCIPLE OF SEIZING THE KNEE OPPOSITE TO THE PRESENTING SHOULDER ILLUSTRATED.

So long as there is any liquor amnii present in the uterus, and often for some considerable time afterwards, the bi-polar method of turning is applicable. But a period arrives when it becomes necessary to pass a hand fairly into the uterus in order to seize a limb. We will now discuss the mode of turning under the more difficult circumstances of loss of liquor amnii, more or less tonic contraction of the uterus upon the child, and descent of the shoulder into the pelvis.

The contraction of the uterus, naturally concentric or centripetal, tends to shorten the long axis of the child's body. (*See* Figs. 39, 40, 41.) The effect is to flex the head upon the trunk, and to bend the trunk upon itself, reducing the ovoid to a more globular form. This brings the knees nearer to the chest, but does not diminish the difficulty of turning.

I need not pause again to discuss minutely the

preparatory measures. It is only necessary here to call to mind that chloroform or opium is especially serviceable, and that it is important to empty the bladder and rectum.

The first question to determine is, *Which hand will you pass into the uterus?* I have given some of the reasons why the left hand should be preferred in Lecture XII. In the majority of cases the child's back is directed forwards; to reach the legs, which lie on the abdomen, your hand must pass along the hollow of the sacrum, and this can hardly be done, the patient lying on her left side, with the right hand, without a most awkward and embarrassing twist of the arm. I need scarcely point out how violent and unnatural a proceeding it would be to pass up the right hand between the child's back and the mother's abdomen, to carry the hand quite round and over the child's body in order to seize the feet which lie towards the mother's spine, and then to drag them down over the child's back. If you attempted this, you would probably get into a difficulty. The child, perhaps, would not turn at all. To avoid this failure, the rule has been laid down to pass your hand along the inside or palmar aspect of the child's arm. This will guide you to the abdomen and the legs. Or the rule has been stated in this way:—Apply your hand to the child's hand, as if you were about to shake hands. If the hand presented to you be the right one, take it with your right, and *vice versâ*.

Rules even more complicated are proposed, especially by Continental authors. Some go to the extent of determining the choice of hand in every case by the position of the child. The fallacy and uselessness of these rules are sufficiently evident from the disagreement among different teachers as to which hand to choose under the same positions. Rules, moreover, which postulate an exact knowledge of the child's position are inapplicable in practice, because this diagnosis is often impossible until a hand has been passed into the uterus; and it is certainly not desirable to pass one hand in first to find out which you ought to use, at the risk of having to begin again and to pass in the other.

The better and simpler rule is this:—*In all dorso-anterior positions, lay the patient on her left side; pass your left hand into the uterus*—it will pass most easily along the curve of the sacrum and the child's abdomen; *your right hand is passed between the mother's thighs to support the uterus externally.*

In the case of *abdomino-anterior positions, lay the patient on her back, and you may introduce your right hand, using the left hand to support the uterus externally.* If the patient is supported in lithotomy position, you can thus manipulate without straining or twisting your arms or body. But it is equally easy to use the left hand internally if the patient is on her back, so that the exception is only indicated to suit those who have more skill and confidence with the right hand.

We will first take a dorso-anterior position. Introduce your left hand into the vagina, along the inside of the child's arm. The passage of the brim, filled with the child's shoulder, is often difficult. Proceed gently, stopping when the pains come on. At the same time support the uterus externally with your right hand. Sometimes you may facilitate the passage of the brim by applying the palm of the right hand in the groin, so as to get below the head and to push it up. This will lift the shoulder a little out of the brim. Or you may adopt a manœuvre attributed to Von Deutsch, but which had been practised by Levret. This consists in seizing the presenting shoulder or side of the chest by the inside hand, lifting it up and forwards, so as to make the body roll over a little on its long axis. This may be aided by pressure in the opposite direction by the outside hand on the fundus uteri, getting help from the bi-polar principle.

Sometimes advantage is to be gained by placing the patient on her elbows and knees. In this position you are favoured by gravity, for the weight of the fœtus and uterus tends to draw the impacted shoulder out of the brim.

The brim being cleared, your hand passes onwards into the cavity of the uterus. This often excites spasmodic contraction, which cramps the hand, and impedes its working. Spread the hand out flat, and let it rest until the contraction is subdued. In your progress you must pass the umbilicus, or a loop of umbilical cord will fall in your way. Take the opportunity of feeling it, to ascertain if it pulsates. You thus acquire knowledge as to the child's life. But you must not despair of delivering a live child because the cord does not pulsate. I have several times had the satisfaction of seeing a live child born where I could feel no pulsation *in utero*. You are now near the arm and hand. They are very apt to perplex. Keep, therefore, well in your mind's eye the differences between knee and elbow, hand and foot, so that you may interpret correctly the sensations transmitted by your fingers from the parts you are touching.

At the umbilicus you are close to the knees. The feet are some way off at the fundus of the uterus applied to the child's breech.

What part of the child will you seize? It is still not uncommon to teach that the feet should be grasped. You will see pictures copied from one text-book to another, representing this very unscientific proceeding. There ought to be some good reason for going past the knees to the feet, which are further off, and more difficult to get at. Now, I know of no reasons but bad ones for taking this additional trouble. You can turn the child much more easily and completely by seizing one knee. Dr. Radford insists upon seizing one foot only, for the following reason:—The child's life is more frequently preserved where the breech presents than where the feet come down first. A half-breech is also safer than cases where both feet come down. The dilatation of the cervix is better done by the half-breech. The circumference of the breech, as in breech presentations, is from twelve to thirteen and a half inches, nearly the same as that of the head; the circumference of the half-breech, one leg being down, is eleven to twelve and a half inches, whilst the circumference of the hips, both legs being down, is only ten to eleven and a half inches.

But a knee is even better than a foot. You determine, then, to sieze *one* knee; which will you choose? The proper one is that which is furthest. The reasons are admirably expressed by Professor Simpson. We have a dorso-anterior position, the right arm and shoulder are downmost; these parts have to be lifted up out of the brim. How can this be best done? Clearly by pulling down the *opposite* knee, which, representing the opposite pole, cannot be moved without directly acting upon the presenting shoulder. If the opposite knee be drawn down, and supposing the child to be alive or so recently dead that the resiliency of its spine is intact, the shoulder must rise, and version will be complete, or nearly so. But if both feet are seized, or only the foot of the same side as the presenting arm, version can hardly be complete, and will, perhaps, fail altogether.

This point is worth illustrating. I have taken Fig. 73 from Scanzoni (*Lehrbuch der Geburtshülfe*, 4th edition, 1867), in order to show you the error in practice which I wish you to avoid. It represents a dorso-anterior position, the right shoulder presenting, and the method recommended by Scanzoni. The operator's left hand is seizing and drawing down the right leg. I have introduced the arrows to indicate the direction of the movements sought to be imparted. You want the shoulder to run up whilst you draw down the leg. Now, drawing on the right leg necessarily tends to bring it towards the shoulder, the line of motion of the leg being more or less perpendicular to that of the shoulder. The body bends upon its side, the leg and shoulder get jammed together, and you have failed to turn.

Contrast this figure with Figs. 74 and 75, which I have designed to show the true method and principle of turning. The arrows, as before, indicate the direction of the movements. By drawing upon the opposite knee to the presenting shoulder, the movements run parallel in directly opposite directions, like the two ends of a rope round a pulley. You cannot draw down the left leg without causing the whole trunk to revolve; and the right shoulder will necessarily rise. To turn effectively, the child must revolve upon its long or spinal axis, as well as upon its transverse axis. Turning, in short, is a compound or oblique movement between rolling over on the side and the somersault.

Fig. 73.

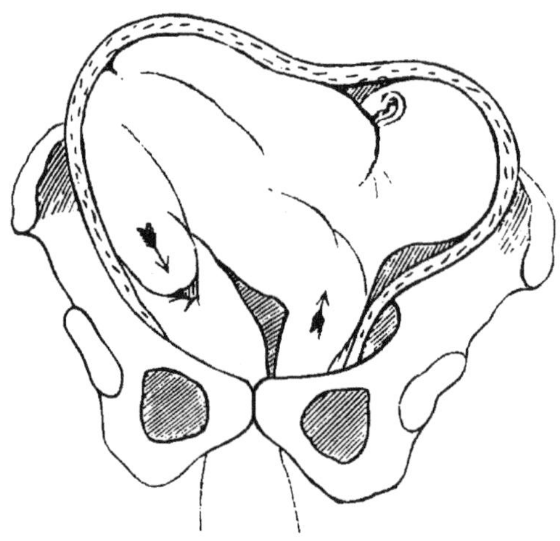

AFTER SCANZONI.

To show the error of attempting to turn by seizing the leg of the same side of the prolapsed shoulder.

If you seize both legs, you mar this process. The only cases in which I have found it advantageous to seize both legs are those in which the child has been long dead. Here the spine has lost its elasticity. The body will hardly turn, and there is nothing to be gained for the child in maintaining the half-breech and preserving the cord from pressure.

The seizure of a foot is not seldom a matter of so

FIG. 74.

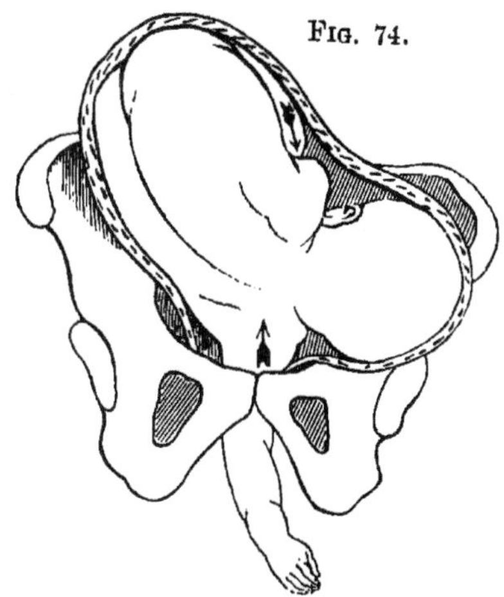

SHOWING THE PRINCIPLE OF TURNING BY BRINGING DOWN THE KNEE OF THE OPPOSITE SIDE TO THE PRESENTING SHOULDER.

The arrows indicate the reverse movements effected. The object is to carry up the right shoulder. By bringing down the left knee this is most surely effected.

FIG. 75.

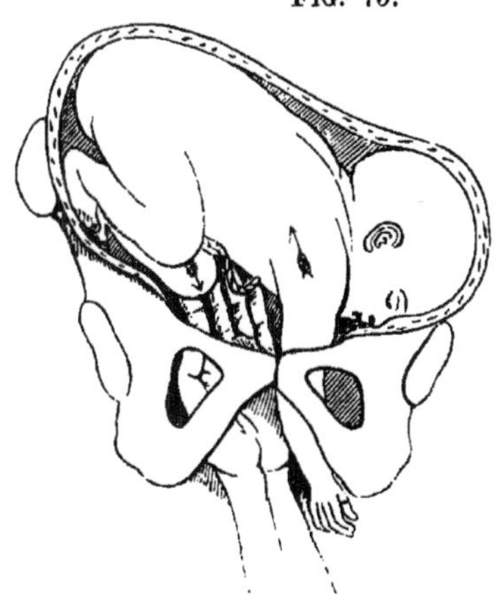

SHOWS TURNING IN PROGRESS.

As the left knee descends, the trunk revolves on its transverse as well as on its long axis, and the right shoulder rises out of the pelvis.

much difficulty that various instruments have been

contrived to attain this object. To draw the foot or feet down, you must grip them firmly—that is, your fingers must be flexed in opposition to the thumb, or two fingers must coil around the ankle. This doubling of your hand takes room. Whereas, to seize a knee, only requires the first joint of the forefinger to be hooked in the ham. Fig. 76 shows Braun's contrivance for snaring a foot. A loop of tape in the form of a running noose is carried by means of a gutta-percha rod, about a foot long, into the uterus, guided by the hand to the foot. When you have succeeded in getting the noose over the ankle, you pull on the free end, and withdraw the rod.

Fig. 76.

BRAUN'S SLING-CARRIER, TO APPLY A LOOP ROUND THE FOOT, OR TO REPLACE THE UMBILICAL CORD.

Hyernaux, of Brussels, has invented a very ingenious instrument, a *porte-lacs*, or noose-carrier, for the purpose. There are many others, but since they are created in order to meet an arbitrary—I might say a wantonly-imposed—difficulty, arising out of an erroneous practice, they need not be described. It is true that it is often convenient to attach a loop to the foot when brought into the vagina, to prevent it from receding before version is complete. But this can be done by the fingers with a little dexterity. The occasions on which it is necessary to seize a foot which can only be barely touched by the fingers are extremely rare.

For these, I think the simple apparatus of Braun, which also serves for the reposition of the umbilical cord, is as efficient as any. The wire-écraseur forms an excellent snare to seize a foot. A loop just large enough for the purpose is made, and guided over the ankle. It can be slightly drawn in to fix the grasp, avoiding, of course, cutting into the limb.

LECTURE XV.

TURNING IN THE ABDOMINO-ANTERIOR POSITION — INCOMPLETE VERSION, THE HEAD REMAINING IN ILIAC FOSSA, CAUSES OF—COMPRESSION OF UTERUS, TREATMENT OF — BI-MANUAL OR BI-POLAR TURNING WHEN SHOULDER IS IMPACTED IN BRIM OF PELVIS.

Turning in abdomino-anterior positions does not differ essentially from turning in dorso-anterior positions. I have already said that the best position for the patient is on her back, and that the right hand may be used. The uterus, as in all cases, is supported externally, whilst you pass your right hand along the inner aspect of the child's arm and behind the symphysis pubis; it proceeds across the child's belly, to seize the opposite knee. Drawing this down in the direction of the arrow in Fig. 77, the presenting shoulder rises out of the pelvis.

There is a feature in the history of turning which has not received the attention it deserves. I have found that, notwithstanding diligent adherence to the rules prescribed, turning is not always complete. The head and part of the chest are apt to stick in the iliac fossa, the trunk being strongly flexed.

Indeed, I believe that complete version is rather the exception than the rule in cases where the liquor amnii is drained off, and the uterus has moulded itself upon the fœtus so as to impede the gliding round of the fœtus.

The complete version which exists as the ideal in the minds of most of those who perform the operation is not often realized. Indeed, it can hardly take place

Fig. 77.

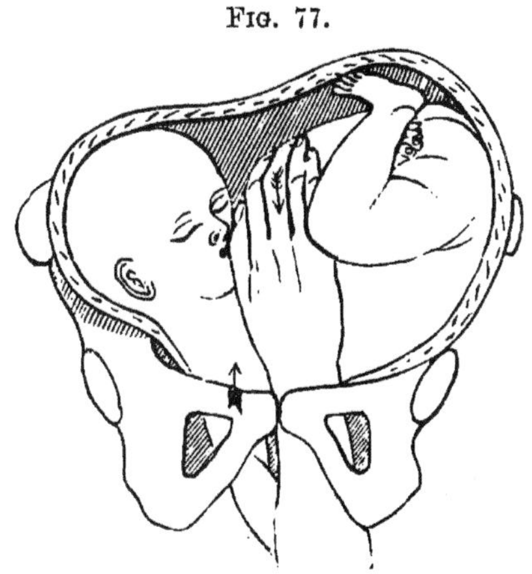

SHOWS TURNING IN ABDOMINO-ANTERIOR POSITION.

The operator's right hand seizes the upper or left knee. As this comes down, the child's body, rotating upon its transverse and long axes, draws the right or presenting shoulder up out of the pelvis.

unless the bi-polar method by combined external and internal manipulation is carefully pursued. The head may commonly be felt throughout the entire process nearly fixed in the iliac fossa, and sometimes the forearm remains in the upper part of the pelvic cavity. The nates and trunk are delivered as much by bending as by version. The process is something between spontaneous version and spontaneous evolution. The two following diagrams (Figs. 78 and 79), taken from memoranda made of a case which occurred to me, will serve to illustrate both this feature of incomplete turning and the importance of the principle of drawing upon the leg opposite to the presenting shoulder.

FIG. 78.

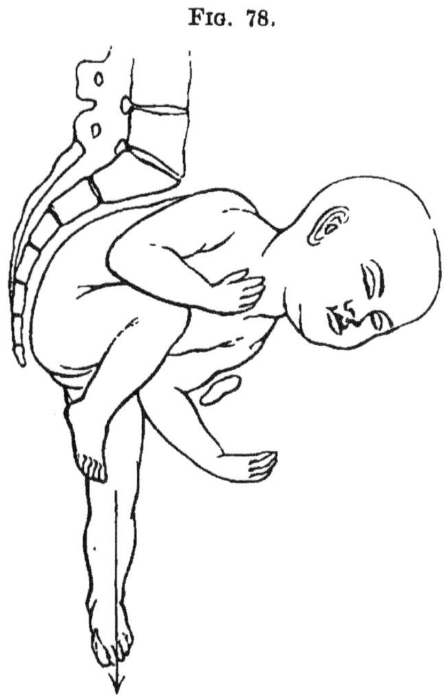

REPRESENTS AN ABDOMINO-ANTERIOR POSITION, LEFT SHOULDER PRESENTING.
Traction was first made upon the left leg, as shown by the arrow. The effect was to bend the trunk and jam the shoulder against the symphysis.

If the head and shoulders rise enough to permit the breech to enter the brim, delivery will not be seriously obstructed. But it not uncommonly happens in extreme cases of impaction of the shoulder in the upper part of the pelvis, that, even when you have succeeded in bringing down a leg into the vagina, version will not proceed: the shoulder sticks obstinately in the brim. In such a case the bi-polar principle must be called into action. It is obvious that if you draw down upon the leg,

FIG. 79.

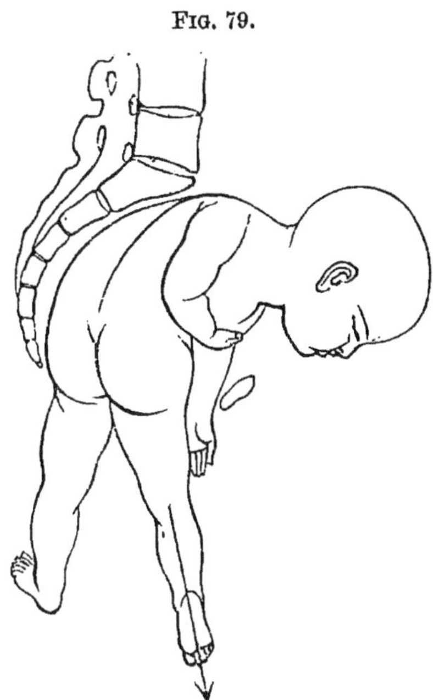

REPRESENTS THE CORRECTION OF THE ERROR COMMITTED IN FIG. 78.

By drawing on the opposite, or right, leg, the trunk was made to revolve on its spinal axis, drawing up the presenting arm from the pelvis, and allowing the breech to descend. Although delivery was now effected, version was not complete, as the head remained in the iliac fossa, and the hand never quitted the pelvic cavity.

whilst you push up the shoulder, you would act at a great advantage. But you cannot get both your hands into the pelvis. Sometimes you may release the shoulder by external manipulation, pressing up the head by the palm of your hand insinuated between it and the brim of the pelvis. In cases of real difficulty, however, this will not answer. You must push up the shoulder by the hand inside. To admit of this, you pass a noose of tape round the ankle in the vagina, and draw upon this. The noosing of the foot is not always easy. To effect it you carry a

Fig. 80.

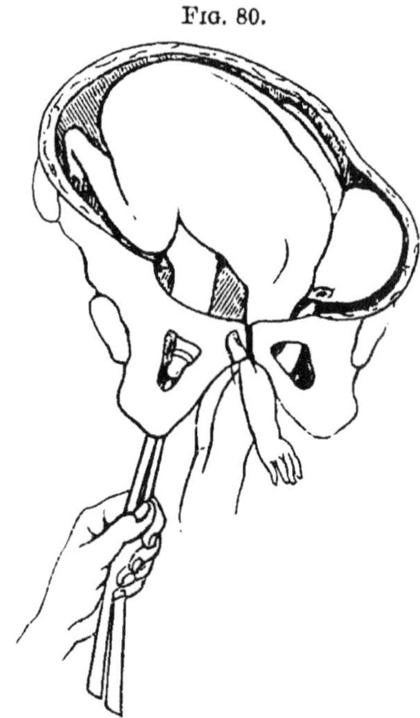

REPRESENTS THE BI-POLAR METHOD OF LIFTING AN IMPACTED SHOULDER FROM THE BRIM OF THE PELVIS, SO AS TO EFFECT VERSION.

running noose on the tips of two or three fingers of one hand up to the foot, held down as low as possible in the vagina by the other hand. Then the loop is slipped up *beyond the ankles and heel*, and drawn tight. Often you will have to act with one hand only in the vagina, the hand outside holding on the free end of the tape ready to tighten the noose as soon as it has got hold. Or, whilst holding the foot with one hand, you may carry the noose by help of Braun's instrument. (*See* Fig. 76). The foot being securely caught, the right hand is passed into the vagina, and the fingers or palm, if necessary, are applied to the shoulder and chest. Now, you will find it difficult to draw upon the tape and to push upon the shoulder exactly simultaneously. There is so little room that, whenever you push, there is a tendency to carry the leg up as well. The most effective movement is as follows:—Pull and push alternately. Presently you will find the leg will come lower, and the prolapsed arm will rise.

In pushing the chest and shoulder, it is not unimportant in what direction you push. You cannot push backwards, or even directly upwards. Your object is to get the trunk to roll over on its spinal axis. Here, then, is an indication to carry out the manœuvre of Levret and Von Deutsch. *Push the shoulder and adjacent part of the chest well forwards*, so as to make them describe a circle round the promontory as a centre.

Various other contrivances have been designed in order to accomplish this end. Crutches or repellers have been made, by which to push up the shoulder instead of by the hand. The objection to these is that you cannot always know what you are doing. But your hand is a sentient instrument, which not only works at your bidding, but constantly sends telegrams to the mind, informing it of what is going on, and of what there is to do.

LECTURE XVI.

TURNING (*continued*) — IMITATION OR FACILITATION OF DELIVERY BY THE PROCESS OF SPONTANEOUS EVOLUTION — EVISCERATION — DECAPITATION — EXTRACTION OF A DETRUNCATED HEAD FROM THE UTERUS.

We have discussed the mode of dealing with those cases of difficult shoulder-presentation in which, the uterus having contracted closely upon the child, the shoulder is more or less firmly, but not immoveably, fixed in the pelvic brim. In the majority of cases of this kind we are justified in attempting to turn, because there is still a prospect of the child being preserved.

But there are cases in which matters have proceeded a stage further, in which the shoulder and corresponding side of the chest are driven deeply into the pelvis—in which, consequently, the body is considerably bent upon itself. Now, this can only occur after protracted uterine action, such as is scarcely compatible with the life of the child. Either the child was already dead at an early stage of labour—a condition, especially if the child were also of small size, most favourable to the carrying out of this process of spontaneous evolution—or the child has been destroyed under the long-continued concentric compression of the uterus.

In the presence of such a case, the first question we have to consider is—Will Nature complete the task she has begun? Will the child be expelled spontaneously? A little observation will soon enable us to determine how far this desirable solution of the difficulty is probable, and when we ought to interpose. If the pelvis is roomy in proportion to the child; if the child is dead, small, and very flaccid; if we find the side of the chest making progress in descent under the influence of strong uterine action possessing an expulsive character, and if the patient's strength be good, we shall be justified in watching passively. But if we find no advance, or but very slow advance, of the side of the chest, the child being large and not very plastic; if the uterus has ceased to act expulsively, and the patient's strength is failing, we must aid delivery. Then comes the second question: In what manner? This must depend upon the circumstances of the case. If a little help *à fronte* to make up for deficient *vis à tergo* promise to be enough, we may imitate the proceeding of Peu, who, in a case in which spontaneous evolution was in progress, passed a cord round the body to pull upon and aid the doubling.

Or we may much facilitate the doubling and expulsion by *evisceration*. This operation consists in perforating the most bulging part of the chest, and picking out the thoracic and abdominal viscera. When this is done, traction upon the body by the crotchet or the craniotomy forceps, and dragging in the direction of Carus's curve, will commonly effect delivery without difficulty. This operation is represented in Fig. 81.

FIG. 81.

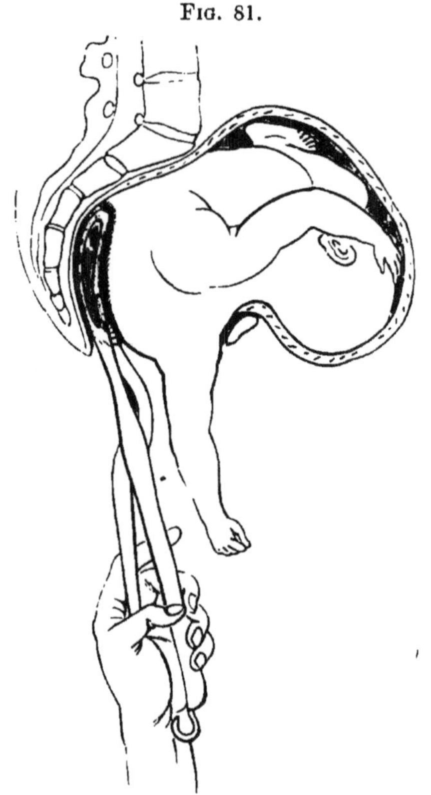

REPRESENTS EXTRACTION AFTER PERFORATION OF THE CHEST, AND DELIVERY IN IMITATION OF SPONTANEOUS EVOLUTION.

Sometimes perforation and evisceration are insufficient in themselves, and another step will be necessary in order to complete delivery. This ultimate step is *decapitation*, an operation of extreme importance, capable of bringing almost instant relief and safety to the mother. It is pointed out by Celsus, and was clearly described by Heister after Von Hoorn.

The recognition, or at least the application, of this proceeding is so inadequate, that I think it useful to state the arguments in favour of it with some fulness.

The late Professor Davis, in his great work,* a work too much neglected by his successors, says:—"It may be considered a good general rule never to turn when the death of the child is known to have taken place." In cases of long impaction he recommends an operation to be performed *upon the child*—namely, "*bisection of the child at the neck.*" Again, he says: "It ought to be an established rule in practice to decapitate in arm-presentations not admitting of the safer performance of turning."

Dr. Ramsbotham† also says:—"It appears to me better practice either to eviscerate or decapitate the fœtus, than to endeavour to deliver by turning, in all cases where the uterus is so strongly contracted round the child's body as to cause apprehension of its being lacerated by the introduction of the hand; because if such a degree of pressure is exerted on it as to render the operation of turning very difficult, the child must have died, either from the compression on its own chest, or on the funis, or on the placenta itself."

Such men speak with an authority that commands respect. Their opinions are not the crude utterances of dogmatism presuming on a little reading and small experience, but conclusions drawn from repeated en-

* "Obstetric Medicine."

† "Medical Times and Gazette," December, 1862.

counters with great emergencies. The justness of the rule thus distinctly expressed by Davis and Ramsbotham is attested by the practice of the most eminent Continental practitioners. Decollation has been advocated and practised by l'Asdrubali (1812), by Paletta,* by Braun of Vienna, by Dubois, by Lazzati of Milan,† and by many others.

Various instruments have been designed to effect decapitation. Ramsbotham's hook is perhaps best known in this country, and it has served as a model for several modifications made abroad. It is named after the first Ramsbotham. It was described and recommended by Professor Davis. It consists of a curved hook, having a cutting edge on the concave part, supported on a strong straight stem, mounted on a wooden handle. Professor Davis also used another instrument, of his own contrivance—the guarded embryotomy knife. It consists of two blades working on a joint like the forceps. One blade is armed with a strong knife on the inner aspect. The other blade is simply a guard; it is opposed to the knife, and receives it when the neck is severed.

A plan sometimes resorted to is to carry a strong string round the neck, and then, by a to-and-fro or sawing movement, effected by cross-bars of wood on the ends to serve as handles, to cut through the parts.

In a discussion on the subject at a recent meeting of the Edinburgh Obstetrical Society, Dr. C. Bell suggested the use of an instrument like that for plugging the nares in epistaxis to carry a cord. Dr.

* "Del Parto per il Braccio," Bologna, 1808.

† "Del Parto per la Spalla," 1867.

Keiller mentioned an instrument designed by Dr. Ritchie, like an écraseur, with a perpetual screw and chain.

Strong scissors have been made for the purpose, which cut through the vertebræ. A good instrument of this kind has been designed by Dr. Mattéi, of Paris. I believe it is a very useful form, as it is sometimes easy to cut through the spine when it is difficult to pass a hook over the child's neck. It resembles the surgical bone-forceps.

It is important to remember that the spine may be divided by piercing the vertebræ with the common perforator, then separating the blades so as to rend or crush asunder the bones. What remains may then be divided by scissors. Failing special instruments, the spine may be divided by strong scissors, or even by a penknife or a Wharncliffe blade.

The favourite instrument in Germany and Italy is Braun's blunt hook or "decollator." I have not tried it, having hitherto used Ramsbotham's cutting hook; but I am inclined to think it is a better instrument than Ramsbotham's. Dr. Garthshore performed the operation with an ordinary blunt hook. It is certainly desirable to do away with the cutting edge, which is not without danger to the mother and the operator. Braun's instrument is twelve inches long, including the thickness of the handle; the greatest width of the hooked part is one inch; the geratest thickness of the stem is from four to five lines. Lazzati introduces a gentle curve into the stem near the hook.

The Operation of Decapitation.—It will be best described as consisting of *three stages.* The first stage is the application of the decapitator and the bisection of the neck; the second is the extraction of the trunk; the third the extraction of the head.

The First Stage.—The patient may lie on her left side. Take Ramsbotham's hook or Braun's decollator. As the instrument should be passed up over the back of the child's neck, it is, in the first place, necessary to ascertain whether the position be dorso-anterior or abdomino-anterior. It is also necessary to determine accurately whether the fœtus is still in great part above the brim lying transversely or obliquely, in which case the head and neck will be in one or other side; or whether, a great part of the chest having descended into the pelvis, the movement of rotation has taken place, in which case the head and neck will be found in front near the symphysis. The observation of the relations of the prolapsed arm and exploration with the hand internally will inform us as to these particulars. The next step is to get an assistant to pull upon the prolapsed arm, so as to bring down the shoulder and fix it well. This brings the neck nearer within reach. The operator then passes his left hand, or two or three fingers, if this be enough, into the vagina, over the anterior surface of the child's chest, until his fingers reach the fore part of the neck. With his right hand he then insinuates the hook, laying flat, as in the dotted outline in Fig. 82, between the wall of the vagina and pelvis and the child's back, until the beak has advanced far enough to be turned over the neck. The beak will be received, guided, and adjusted by the fingers of the left hand. The instrument being *in situ*, whilst cutting or breaking through the neck, it is still desirable to keep up traction on the prolapsed arm. In using Ramsbotham's hook, a sawing movement must be executed, carefully regulating your action by aid of the fingers applied to the beak. If Braun's decollator be used, the movement employed is rotatory, from right to left, and at the same time, of course, tractile. The instrument crushes or breaks

Fig. 82.

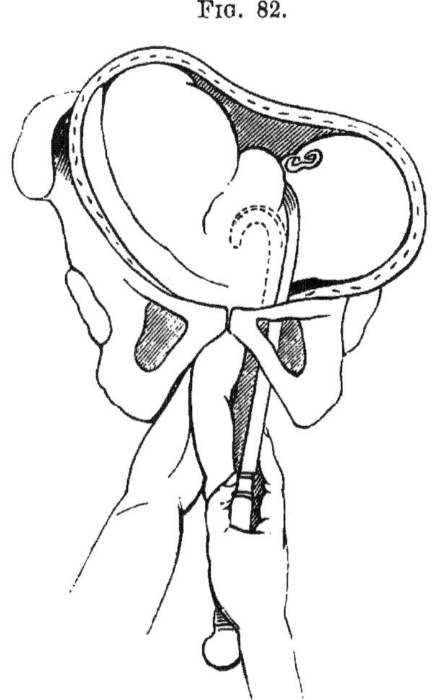

REPRESENTS THE FIRST STAGE OF DELIVERY BY DECAPITATION.

The dotted outline shows how the hook (Ramsbotham's) is introduced, lying flat upon the back of the child's neck, the beak being then turned over the neck, and meeting the fingers of the left hand on the anterior aspect.

through the vertebræ. When the vertebræ are cut through, some shreds of soft parts may remain. These may be divided by scissors, or be left to be torn in the *second stage* of the operation—*the extraction of the trunk.*

The wedge widening above the brim, that hitherto

obstructed delivery, is now bisected, divided into two lesser masses, each of which separately can readily be brought through the pelvis. By continuing to pull upon the prolapsed arm the trunk will easily come through, the head being pushed on one side out of the way by the advancing body. (See Fig. 83.)

FIG. 83.

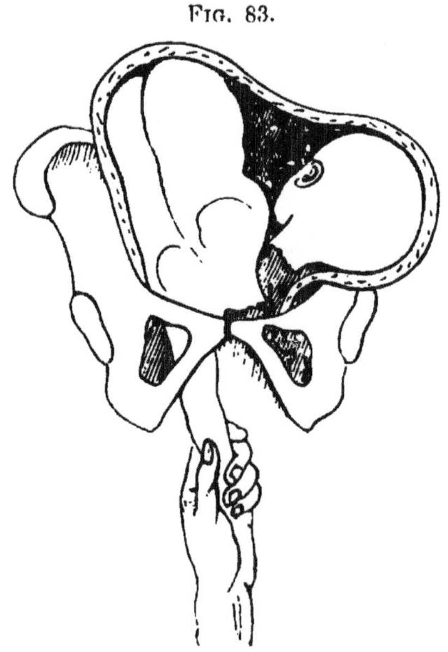

REPRESENTS THE SECOND STAGE OF DELIVERY AFTER DECAPITATION, OR EXTRACTION OF THE TRUNK BY PULLING ON THE PROLAPSED ARM.

The head, no longer linked to the body, is pushed out of the way as the trunk descends.

The Third Stage: The Extraction of the Head.—The problem how to get away a detruncated head left behind in the uterus is not always easy of solution. In the case before us, the child having probably been dead many hours, the bones and other structures have lost all resiliency, the connections of the bones are broken down by decomposition, and the whole becomes a plastic mass, easily compressible. Such a head will sometimes be expelled spontaneously. I have taken away a head, under the circumstances under discussion, by seizing it with my fingers. On the other hand, I have on several occasions been called in to extract a head which resisted ordinary means. There are four modes of action. The crotchet, the forceps, the craniotomy-forceps or the cephalotribe may be used. If the *crotchet* can be passed into an orbit, or into the cranial cavity, getting a good hold, this plan may answer. The objection to it is the difficulty of getting such a hold, and the risk of the point slipping and rending the soft parts of the mother. The head, being loose, rolls over in the uterus when an attempt is made to seize it. I therefore discard the crotchet.

The *forceps* is better adapted. If the head can be seized, which is not always easy, for it is apt to escape high above the brim, and roll about when touched by the blades, extraction is not difficult. Care must, moreover, be taken to seize the head in such a manner that the spicula resulting from the severance of the vertebræ shall not drag along or injure the mother's soft parts.

I prefer the *craniotomy-forceps* as being much the most certain and safe. In order to obtain a hold, it is generally necessary first to perforate. The free rolling of the head when pressed by the point of the perforator tends to throw this off at a tangent, missing the cranium, and endangering the soft parts of the mother. To obviate this difficulty, the head must be firmly fixed down upon the pelvic brim by an assistant, who grasps and presses the uterus and head down by both hands spread out upon them. The operator then, feeling for the occiput with two fingers of his left hand, and guided by them, carries up the perforator with his right hand, taking care that the point shall strike the head as nearly perpendicularly as possible. He then,

Fig. 84.

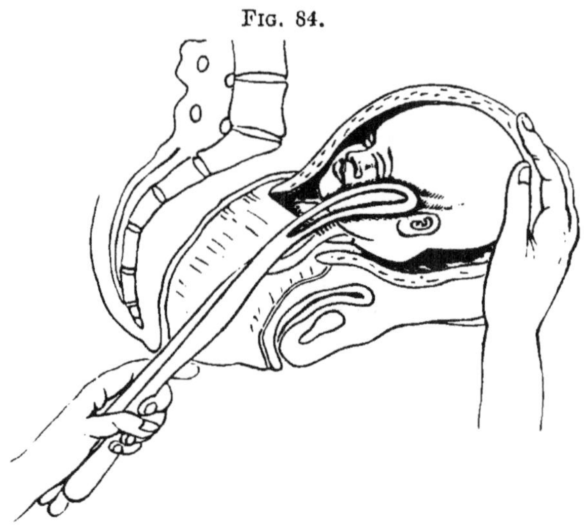

REPRESENTS THE THIRD OR LAST STAGE OF DELIVERY AFTER DECAPITATION.
The head is seized by the craniotomy-forceps, and extraction is made in the direction of Carus' curve.

partly by a drilling, screwing, boring motion, partly by pushing, perforates the cranium. The drilling movement avoids the necessity of using much pushing force, and thus lessens the risk of the instrument slipping.* When a sufficient opening is made into the

* It is in perforating under such circumstances that the vice of perforators curved in the blades is most apparent. Such instruments can hardly strike in a true perpendicular, and the point is almost certain to glide off. The superiority of a straight powerful perforator like Oldham's will not be disputed by any one who has had to perforate under difficulties. The trephine-perforators are not applicable to the case under discussion, owing to the difficulty of fixing the head.

cranium, the craniotomy-forceps is applied, one blade inside, the other outside; the blades are adjusted and locked; and traction made in the orbit of Carus' curve commonly brings the head away without further difficulty. During extraction, the fingers of the left hand should be kept upon the skull at the point of grasp by the instrument, guarding the soft parts from injury by spicula, and regulating the force and direction of traction. In cases where there was difficulty in extracting the trunk, Dr. D. Davis used a double-guarded crotchet, the two blades of which, fixing themselves in the trunk, extracted like a forceps.

If there be any likelihood of difficulty in extraction, there is a last and an effectual resource in the *cephalotribe.* When the head is left behind after turning in contracted pelvis, the cephalotribe to crush down the head is invaluable.

LECTURE XVII.

TURNING IN CONTRACTED PELVIS AS A SUBSTITUTE FOR CRANIOTOMY — HISTORY AND APPRECIATION — ARGUMENTS FOR THE OPERATION: THE HEAD COMES THROUGH MORE EASILY, BASE FIRST—THE HEAD IS COMPRESSED LATERALLY—MECHANISM OF THIS PROCESS EXPLAINED—LIMITS JUSTIFYING OPERATION—SIGNS OF DEATH OF CHILD—ULTIMATE RESORT TO CRANIOTOMY IF EXTRACTION FAILS—THE INDICATIONS FOR TURNING IN CONTRACTED PELVIS—THE OPERATION.

WE now come to a long-contested and still undecided question in obstetric practice—*Is turning ever justifiable as a means of delivery in labour obstructed by pelvic deformity?*

The next alternative in the descending scale of operations is a transition from conservative to what may be distinguished as sacrificial midwifery, involving the destruction of the child. It is obviously a matter of exceeding interest to cultivate any operation that shall hold out a reasonable hope of safety to the child, without adding unduly to the danger of the mother. So much may be conceded on both sides. The question, then, may be set forth as follows:—Do cases of dystocia from pelvic contraction occur in which the child can be delivered alive by turning, and without injury or danger to the mother, which must otherwise be condemned to the perforator? And, not to blink in any way the serious character of the inquiry, it is necessary to append this secondary question to the first—namely:

Assuming that such cases do occur, can they be diagnosed with sufficient accuracy to enable us to restrict the application of turning to them? And if we err by turning in unfitting cases, what is the penalty incurred?—how can we retrieve our error?

These questions I will endeavour to illustrate, if not to answer, by the light of the writings of others, and my own experience and reflections.

The choice of an operation in obstetrics will, in many cases that fall within the debateable territory claimed by two or more rival operations, be determined by the relative perfection of these operations, and by the relative skill in them possessed by the individual operator. This law has governed the history of the progress of the art of delivering women in whom labour was obstructed by disproportion.

"The Fluctuations in Opinion that have prevailed among Practitioners of Midwifery with reference to the Performance of Turning, and the Application of the Forceps in Cases requiring Artificial Delivery on account of Deformity of the Pelvis," is the title of a most interesting memoir by Dr. Charles West.* I refer to it as the most important and most instructive epitome of this subject.

The operation of extracting a child through a contracted brim has no doubt often been performed as a

* "Medical Gazette," 1850.

matter of assumed necessity, as, for example, when the shoulder has presented; and contraction of the pelvis is certainly a cause of shoulder-presentation. The observation of such cases, a certain proportion of which terminated successfully for the child, could not fail to suggest the deliberate resort to the operation in cases of similar contraction where the head presented.

Before the forceps was known, and before the instruments for lessening and extracting the head had been brought to any degree of perfection, turning was commonly resorted to in almost all cases of difficult labour. Thus Deventer, who wrote in 1715, as well as La Motte, declaimed against the use of instruments, and recommended turning by the feet in all cases of difficult cranial presentation. The consequence was that the art of turning was cultivated very successfully by some of the followers of Ambroise Paré. It appears to me evident that, in the early part of the last century, turning was better understood and more skilfully performed than it was at the beginning of the present century; and it is equally evident to me that, by turning, many children were saved under circumstances that are now held to justify their destruction. Of course this gain was not achieved without a drawback. If children were sometimes saved, many mothers were injured or lost by attempts to turn under circumstances which are now encountered successfully by the forceps or by craniotomy.

As instruments were improved, the choice of means was extended. The forceps first contested the ground. The contest, indeed, was for exclusive dominion. The reputed inventor of the forceps, Hugh Chamberlen, did not hesitate to accept the challenge of Mauriceau to attempt to deliver a woman with extreme pelvic contraction by means of his instrument, feeble and imperfect as it was. As science advanced, the contest was better defined. As the obstruction to delivery was due to contraction of the pelvic brim, and the problem was how to extract a live child arrested above the brim, it is obvious that a short single-curved forceps must fail. It was only when the long double-curved forceps was designed, that the knowledge and the power arose which enabled the obstetric surgeon to bring another means into competition with turning, for the credit of saving children from mutilation.

It is, then, from the time of Smellie and Levret, who perfected and used the long forceps, that the real interest of the inquiry dates. It is not a little remarkable that amongst those who have most distinctly recognised the value of the long forceps have been found the advocates for turning in contracted pelvis. The following words, written by Smellie in 1752, challenge attention now:—"Midwifery is now so much improved that the necessity of destroying the child does not occur so often as formerly; indeed, it never should be done, *except when it is impossible to turn or to deliver with the forceps;* and this is seldom the case but when the pelvis is too narrow, or the head too large to pass, and therefore rests above the brim."

Pugh, of Chelmsford (1754), who advocated the long forceps, says:—"When the pelvis is too small or distorted, the head hydrocephalic or very much ossified, or its presentation wrong. . . . provided the head lies at the upper part of the brim, or, though pressed into the pelvis, it can without violence be returned back into the uterus, the very best method is to turn the child and deliver by the feet." He then goes on to lay down the conditions which would induce him to prefer the curved forceps, and states that, as the result of these two modes, "I have never opened one head for upwards of fourteen years."

Has not midwifery retrograded since his time? Perfect (1783), who used the long forceps, delivered a rickety woman whose conjugate diameter measured three inches, the head presenting, and brought forth the first living child out of four, the three first having been extracted after perforation. La Chapelle (1825) advised and practised the method. She relates that out of fifteen children extracted by forceps (long) on account of contracted pelvis, eight were stillborn, seven alive; and that out of twenty-five delivered footling, sixteen were born alive, and nine dead.

It is not less remarkable that it is amongst those who reject the long forceps that the strongest opponents of turning in contracted pelvis are to be found. This is the more astonishing when we reflect that this school, rejecting the two saving operations, has nothing to propose but craniotomy for a vast number of children that claim to be brought within the merciful scope of conservative midwifery.

Denman, who used the short forceps exclusively, was, upon the whole, adverse to the operation, although he relates a striking case in illustration of its advantages. He delivered a woman of her eighth child alive at the full period, all her other children having been stillborn. "The success of such attempts," he says, "to preserve the life of a child is very precarious, and the operation of turning a child under the circumstances before stated is rather to be considered among those things of which an experienced man may sometimes avail himself in critical situations, than as submitting to the ordinary rules of practice."

Those who have studied the history of obstetric doctrine cannot fail to see that this dread of encouraging enterprise in practice lest disaster should result from unskilfulness, has cramped the teaching, obstructed the progress of knowledge, and enforced a slavishly timid, yet barbarous, practice which still persists down to the present time. That the precepts and practice of Smellie and his immediate disciples were infinitely more scientific and successful than those which prevailed in the time of Denman, and in the first half of the present century, cannot be doubted. Possibly the cautious teaching of Denman and many of his successors was justified greatly by the general imperfection of medical education. They had, as we now have, to teach according to the average capacity and trustworthiness of their pupils. They taught men with the same feeling of reserve with which we should still teach midwives. But surely the day is past for all this. We may safely venture to teach men of a higher standard upon more liberal principles. I am not aware that a similar reticence or restraint has at any time, to a like extent, gagged the teachers of medicine or surgery proper. May we not see in this fact a striking testimony that the practice of obstetrics demands, even more than medicine or surgery, steadiness yet promptitude in judgment, courage under difficulties, and physical skill?

On the other hand, the forceps has by some been held to be of superior efficacy to turning in contraction of the pelvis; that is, whilst certain lesser degrees of contraction may be dealt with by turning, the forceps claims the preference in more advanced degrees of contraction. It is needless to say that those who advocate this preference rely upon a very powerful forceps. It is accordingly in Germany and France especially that the claim for the superiority of the forceps is contended for.

Stein (1773), Osiander the elder (1799), preferred the forceps. Boër was opposed to turning. In France, Baudelocque maintained the same doctrine as Stein and the elder Osiander; and the recent experiments of Joulin, Chassagny, and Delore with the "appareils à traction," by which a powerful extracting force is added to the forceps, enabling it to bring a head through a greatly contracted passage, seem to strengthen the comparative claim of this instrument.

We will now discuss the question—What is the penalty incurred, or how can we retrieve our error, if we turn and fail to bring the head through the too-contracted brim? Undoubtedly, the patient will have to go through a second operation. We are driven to perforate after all. We have tried to save the child, and have failed. Is the mother imperilled by this attempt and failure? This also must be answered by experience. Of course, the mother may suffer if we persevere in dragging the child too long and too forcibly. But we have a right to assume that the attempt is controlled by skill and discretion. The amount of force that can be safely endured is very great—far greater than those who have never seen the operation would readily credit. The violence to which the soft structures are subjected seems to be small in proportion to the traction-force exerted. There appears to be some saving or protective condition. This, I think, is found in the mechanism of the process. I refer to Lecture V. for an illustration and description of the mechanism of labour in contraction of the pelvis from projection of the promontory. This projecting promontory forms the centre of rotation around which the head must revolve in order to enter the pelvic cavity. The side of the head applied to this point scarcely moves at all. The promontory catches the fœtal skull in the fronto-temporal region. If the coarctation be decided, the skull where it is caught bends in. All the onward movement is effected by the opposite or pubic side of the skull sweeping in a circle, which I have called "the curve of the false promontory," until the equator or greatest circumference has passed the plane of the brim, when the whole head slips into the cavity with a jerk. Now, injurious pressure is avoided on the pubic side by the smoothness and flatness of the inner surface of the pelvic brim, and by a gliding movement of the soft parts intervening between the head and the bony canal. Injurious pressure is avoided over the promontory by the yielding or moulding of the head. The temporal and parietal bones will bend in, even break. Children have been born alive after this bending or breaking. Sometimes a large cephalhæmatoma forms at the point of depression. In other cases the child perishes. The observation of these cases shows that the mother will bear with safety an amount of pressure which is sufficient to kill the child.

What follows? This obvious corollary: that the mother will safely bear that lesser degree of pressure which is required to bring through a living child.

The operation, then, is justified in cases of contraction that admit of the passage of a living child. It is further justified in cases of contraction to a certain, though small, degree of contraction beyond this, which admits of the passage of a dead child. We have here, perhaps, carried the experiment to the verge of what is justifiable. Beyond this, there being no possibility of getting a child, live or dead, through the pelvis, it would of course be better not to go. And if all the conditions of the problem could be precisely ascertained beforehand, we should not go beyond this. But, whilst calculating upon an average or standard head, we may encounter a head above the standard in size or hardness, and thus, in our endeavour to save the child, we may find ourselves in a difficulty. The extrication is by perforation. By lessening the head, it is brought within the capacity of the pelvis. This is, indeed, an acknowledgment of defeat; it is beating a retreat. The justification, however, is that we accomplish in the end exactly that which those who reject the operation accomplish, namely, the safety of the mother. We have tried to do more; to save the child as well.

Is there any great difficulty or danger in perforating after turning? I believe not. The child's body is drawn well over to one side by an assistant, so as to facilitate the access of the operator's guiding fingers and the perforator to the child. The best place to perforate is in the occiput; but if that part be not easily struck, the perforator may be run up through the base of the skull. An opening into the cranium being made, the crotchet is passed into it, and the discharge of brain facilitated. Then, resuming traction on the trunk cautiously, the skull will probably collapse enough to pass easily. If not, the craniotomy forceps can be applied; or, better still, the cephalotribe, to crush up the base of the skull. Now, under the postulates of the case, this late recourse to craniotomy must not be considered as a severe or hazardous addition to the risks of the woman. The turning has been performed early in labour—that is, before the liquor amnii has all drained away, whilst the child is still freely moveable, and before there is any serious exhaustion of the mother. Under these circumstances the turning, especially if conducted, as it commonly may be, on the bi-polar principle, is not necessarily a long or a severe operation. If we fail in extraction, which is soon ascertained by observing that the head makes no advance, but that its globe expands broadly above the brim of the pelvis, perforation can be performed in good time. In short, the safety of the mother is secured by carrying through both operations whilst her strength is good. If exhaustion had set in, we should not have turned at all, but have proceeded to craniotomy in the first instance. To these considerations must be added the result of experience, which is to the effect that the retrieval by the secondary operation of craniotomy is successful.

What is the chance of saving the child? Dr. Churchill urges that "the life of the child is not secured, and its chance but little increased, even if our estimate of the pelvic diameters be accurate; for, if in turning with an ordinary-sized pelvis rather more than one-third of the children are lost, the mortality will be surely much increased if its diameter be reduced more than one-fourth." I will not stop now to press the pre-

T 2

liminary objection I entertain to submit the decision of this or any other question in obstetric practice to *à priori* arguments drawn from statistics. It would not be difficult to prove that the statistics employed by Dr. Churchill are a confused heap of incongruous facts, and that rules to guide practice drawn from them must be stultified by endless fallacies. It is enough to state that the operation is not recommended by any one when the pelvis is contracted more than one-fourth—that is, below three inches—therefore the argument, statistical or other, is beside the question. I am not able to state or to estimate the proportion of children saved or lost under the operation. It is enough to justify the operation if we save a child now and then. I believe, however, that, exercising reasonable care in selection of cases and skill in execution, more than one-half of the children may be saved. And to save even one child out of twenty is something to set against the deliberate sacrifice of all.

Not assenting to the proposition that one child out of every three is lost by turning where there is no disproportion, I do not doubt that a much larger proportion would be lost under turning where the conjugate diameter of the pelvis is less than 3″. But, as I have said, the comparison is gratuitous, for no one, I believe, recommends turning in this latter case unless the child be premature. Experience here again corrects the foregone conclusion deduced from statistical reasoning. The risk to the child is considerably less than might be fairly anticipated. It is a matter of observation that *in cases of moderate contraction the funis is safer from compression than in cases of normal pelvis.* I have found the cord commonly fall into the side of the pelvis towards which the face looks, and there it is protected in the recess formed by the side of the jutting promontory (*see* Fig. 85, p. 281); so that if the soft parts are sufficiently dilated not to compress the cord against the child's face, and if the labour can be completed under 5″, or even a little more, the child has a very good chance indeed. This proposition is especially true in the case of premature labour with contracted pelvis. In this case the child may, in the great majority of cases, be saved by turning. I have in this way saved many children who still survive to parents who would otherwise be childless. It deserves, I think, to be laid down as a rule in practice that *where the conjugate diameter measures from* 2·75″ *to* 3″, *delivery by turning should be the complement to the induction of labour at seven or eight months*—at least, I have acted on this rule with the happiest results.

Since the design of the proceeding is to save the child, it is obviously useless if the child is dead. How do we know when a child is dead? It is by no means easy to acquire certain knowledge of this fact. Nothing is more common than to read in clinical records "that the pulsations of the fœtal heart being no longer audible with the stethoscope," or, "the pulsation in the cord having ceased," or "meconium having escaped," the death of the child was assumed, and the perforator was used without hesitation. I will not dispute that these are presumptive evidences of death, but I have too often experienced the satisfaction of seeing a child resuscitated after I had ceased to feel the pulsation in the cord, and after the free escape of meconium, to abandon the hope of saving the child without more certain evidence. This is found in great mobility and crackling of the cranial bones; the caput succedaneum falling into loose skin-folds; the coming away of epidermis and hairs.

So long as there is tonicity, rigidity, or firmness of the limbs, life is present; but flaccidity is not a certain sign of death. A sign of threatening imminent death is a twitching or convulsive movement of the leg held in your hand. This indicates an attempt at inspiration, made to supplant the suspended placental circulation. When this is felt, it is a warning to accelerate delivery, and to excite aërial respiration.

The value of turning in moderate degrees of pelvic contraction rests greatly upon the truth of the following proposition:—*The head will come through the pelvis more easily if drawn through base first than if by the crown first.* Baudelocque affirmed this proposition.* He said:—"The structure of the head is such that it collapses more easily in its width, and enters more easily when the child comes by the feet, if it be well directed, than when it presents head first." Osiander had maintained the same opinion. Hohl (1845) also pointed out that the bones overlapped more readily at the sutures when the base entered first. Simpson (1847) insisted strongly upon the truth of this proposition, and illustrated the mechanism of head-last labours with much ingenuity. The proposition has, however, been disputed, and that by Dr. M'Clintock†. He says:—"I do not believe that the diameters of the head are more advantageously placed with regard to those of the pelvis,

* "L'Art des Accouchements."

† "Obstetrical Transactions," vol. iv., 1863.

nor can I believe that the head is more compressible when entering the strait with its base than when it does so with its vertex, till this be demonstrated by direct experiment."

It is also contested by Professor E. Martin, of Berlin.* He especially insists that when the vertex presents, moulding may go on safely for hours; but that if the base come first the moulding must be effected within five minutes to save the child.

I venture to submit that I have made such clinical observations as are equivalent to direct experiments. In the first place, let me state a fact which I have often seen. A woman with a slightly contracted pelvis, in labour with a normal child presenting by the head, is delivered, after a tedious time, spontaneously or by the help of forceps; the head has undergone an extreme amount of moulding, so as to be even seriously distorted. The same woman in labour again is delivered breech first; the head exhibits the model globular shape, having slipped through the brim without appreciable obstruction. For examples see my outlines of heads.†

In the second place, I have on several occasions been called to an obstructed labour, in which the head was resting on a brim contracted in the conjugate diameter. Of course, Nature had failed; the *vis à tergo* was insufficient. I have tried the long double-curved forceps, trying what a moderate compressive power, aided by considerable and sustained traction, would do to bring the head through, and have failed. I have then turned, and the head coming

* "Monatsschr. f. Geburtsk," 1867.
† "Obstetrical Transactions," 1866.

base first has been delivered *easily*. Upon this point I cannot be mistaken; and I think this greater facility can be explained. Dr. Simpson has illustrated by diagrams how the head, caught in the conjugate at a point below its bi-parietal diameter, is compressed transversely as traction-force is applied below, causing the mobile parietals to collapse and overlap at the sagittal suture. And surely no one can doubt that the traction-power, and, therefore, the compressing power, acquired by pulling on the legs and trunk, is infinitely greater than can be exerted by the strongest forceps. But there is another circumstance in the clinical history of head-last labours in narrow conjugate which affords a remarkable illustration of this proposition. *The head is rarely, or never, seized in its widest transverse diameter; it is seized by the conjugate at a point anterior to its greatest width—that is, in the bi-temporal diameter;* the bi-parietal and occiput commonly finding ample opportunity for moulding in the freer space left in the side of the pelvis behind the promontory. The head, in fact, fits or moulds into the kidney-shaped brim wherever there is most room. I have given illustrations of this point also in the memoir referred to.* I think, therefore, it may be taken as demonstrated, that the head coming base first passes the contracted brim more easily than coming crown first; and if the head comes through more easily, it may be expected that the child will have a better prospect of being born alive.

Can we define with any precision the conditions as to degree of pelvic contraction that are compatible

* "Obstetrical Transactions," 1866.

with the birth of a living child? The question is not easy to answer; nor is it important to be able to answer it very precisely. The great fact upon which the justification of the operation rests is this: many children have been delivered by it alive, with safety to the mother. We know accurately only one element of the problem—namely, the degree of con-

Fig. 85.

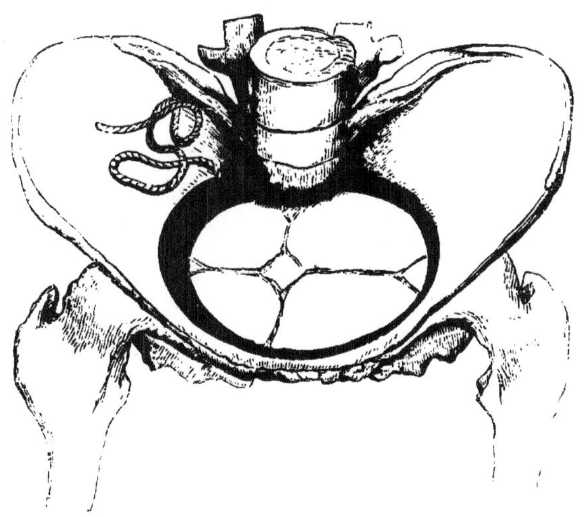

REPRESENTS THE HEAD ENTERING A CONTRACTED BRIM, BASE FIRST.

It is nipped in the small transverse diameter, the greater or bi-parietal diameter and the occiput finding room in the side of the pelvis. The cord lies in the side of the pelvis to which the face is directed, and is protected by the promontory.

traction of the pelvis. The other element, the relative size and hardness of the fœtal skull, we can but estimate. *We must assume, in many cases, a standard head.* With this assumption the practical question is reduced to this: *What is the extreme limit of pelvic contraction justifying the attempt to deliver by turning?* In other words, this means: What is the narrowest pelvis that admits of the passage of a normal head? This is answered chiefly by experience. It is not to be answered by *à priori* reasoning like that urged by Dr. Fleetwood Churchill, who says, even in his last edition:* — "The bi-mastoid diameter in the six cases measured (by Dr. Simpson) varied from $2\frac{6}{8}$ to $3\frac{2}{8}$ inches, and a living child can pass through a pelvis of $3\frac{2}{8}$ inches antero-posterior diameter, with or without the forceps. With a pelvis of this size, then, the operation is unnecessary; and if the antero-posterior diameter be less than $2\frac{6}{8}$ inches, the operation would be impracticable. These, then, are the limits of the operation; for us to attempt to drag a child through a smaller space would be unjustifiable."

To this statement of the case serious objections may be taken. The proposition that a living child can pass through a pelvis with an antero-posterior diameter measuring 3·25″, with or without the forceps, can only be accepted with considerable qualifications. I claim to speak with the confidence drawn from large experience, when I say that a head of standard proportions and firmness will hardly ever pass a conjugate reduced to 3·25″ without the forceps, and very rarely indeed with the forceps—that is, alive. I might even extend the conjugate to 3·50″, and affirm the same thing. The compressive power of the forceps, unless very long sustained, is not great, rarely great enough to reduce a bi-parietal diameter of 4·00″ to 3·50″ without killing the child. My opinion, then, is, that a standard head, especially if it happen to be a female head, which is more compressible than a male one, *may be* drawn through a conjugate of 3″,

* "Theory and Practice of Midwifery," 1866.

but not with much prospect of life; and that the proper range of the operation of turning is from 3·25″ to 3·75″, at the latter point coming into competition with the forceps. I believe no one advocates resort to turning when the conjugate measures less than 3″.

A correlative proposition to the foregoing is the following:—*Compression of the head in its transverse diameter is much less injurious to the child than compression in its long diameter.* The truth of this is attested or admitted by most authors who have considered the point. It is insisted upon by Radford, Ramsbotham, and Simpson. It is confirmed by the observation of the form which the head assumes under moulding in natural labour, which, as I have shown, is effected by the lengthening of the fronto-occipital diameter and the shortening of the transverse diameter.*

Now, it is an almost necessary consequence that when the head, arrested on a contracted brim, is seized by the forceps, it is seized by its fronto-occipital diameter, and to the longitudinal compression is added the increased obstruction to the entry of the head into the narrowed conjugate caused by the lateral bulging.

The Indications for the Operation.

Assuming a standard head whose base, unyielding, measures 3″, this is obviously the limit beyond which the operation would be useless; for although the head is caught in the bi-temporal diameter, a little in front of the greatest transverse or bi-parietal diameter, the base must be exposed in its full width to the

* "Obstetrical Transactions," vol. vii., 1866, and "Medical Times and Gazette," vol. ii., 1867.

narrowed strait. Even if the side of the head be indented by the promontory, no important degree of canting or obliquity of the base can be counted upon. If the head should fortunately be undersized or unusually plastic, there is a fair prospect of the child being drawn alive through a conjugate diameter measuring 3·00″.

But, generally, from 3·25″ to 3·50″, or even a little more, is the working range for a child at term. The great majority of those who advocate the operation insist upon this amount of space. It is very important to have a fair oblique or sacro-cotyloid diameter on one side; for if the ileo-pectineal margin of the brim incline rapidly backwards, the occiput will not find room.

The operation is also indicated if, the conjugate diameter being 3·50″ or more, the forceps have failed.

Velpeau (1835), Chailly (1842), Edward Martin, and others, advise the operation in cases of unequally contracted pelvis where there is more room in one side of the pelvis than in the other—when the thicker or occipital end of the head is not already engaged in this larger side.

I have already shown that the head is always nipped in its small or bi-temporo-frontal diameter, which generally measures about 3″, and is more compressible than the bi-parietal diameter. *The mark of pressure or indentation against the jutting promontory is always seen at one end of this short diameter whenever there has been obstruction in delivery.* It follows, then, that for the operation to be successful there ought to be room enough on one side of the pelvis to receive the occiput or big end of the head.

The operation may also be performed as the complement to the premature induction of labour where the conjugate measures from 2·75″ to 3·50″. Indeed, this I believe to be one of its most valuable applications.

The next condition is, that there be reasonable presumption that the child is alive.

The cervix should be dilated enough to admit the fingers pointed in a cone, and dilatable enough to yield with readiness under the extraction of the trunk. In this, as in most cases where the head cannot press fairly upon the cervix, we are not to expect complete spontaneous dilatation.

The membranes should be intact, or there should be enough liquor amnii present to permit of the ready version of the child.

The *contra-indications of the operation are*:—

1. A conjugate diameter narrowed to less than 3″.
2. Firm and close contraction of the uterus round the child, compressing it into a globular shape.
3. Impaction or very firm setting of the head in the brim of the pelvis.
4. Marked exhaustion or prostration of the mother.
5. Death of the child.

As Hohl has remarked, the sudden emptying of the uterus of a woman far gone in prostration, acting as a new shock, is apt to increase the *collapsus post partum*.

The Operation.—The preparatory steps are the same as for the ordinary operation of turning. As the conditions postulated admit of bi-polar action, it is important to avail ourselves of a means that so greatly lessens the force necessary to use, and which further enables the operator to bring a leg and the breech through a cervix that would not permit the passage of his hand. Chloroform will be useful chiefly during extraction.

If exploration by the whole hand in the pelvis satisfy us that the pelvis is symmetrical—that is, that there is equal and sufficient space for the big end of the head in either side—we turn according to the ordinary rules. Finding the head in the first position, or with the occiput to the left ilium, depress the breech towards the right with the right hand externally; push the head across to the left iliac fossa with the fingers of the left hand passed through the os uteri, and seize the further knee. Extraction must be performed at first slowly, so as to allow the half-breech to dilate the cervix. This is especially a case where hurry is misplaced. The extraction should go on slowly whilst the trunk is passing. As soon as the funis is felt, draw down a loop, and direct it towards the posterior wall of the pelvis. So long as it pulsates freely, do not hurry. But if the pulsations flag, lose no time in liberating the arms. The pelvic contraction makes this a little more difficult than under ordinary circumstances. I refer for the description of this proceeding to Lecture XIII. As soon as the arms are liberated, the real difficulty begins: the extraction of the head. Sometimes the head is delayed by being encircled by the imperfectly dilated os uteri. This is an unfortunate complication, since compression at this point is likely to stop the circulation through the cord. To avoid this risk, it is necessary not to hurry the trunk through the cervix. It is above all things necessary to draw at first as much backwards as possible, so as to make the head revolve round the jutting sacral promontory until it clears the strait, when the head can enter its natural orbit, the curve of Carus. Then, traction is changed to the direction of the pelvic outlet. Traction is effected by holding the legs with one hand, and the nape of the neck with the other. Commonly, the force thus obtained is enough; but sometimes more is wanted. This is obtained by crossing a fine napkin or silk handkerchief over the neck, and bringing the ends in front of the chest, and drawing upon them, as in

FIG. 86.

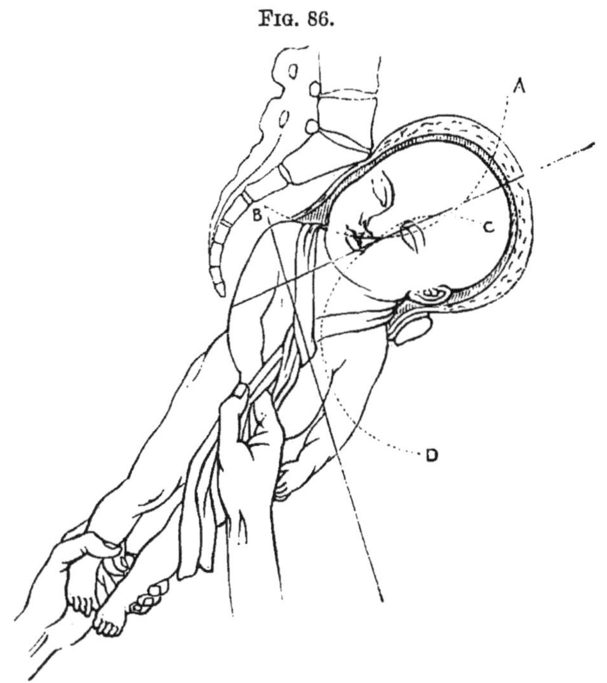

SHOWS THE MODE OF EXTRACTION AFTER TURNING WHEN HEAD IS JAMMED IN THE CONJUGATE DIAMETER.

The right forehead fixed against the jutting promontory is the centre of revolution. The left side of the head, resting on the pubes, sweeps round in the orbit of the false promontory A B. To favour this first movement, traction is made well backwards. As soon as the equator of the head-globe has slipped through the conjugate, the head enters the true orbit C D, revolving round the pubes.

Great assistance in extraction may be gained, and traction-force economised, by getting an assistant to press firmly upon the vault of the head through the abdominal walls, thus helping to push the head through the strait. This proceeding was advised by Pugh and Wigand, and quite recently Dr. Strassmann* has insisted upon its utility. The possibility of deriving advantage from it should be borne in mind in all cases of head-last labour.

Where the pelvis is unequally contracted, one half being smaller than the other, the object is to throw the big or occipital end of the head into the larger half. Professor E. Martin describes three modes of accomplishing this.

1. *A suitable position of the woman.* Let her lie on that side towards which the forehead is directed. The fundus uteri will gradually sink with the pelvic end of the child to this side; the spine draws the occiput to the opposite side of the pelvis, and the forehead sinks more deeply towards the brim. Martin refers to a case in which he successfully carried this plan into execution, the pelvis measuring three inches only.

2. *The forceps* is a means of releasing the posterior transverse diameter of the head when imprisoned in the pelvic conjugate. This explains the frequent easy extraction when a little traction has been made. Martin admits that we must not be sanguine as to the success of this plan. We must be prepared, he says, to perforate, if there be evidence of exhaustion. My own experience is decidedly adverse to it if the contraction is at all marked.

3. *Turning by the Feet.* How is this to be done?

* "Monatsschr. f. Geburtsk," June, 1868.

In consequence of the well-known law that, in complete foot presentation, the foot that is drawn down always comes under the pubic arch, if the fœtus is not abnormally small, or the pelvis too large, if we draw down the right foot, the child's back, and also its occiput, will come into the right half of the uterus, and *vice versâ.* If, therefore, the right half of the pelvis is the larger, seize the right knee; if the left side is larger, seize the left knee.

Hohl and Strassmann doubt the possibility of securing this result. If it happens, it does so by accident. I believe, however, the rule and the practice are good and feasible. But the success of the operation is not necessarily imperilled, if even the occiput should fall into the narrower half of the pelvis. I have saved children when this has happened, and Strassmann relates some striking cases* in proof of this proposition.

To determine which side of the pelvis is the more contracted, attention to the following points will help:—

1. If the woman walks straight, and the legs are of equal length, the defect in symmetry will be but slight; but the presumption is that the right side is the larger.

2. If the woman has one hip affected, or one leg shorter than the other, the corresponding side of the pelvis will be the smaller.

3. You may measure and compare the two half circumferences of the pelvis externally from the crest of the sacral spine to the symphysis pubis.

4. The hand in the pelvis may take a very close estimate of the relative space in the two sides.

* "Monatsschr. für Geburtskunde," June, 1868.

LECTURE XVIII.

CRANIOTOMY—THE INDICATIONS FOR THE OPERATION—THE OPERATION—TWO ORDERS OF CASES—PERFORATION SIMPLE, AND FOLLOWED BY BREAKING-UP OR CRUSHING THE CRANIUM, AND EXTRACTION—EXPLORATION—PERFORATION—EXTRACTION BY CROTCHET, BY TURNING—DELIVERY BY THE CRANIOTOMY-FORCEPS—USE AS AN EXTRACTOR—AS A MEANS OF BREAKING UP THE CRANIUM—USE IN EXTREME CASES OF CONTRACTION—DELIVERY BY THE CEPHALOTRIBE—POWERS OF THE CEPHALOTRIBE—COMPARISON WITH CRANIOTOMY-FORCEPS—THE OPERATION—DR. D. DAVIS'S OSTEOTOMIST—VAN HUEVEL'S FORCEPS-SAW—DELIVERY BY THE AUTHOR'S NEW METHOD OF EMBRYOTOMY BY THE WIRE-ÉCRASEUR—INJURIES THAT MAY RESULT FROM CRANIOTOMY.

WE have lingered long on the border-land between Conservative and Sacrificial Midwifery, unwilling to abandon the hope of saving mother and child; striving to set back, as far as the dictates of science and the resources of art will enable us, those limits where the death, certain or probable, of child or mother must be encountered. We must at length pass the boundary; we must lay aside the lever, the forceps, and turning, and take up the perforator, the crotchet, the craniotomy-forceps, the cephalotribe — instruments the use of which is incompatible with the preservation of the child's life. A law of humanity hallowed by every creed, and obeyed by every school, tells us, where the hard alternative is set before us, that our first and paramount duty is to preserve the mother, even if it involve the sacrifice of the child. As, therefore, we have striven to give the highest possible perfection to the forceps and turning in order to save mother and child, so it now behoves us to exhaust every effort in perfecting the means of removing a dead child, in order to rescue the mother from the Cæsarian section, that operation which the late Professor Davis justly called "the last extremity of our art, and the forlorn hope of the patient."

The Indications for Craniotomy.

These are generally:—1. Such contraction of the pelvis or soft parts as will not give passage to a live child, and where the forceps and turning are of no avail. Contraction of this kind may be due to contraction or distortion of the pelvis, which is most frequent at the brim; to tumours, bony, malignant, or ovarian, encroaching upon the pelvic cavity; to growths, fibroid or malignant, in the walls of the uterus; to cicatricial atresia of the cervix uteri or vagina; to extreme spastic contraction of the uterus upon the child, forbidding forceps or turning. Craniotomy and cephalotripsy are the means of effecting delivery in cases where labour at term is obstructed from disproportion, the pelvic contraction ranging from 3·25″

as a maximum to 1·5″ as a minimum. If labour occur at seven months, these means may be applied to contraction measuring even less than 1·50.″ We are not hastily to assume, because a woman has been delivered on previous occasions by natural powers or by forceps, that it is therefore unnecessary to resort to craniotomy. It is, indeed, ample reason to pause and to examine anxiously. But it is a matter of experience that some women bear children with a constantly increasing difficulty. This may be from two causes—1. Advancing pelvic contraction; 2. Increasing size of the children. I can affirm the reality of the first cause from repeated observation. I have had to record the histories of many women whose first labours may have been natural, and whose subsequent ones exhibited difficulties increasing in a kind of accelerated ratio, rising from the forceps to turning and craniotomy. The second cause may be independent of, or aggravate, the first. D'Outrepont says he has constantly observed that in fruitful women whose first children were small, subsequent ones became bigger and bigger. On the other hand, Dr. Matthews Duncan contends that the maximum weight of children is found in women aged from 25 to 29, and that the weight afterwards falls. But I have known many exceptions to this.

Professor Elliot makes the apposite observation, that "the same degree of deformity admits of varying results in successive pregnancies." For example, the child may at the time of labour have attained a different degree of development: it may present differently; there may be accurate dip of the head in one labour, and not in the rest.

2. *Certain cases where obstruction to delivery is due to the child*—as some cases of face-presentation; some cases of locked twins in which the lessening one head is necessary to release the other; excessive size of head, as from hydrocephalus; cases where there is obstructed labour, the head presenting, and the child is dead.

3. *Conditions of danger to the woman, rendering it expedient to deliver her as speedily as possible*, and where craniotomy is the quickest way, involving the least violence. Amongst these are *some* cases of convulsions; *some* cases of hæmorrhage; great exhaustion; some cases of rupture of the uterus; and generally where, delivery being urgently indicated, the cervix uteri is not sufficiently dilated to admit of other operations.

An important question is—At what stage of labour shall we begin? As most of the dangers flow from exhaustion, it is obviously proper to begin as soon as the indication for the operation is clear. On the Continent especially, it is still urged by many that we should wait until the child is dead. Now, if it be admitted, and the conditions of the case involve these postulates—1. That the child cannot come through alive; 2. That the operation is undertaken in order to save the mother—waiting till the child is dead is opposed both to reason and to humanity. It seems a refinement of casuistry to distinguish between directly destroying a child, and leaving it exposed to circumstances which must inevitably destroy it; and it is risking the very object of our art to wait for this lingering death of the child until the mother's life is also imperilled. If, then, we have clearly determined by our knowledge of the patient, by exploration, by trial, that the child cannot come through the pelvis by spontaneous action, by forceps, or by turning, it is our duty at once to adopt the best means of securing the safety of the mother. There is no need to wait for the far advance of labour. We should not wait long after the rupture of the membranes. It would, in the majority of cases, be useless to wait for complete dilatation of the cervix uteri. It is one of the necessary results of contracted brim that the cervix uteri dilates slowly and imperfectly. The head-globe resting by two points on the contracted brim cannot bear upon the cervix. It is not, therefore, often desirable to wait for more opening than is enough to admit two or three fingers to guide the perforator. When the head collapses and comes down into the pelvis, it bears upon the cervix, which yields gradually.

Although it is a good general rule to perform every operation as early as the indication for it is clearly recognised, it is not desirable, in minor degrees of contraction, to arrive at once at the conclusion that perforation is necessary. Some time and opportunity should be given to Nature. The head may be small or plastic, and occasionally even a full-sized head will, under continued action of the uterus, become so moulded as to admit of delivery either spontaneously or by aid of the forceps.

Perforation should be the first step of all operations for lessening the bulk of the head. The necessary condition for full collapse of the bones of the head is that the support given by the brain and the integrity of the cranial vault should be broken down. Until this is done, you may obtain, with considerable expenditure of time and force, some amount of moulding or alteration of form, but no diminution of bulk. It is astonishing what resistance to compression the unopened head possesses. The most powerful forceps, and even cephalotribes, may be bent in the attempt to crush it in. Whereas, break the arch of the cranial vault, allow the contents to escape, and very moderate compression will cause collapse, more or less complete. It is remarkable that not a few Continental obstetrists practise cephalotripsy without perforating.

Craniotomy, being used as a general term to include all the proceedings for reducing the bulk of the head, may be divided into three principal orders:—

The *first* includes those cases of minor disproportion in which perforation is enough to allow of such an amount of collapse of the head under the natural forces as will effect delivery.

The *second* order includes those cases of major disproportion in which perforation must be supplemented by breaking-up, removal of parts of the cranium, or by crushing down, and followed or not by extraction.

The *third* includes those cases of extreme disproportion in which the head has to be removed in sections, as by my New Method of Embryotomy.

The *preparations* are generally the same as for other operations.

Position.—The patient lies on her left side, with her knees well drawn up, near the edge of the bed, and with the head supported on a *low* pillow, directed towards the middle of the bed.

Exploration.—The left hand of the operator is introduced, if necessary, into the pelvis, so as to explore thoroughly the shape and dimensions of the brim, and the relations of the head and cervix uteri. Three points especially should be clearly made out: First, the projection of the promontory, which in extreme cases has been mistaken for the head; secondly, the head; thirdly, the os uteri. The finger passed inside the os should be made to sweep all round the circumference of the head.

Perforation.—The point to be selected for perforation is the most centrally presenting. It is easier to strike; it offers a better resistance to the point of the instrument. The opening made allows a freer exit to the contents of the skull, and affords greater facility for the introduction of the crotchet or the blades of the craniotomy-forceps, which have to follow. Two most essential things to be attended to are: That an assistant shall support the uterus and child externally, pressing them firmly down towards the pelvis, so as to fix the head upon the brim, and obviate the retreat or rolling of the head under the impact of the instrument. The other thing is, to take care that the instrument shall strike the head perpendicularly. If it strike at an angle, the point will be apt to fly off at a tangent, at the risk of wounding the mother.

Sometimes in cases of great deformity the uterus is so twisted from its normal direction that reposition is necessary before the os can be brought near the centre of the brim to allow of safe perforation.

Two fingers of the left hand then are passed up to the head, keeping the os uteri at their back; the instrument is run up in the groove formed by the fingers; the point having struck the part desired, the perforation is effected by a movement combining boring and pushing. When the skull is pierced, push the blades in up to the shoulders; then open the blades to enlarge the aperture, turn the handles at right angles to the first position, and open the blades again, so as to make a free crucial opening. This breaks the continuity of the arch; allows free discharge of brain, and ample entry for the crotchet or craniotomy-forceps.

Now you may wait a little, to afford opportunity for spontaneous compression and collapse, or you may at once pass in the crotchet. This should be carried in as deeply as possible, and moved freely round in all directions, to break up the tentoria and the brain. This proceeding greatly facilitates the evacuation and collapse of the skull.

If the disproportion is not great, and the powers of the patient are good, it commonly happens that uterine action sets in as soon as the bulk of the head is a little diminished, and the compression and propulsion resulting will often suffice to expel the child. Reasonable opportunity should be given for this spontaneous process.

Should no advance be made, the case falls into the second order, and we must proceed to extraction, or artificial compression of the skull. Some operators advise to pass a catheter or other tube into the skull, to inject a stream of water through it, in order to wash out the brain. Here the cephalotribe comes in admirably, quickly reducing the bulk of the head, rendering it plastic, and thus facilitating its extraction.

Extraction may be accomplished in several ways.

1. By the *crotchet.* This instrument is generally

preferred by the Dublin school. It was very naturally resorted to in preference to bad craniotomy-forceps; and some practitioners of great experience, who have acquired exceptional skill in the management of it, accomplish delivery by its aid in cases of extreme disproportion. Until I had contrived a good craniotomy-forceps, I myself trusted to it entirely. I am now satisfied that, for safety and expeditiousness in extracting a head, it is very inferior to the craniotomy-forceps. The way of using it is as follows:—Two fingers of the left hand guide the end of the crotchet into the hole in the skull. The ends of the fingers are then passed up outside the skull, to serve as a guard and support to the sharp point of the crotchet, which is fixed into the bone inside. The part to which the crotchet is first applied is not perhaps very important, since, if there be any great resistance, the part will be broken away, and the instrument will have to find fresh hold. This may have to be repeated several times, pieces of the parietal, occipital, or frontal bones being successively torn out. Whenever a piece of bone is detached, it is wise to remove it altogether, which may be generally done with the fingers. By-and-by—for the process is apt to be tedious—when the cranial vault is much broken up, if a good hold can be obtained in the occipital bone or in the foramen magnum, collapse or falling in of the skull takes place, and extraction is successful. In very difficult cases, when the vault is well broken up, it is better to take hold in the orbital region, fixing the point of the crotchet either inside the skull, under the sphenoid on one side of the sella turcica, or in the eyeball. In this way the base is brought into the brim edgewise or end on.

In the last century it was a recognised plan to perforate and leave the evacuation of the brain and the compression of the skull to the action of the uterus. The process was usually slow and tedious; commonly, some degree of decomposition had to take place before the bones would collapse sufficiently, and exhaustion, or even inflammation of the uterus, and fatal prostration, would sometimes ensue. The late Professor Davis records that, in his time, this mode of proceeding was still followed by the disciples of one school in London, and that he was often called in to witness the most disastrous consequences. Indeed, when it is considered that craniotomy is not often performed except at an advanced period of labour, and after much suffering has been endured, it will hardly appear justifiable to throw upon an enfeebled system a task entailing further exhaustion, and under which it may sink. It is our duty to relieve Nature, and not to leave her to struggle through unaided.

Dr. Hull says it was Dr. Kelly who first advised and practised the method of waiting twenty-four hours before using the crotchet, in order to allow the head to collapse and settle in the pelvis. Dr. Osborn advocated the same practice, and went so far as to assert that delivery by the crotchet might always be effected.

2. *Delivery by turning.*—When the cranial arch is broken the bones will readily collapse, if the skull be drawn through the contracted brim base first. In certain cases turning is a very efficient method of

completing delivery. The torn scalp during extraction is drawn over the aperture made by perforation, and sheaths the jagged edges of the bones. This plan will be generally inferior to the use of the craniotomy-forceps. The child, being dead, does not always lend itself readily to turning. It may be necessary to pass the hand into the uterus, which is moulded upon the child, and through a brim so contracted as to oppose considerable difficulty. Turning, however, must be regarded as a valuable resource in certain cases of exceptional emergency.

3. *Delivery by the Craniotomy-forceps.*—The use of this instrument is twofold:—It will seize and extract the head; it will seize and remove portions of the bones of the cranial vault. The first use is adapted to minor degrees of disproportion.

What part is best to seize? If the head is found to collapse well, and the disproportion is not great, it is enough to seize the forehead, which, being generally directed to the right ilium, is the easiest to do. But if there is any great difficulty, it is better to quit the forehead and seize by the occiput. The head will not come down well, face presenting, unless the vault and occiput are in a condition to be crushed in against the base. In this proceeding compression of the skull is effected by its being drawn through the narrow passage formed by the soft parts supported by the pelvis. The head must, therefore, be ductile enough to admit of the necessary compression and elongation. If the skull be too unyielding, or the passages too small for this process, a totally different principle must guide us. Portions of the vault must be removed, and then we get the most remarkable advantage.

Dr. Osborn contended that by canting the base of the skull, so as to bring it edgewise into the brim, it was quite possible to deliver a full-sized child through a conjugate diameter measuring an inch and a half

FIG. 87.

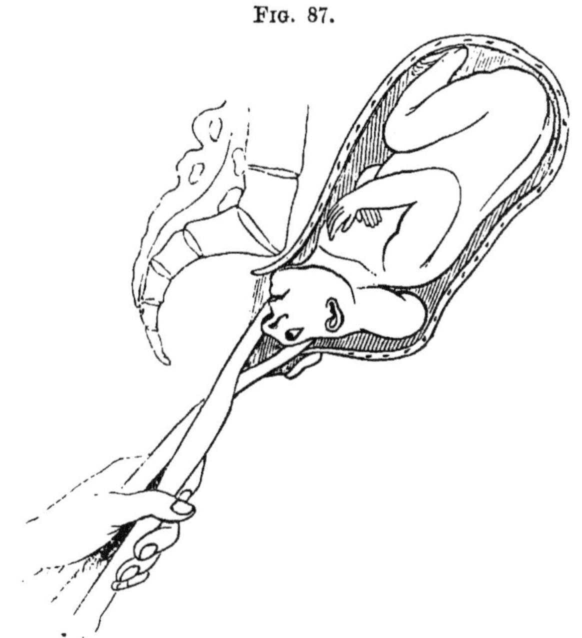

SHOWS THE BONES OF THE CALVARIUM REMOVED, AND THE BASE OF THE SKULL, GRASPED BY THE CRANIOTOMY-FORCEPS, DRAWN THROUGH THE CONTRACTED BRIM EDGEWISE, FACE FIRST.

only. His contention was hotly disputed by Dr. Hüll. Dr. Burns came to the same conclusion as Dr. Osborn, and showed that by removing the calvarium, reducing the skull to its base, and bringing it through as in a face-presentation, nothing was opposed to the conjugate but the distance from orbital plates to chin, which is rarely much more than an inch. Thus an inch and a half to an inch and three-quarters conjugate diameter, with a transverse diameter of three inches, is enough; and degrees of contraction beyond this, requiring the Cæsarian section, are rare indeed. This question has been investigated and illustrated anew by Dr. Braxton Hicks.* He describes very fully the mechanism of the proceeding. Having removed the calvarium, he grapples the orbit with a small blunt hook, the hook of which is hard, and the stem soft, so as to admit of easier adaptation. The face is then gently drawn down, turning the chin forwards, as occurs in ordinary face-labour. A fresh hold, in the mouth or under the jaw, is then taken for traction. The evidence given by Dr. Hicks of the efficacy of this proceeding is conclusive. I, however, prefer the craniotomy-forceps. The proceeding I practise is as follows:—I pass the inner or small blade into the cavity of the skull as usual, then the outer blade in between the portion of bone to be removed and the scalp. Then, a considerable piece of parietal or occipital bone being seized, by a sudden wrench is broken, and then cautiously torn away under the guidance and protection of the left hand in the vagina. If the distortion is not extreme, it may be enough to break away two or three pieces, say an angle of each parietal and of the occipital. This destroys the arch of the calvarium, so that the remains of the walls easily fall in upon the base, forming a flat cake, when the head comes to be compressed in the chink of the brim. When enough has been taken away to admit

* "Obstetrical Transactions," vol. vi., 1865.

of this flattening in, the blades of the forceps are made to seize the forehead and face, the screw working at the ends of the handles helping to crush in the frontal bones and to secure an unyielding hold. The

FIG. 88.

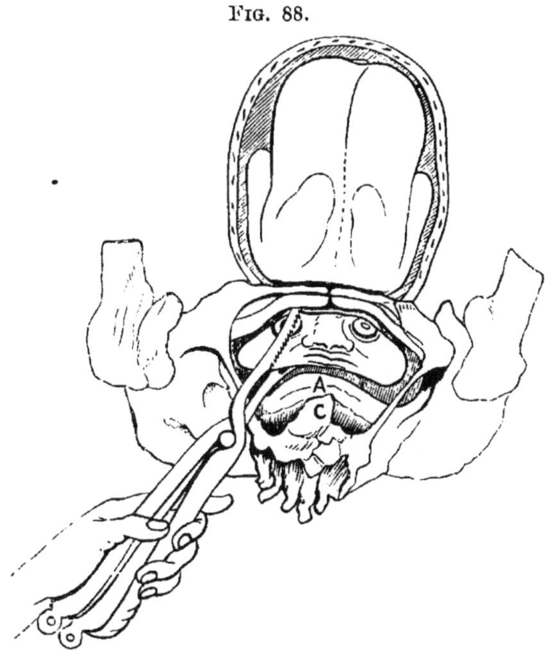

SHOWS THE REMAINS OF THE SKULL DRAWN THROUGH THE CHINK OF THE BRIM, FLATTENED LIKE A CAKE.

The calvarium being removed, the head resembles that of an anencephalous fœtus. A, projecting promotory of the sacrum; C, coccyx.

craniotomy-forceps, in fact, here acts like, and fulfils the chief function of, the cephalotribe. Then traction is made, carefully backwards at first, in the course of the circle round the false promontory. As the face descends it tends to turn chin forwards, and this turn may be promoted by turning the handles of the instrument. It is not necessary that this turn should take place, for the case differs entirely from that of the normal head. There is no occiput to roll back upon the spine between the shoulders. The head comes through flatwise like a disc by its edge.

If the pelvic deformity be very decided—say to 2·50″ or 2·00″ or under—it will be wise to take away the greater part of the frontal, parietal, squamous, and occipital bones before beginning traction.

By adopting this method, I entirely agree with Osborn, Burns, and Hicks, that a full-sized head may be delivered with safety to the mother through a pelvis measuring even less than 2·00″ in the conjugate, provided there be 3·00″ in the transverse diameter. I go further, and declare that it is perfectly unjustifiable to neglect this proceeding, and to cast the woman's life upon the slender chance afforded by the Cæsarian section.

The late Professor Davis, relying upon his method of embryotomy, used these words:—"There are few pelves with superior apertures so small as not to furnish from 1″ to 1½″ in the conjugate diameter, or at least of antero-posterior diameters across some part of the brim. In any such cases it would be the practitioner's duty to avail himself of the use of the osteotomist, and to undertake delivery by the natural passages. It will have the effect of reducing almost to zero the necessity of having recourse to the Cæsarian section."

4. *Delivery by Cephalotripsy.*—What are the relative powers of cephalotripsy and of craniotomy as just described? I doubt much whether cephalotripsy can carry the possibility of safe delivery at all beyond the point attained by craniotomy. It nevertheless

possesses considerable independent advantages under many circumstances, and may lend much help to other proceedings.

The Powers of the Cephalotribe.—The all-essential point is that it shall be able to compress and even crush down the base of the skull. A secondary property which it is desirable to possess is that of holding during extraction. The crushing power can be attained in sufficient perfection, and with a gain in the facility of handling, if the instrument be made much less formidable in bulk than are most of the Continental cephalotribes. Three good modifications have been constructed here. Sir James Simpson's is the best known. He insists upon a pelvic curve in the blades as being less likely to slip than straight blades. Dr. Kidd's, of Dublin,* is the best type of a straight-bladed cephalotribe. Dr. Kidd insists strongly upon the advantages of long straight blades on the following three grounds:—First, straight blades admit better of the head being rotated whilst in the grasp; secondly, they are easier to introduce; and lastly, they hold more securely. Dr. Braxton Hicks has modified Sir James Simpson's cephalotribe, producing a very handy and efficient instrument. He preserves a moderate pelvic curve, and adapts a very convenient screw to the handles as a crushing power. I believe that to seize a head above the brim, as is necessarily the case where crushing is required, the blades should be curved; but this curve should be very slight, otherwise the inconvenience in rotating or shifting the relation of the instrument to the pelvis referred to by Dr. Kidd will be felt.

* See "British Medical Journal," October, 1867.

When the instrument is applied to the perforated head, it may be made to completely crush the base, flattening the head sideways, or doubling up the base; or, by the slipping of one of the blades inwards a little, the base is tilted edgewise, and the skull is flattened by the pressing inwards on to the base of the squamous and parietal bones. Under either of these proceedings the head can be so flattened as to allow the blades to meet, and, as the instrument then measures only 1·50″, the obstacle is reduced to that degree. It is generally desirable to repeat the crushing, which is done by taking a fresh hold in a different direction, and then compressing again. Two crushings will generally be enough.

A distinctive advantage of the cephalotribe was pointed out by Curchod (Berlin, 1842). It is, that the plasticity effected by crushing so modified the form of the head that it was easily moulded to the form of the pelvic brim.

1. What are the limits of application of the cephalotribe? The maximum, of course, is not difficult to determine. It may be usefully employed in almost any case of minor disproportion. But what is the least amount of space admitting of its use? This must depend somewhat upon the form and size of the particular model adopted. In a discussion held at Berlin, the majority of the speakers thought a minimum of 54 mm. = 2·0″ conjugate diameter, was necessary. Lauth* says the application begins at 8 mm., or about 3 inches, at which point the forceps and turning are not available, and ends at 5·0 mm., or a little under 2·0″. But Pajot goes beyond this, and

* "De la Céphalotripsie." Par J. F. Ed. Lauth. Strasbourg, 1863.

contends that it ought to be applied where there is only 1·25″ conjugate diameter. Credé thinks it should be used if only there is rôom enough to apply it. Dr. Hicks has applied it where there was $1\frac{3}{4}$ inches. I have recently used it with perfect success in a case of extreme rickety deformity, at St. Luke's Workhouse, aided by Messrs. Harris, Rogers, and Sison, in which the conjugate certainly did not exceed 1·50″. In this case, after the first crushing, I removed some pieces of the cranial vault which had cropped up, by means of my craniotomy-forceps. The delivery was completed, without hurry, in an hour. I have arrived at the settled conviction, that cephalotripsy is quite practicable with a pelvis measuring an inch and a half in the conjugate diameter; and that the risk to the mother is inconsiderable compared with that attending the Cæsarian section.

Other conditions are—

2. The os uteri must be sufficiently dilated; but this can be readily effected by the caoutchouc water-bags.

3. The head must be previously perforated. Abroad, sometimes the ordinary forceps is put on to hold the head during perforation; or the blades of the cephalotribe are first adjusted, and then perforation is effected. But this is sometimes not feasible for want of room, and is never necessary.

Position.—The patient may lie on her left side, as in other obstetric operations.

Operation.—The rules laid down for the long forceps will generally apply to the application of the blades, and it is equally unnecessary in either case to have an assistant or a "third hand." The lower or posterior blade is passed first, guided by the left hand passed well into the pelvis if possible. This blade is passed along the hollow of the sacrum until the point approaches the brim and touches the head-globe, when the handle is raised, and the point, turning into the left ilium or to the left sacro-iliac synchondrosis, travels over the head. It is passed high up, for the point of the instrument must get beyond the base of the skull. This being *in situ*, the second or anterior blade is introduced also at first in the hollow of the sacrum, crossing the handle of the first blade. When the point approaches the brim, the handle is lowered and carried backwards, and the point rises over the head-globe into the right ilium, or opposite the right cotyloid cavity, when it falls into opposition with the first blade. Being locked, the screw is turned slowly and steadily, the hand in the vagina taking note of the work done. If spicula crop out of the scalp, they should be picked away by the fingers or the craniotomy-forceps. Indeed, the removal of portions of the cranial vault in this way much facilitates the subsequent transit of the head. When the base is crushed in the direction first seized, you may use the instrument as a tractor. If there be any marked resistance, it is better to take off the blades, to re-apply them in the opposite oblique diameter, and repeat the crushing; then, rotate the head by turning the handles about a quarter of a circle, to bring the flattened head into relation with the transverse diameter of the brim before extraction, so as to bring the head, flattened like a disc, to correspond with the chink of the pelvic brim. It is not, indeed, always necessary to give this rotation. I found, in the extreme case referred to above, that the necessary adap-

Fig. 89.

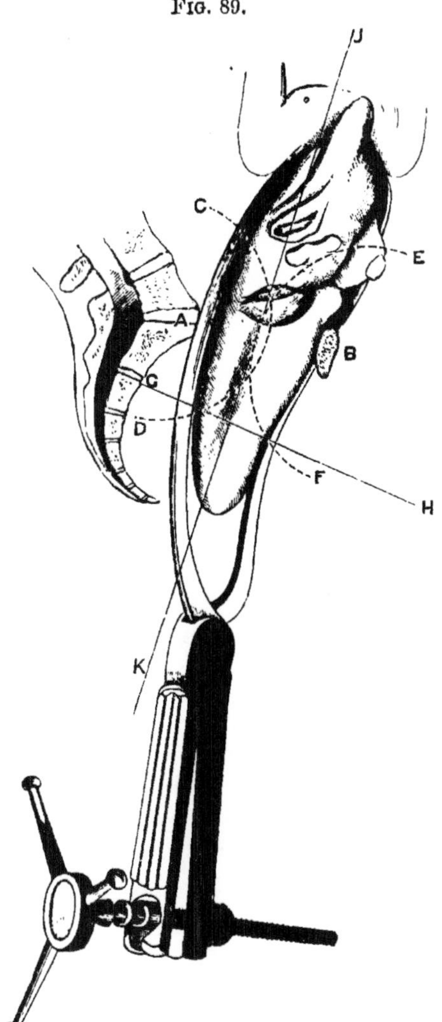

DR. HICKS'S CEPHALOTRIBE SEIZING THE PERFORATED HEAD.

This drawing is based upon a plaster-cast of a head embraced by Dr. Kidd's cephalotribe, exhibited at the Leeds Meeting of the British Medical Association, August, 1869. I have substituted Dr. Hicks's instrument. The head is seized somewhat in the right oblique diameter of the pelvis. It is partly crushed in; but the base is mainly adapted to pass the narrow brim by being canted. A is the projecting promontory, the centre of C D, the false curve which the head must first take; B, the symphysis, the centre of the true or Carus' curve, E F, which the head must enter and follow to emerge; G H, axis of the outlet; J K, axis of inlet.

tation took place almost spontaneously. Extraction may be made by the cephalotribe, taking care to allow time for the dilatation of the cervix uteri and vulva.

The late Professor Davis did not use the cephalotribe, but in extreme cases he cut away the head piecemeal by his osteotomists, and seized and extracted the trunk by a double sharp body-crotchet.

Pajot, of Paris,* has practised a method analogous to that formerly employed in this country in craniotomy. He performs what he describes as "céphalotripsie répétée sans tractions"—that is, he first crushes the base by one operation; he then gently tries to effect a slight rotation of the instrument so as to bring the crushed sides of the head into relation with the contracted diameter. If there is any resistance, he desists, and leaves the case for two or three hours for the uterus to mould the crushed head to the brim. He then repeats the crushing, and again gives two or three hours to Nature. One or two crushings suffice for the trunk. (See also "Osservazioni di Cefalotrissia," by Dr. Chiara, Turin, 1867, for a good case in illustration.) Pajot places this method in distinct competition with the Cæsarian section. The cases related by Pajot lend weight to his recommendation; but I cannot help thinking that the operation may and generally ought to be finished at one sitting. Those who advocate removing the cephalotribe before extracting, or who leave the crushed head and trunk to be expelled, do not sufficiently take into account the great extent to which the head compressed in the cephalotribe will expand again when the compression is removed. The

* "Archives Gén. de Méd.," 1863.

resiliency retained is considerable. A head flattened in the grasp of the cephalotribe may not measure more than one-and-a-half inches, yet, in taking off the blades, it will spring out to more than two inches. Why not, then, keep the blades on?

When the head is extracted, there may be some trouble with the shoulders and trunk. The shoulders will generally be disposed obliquely in the brim—that is, one will be anterior to the other. By keeping up traction on the head backwards, this anterior shoulder will be brought a little down, so that a finger or the blunt hook or crotchet can be fixed in the axilla to pull it through. When this is done, the head is dragged down forwards, so as to enable the same manœuvre to be repeated with the posterior arm. If this cannot be readily done, it is a good plan to crush in with the cephalotribe. Dr. Davis seized the trunk with his double body-crotchet. If turning has been practised after perforating or cephalotripsy, the arms fall in upon the crushed head and offer no obstruction.

To save the assistants the ghastly sight of the mangled head, wrap a napkin round it as soon as it is born. If traction is necessary in delivering the trunk, it is easier to hold when so treated.

4. *Delivery by the Forceps-saw.*—This instrument, introduced by Van Huevel in 1842, may be said to be the distinctive feature of the Belgian school. It is figured in the Obstetrical Society's Catalogue of Instruments, 1867. Dr. Hyernaux, who had been assistant to Dr. Van Huevel at the Maternité of Brussels, in his "Manuel Pratique de l'Art des Accouchements" (Bruxelles, 1857), rejects in its favour all crotchets and cephalotribes as comparatively

dangerous and inefficient. It is therefore used in all cases where embryotomy is indicated. It consists of a powerful long forceps with the pelvic curve, the blades of which are grooved along the inner aspect in order to carry a chain-saw. When the head or other part of the child is seized by the forceps, this chain-saw is worked up from the point whence the blades spring, by means of cross handles attached to the two ends. Thus travelling up the grooves, the saw crosses the head and cuts through it. For extraction, Van Huevel contrived a pair of forceps toothed on one blade to seize the most convenient part of the child. Notwithstanding the formidable and complex appearance of the forceps-saw, it seems to have made its way into use. Professor Faye, of Christiania, a man of singular judgment and ability, says it is the only instrument fitted to cut through any part of the fœtus; but he appears not to be aware of the power of the wire-écraseur. He has simplified the instrument considerably, and extols it as safe, easy, and effectual. It is also used in Germany; and in Italy Dr. Billi has modified it and introduced it into practice. We cannot refuse to lend favourable consideration to an instrument so recommended. It is certainly a new power; and its claims to compete with or to displace other methods of facilitating delivery by embryotomy should be tested in practice. It appears to me, however, who have not yet used it, that it is more especially adapted for those minor degrees of pelvic contraction which can be dealt with satisfactorily by perforation and the craniotomy-forceps; and that in extreme cases, where the conjugate diameter is 2″ or less, where the craniotomy-forceps is still available, and in which the cephalotribe can do good service, the forceps-saw could hardly answer, owing to the size of the blades and the necessity of getting them to lock accurately in order to work the chain. It is capable of being most useful in dividing the neck or other part of the body in cases of impaction of shoulder-presentation.

5. *Delivery by the Author's New Method of Embryotomy.*—I have now to describe a new method of embryotomy, designed by myself, to effect delivery in the most extreme cases of pelvic contraction. It had long appeared to me that, if the problem, how to break up and extract such a body as the mature fœtus through a chink measuring an inch wide and three or four inches long, were proposed to a skilful engineer, he would find a solution. It did not seem to me that we were necessarily restricted from the use of new instruments. I thought I saw in the wire-écraseur the means of effecting the object in view. I had found no great difficulty in snaring an intra-uterine polypus of considerable size with a wire-loop passed through a cervix uteri whose aperture was much smaller than the tumour, guided only by one or two fingers. Why should not the fœtal head be seized in a similar manner and cut in pieces? I performed several experiments with a very diminutive and delicate rickety pelvis, measuring an inch in the antero-posterior diameter, and scarcely more in the sacro-cotyloid diameter, and I will now repeat the operation before you.*

The best instrument is Weiss's écraseur, which has an

* I also demonstrated this operation at the meeting of the Obstetrical Society of the 2nd June, 1869.

Archimedean screw and a windlass, admitting the use of a loop of any size. As in cephalotripsy, but not so

FIG. ?0.

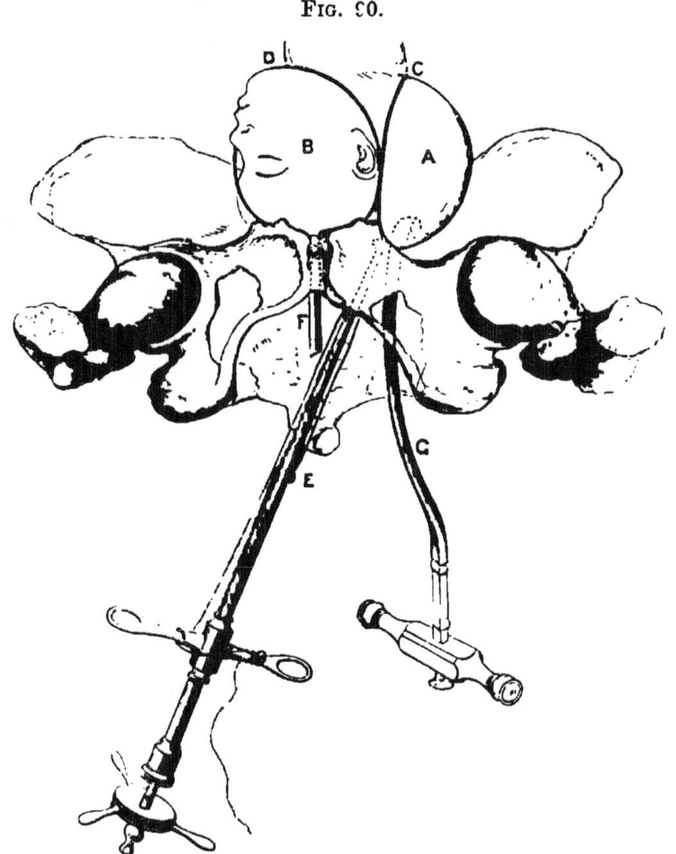

E, stem of écraseur carrying loop of wire over occiput. F, stem of écraseur carrying loop of wire over lateral segment of head. G, crotchet, the point of which is passed into the opening in the cranium made by the perforator, and held by an assistant to steady the head whilst the wire is being applied. A, the occipital segment of the head seized by the wire-loop at C, which buries itself in the head. B, a lateral segment of the head seized by the second application of the wire-loop at D.

urgently, it is desirable, first of all, to perforate the head. The wire cuts through the skull more easily if this be done. In doing this, the head is firmly supported against the brim by an assistant. The crotchet is next passed into the hole made by the perforator, and held by an assistant so as to steady the head. A loop of strong steel wire is then formed large enough to encircle the head. The elasticity of the wire permits of the loop being compressed by the fingers so as to make it narrow enough to slip through the cervix uteri and the chink of the pelvic brim. The loop is thus guided over the crotchet to the left side of the uterus, where the occiput lies. The compression being removed, the loop springs open to form its original ring, which is guided over the occiput, embracing all the posterior segment of the head, as in Fig. 90. The screw is then tightened. Instantly, the wire is buried in the scalp; and here is manifested a singular advantage of this operation. The whole force of the necessary manœuvres is expended on the fœtus. In the ordinary modes of performing embryotomy, as by the crotchet especially, and in a lesser degree by the craniotomy-forceps and cephalotribe, the mother's soft parts are subjected to pressure and contusion. The child's head, imperfectly reduced in bulk, is forcibly dragged down upon the narrow pelvis, the intervening soft parts being liable to be bruised, crushed, and even perforated. And this danger, obviously rising in proportion to the extent of the pelvic contraction, together with the bulk of the instruments used, deprive the mother, in all cases of extreme contraction, of the benefit of embryotomy, leaving her only the terrible prospect of the Cæsarian section. When the posterior segment of the head is seized in the wire-loop, a steady working of the screw cuts through the head in a few minutes. The loose segment is then removed by the craniotomy-forceps.

In minor degrees of contraction, the removal of the occipital segment is enough to enable the rest of the head to be extracted by the craniotomy-forceps. But in the class of extreme cases in which this operation is especially useful, it is desirable still further to reduce the head, by taking off another section. This is best done by re-applying the loop over the anterior side of the head as seen in B, Fig. 90. The wire seizes under the lower jaw beyond the ear. When the screw is worked, the wire has to cut through the base of the skull, dividing the sphenoid bone. The segment thus made is removed by the craniotomy-forceps.

The small part of the head still remaining attached to the trunk offers no obstacle. It is useful as a hold for traction. The craniotomy-forceps now seizes this firmly, and you proceed to deliver the trunk. If the child be well developed, this part of the operation will require considerable skill and patience. An assistant draws steadily on the craniotomy-forceps, directing traction to one side, so as to bring a shoulder into the brim. The operator then hooks the crotchet into the axilla, draws it down, and with strong scissors amputates the arm at the shoulder. This proceeding is then repeated on the other arm. Room is thus gained to deal with the thorax. You perforate the thorax. Introduce one blade of a strong pair of scissors into the aperture, and cut through the ribs in two directions. Then, by the crotchet, eviscerate the thorax and abdomen, until the trunk is in a condition to collapse completely. This done, moderate traction will complete the delivery.

I have imagined a proceeding by which the arms can be amputated even more easily. A curved tube, shaped like Rainsbotham's hook, may be made to carry a strong wire under the axilla, and the end being brought out, and the tube removed, the wire can be attached to the écraseur, which then cuts through the limb with ease and security. Decapitation may be conveniently performed in the same way.

This operation is particularly adapted to extreme cases of narrowing of the pelvic brim from rickets, in which there is commonly left a moderate amount of space at the outlet for manipulation. Indeed, I believe, a case of rickety deformity will rarely be found so great as to compel resort to the Cæsarian section. No doubt the operation I have recommended is more difficult, demands more skill and richness of resource than the Cæsarian section—an operation which cuts the Gordian knot with despotic simplicity, not perhaps unpleasing to the operator, but certainly full of extremest peril to the mother.

The operation is, I freely admit, less practicable in extreme cases of osteomalacic deformity. Here the pelvis is deeper than in rickets; and the deformity bearing in an aggravated degree upon the outlet, leaves insufficient room for manipulation. Where two fingers can barely pass between the tuberosities of the ischia, it will be scarcely possible to guide the écraseur through the pelvis, and to get the loop over the head. But in these cases, as I have already stated, the bones will often open up, under pressure applied within. Professor Lazzati tells me he relies upon this dilatability in all cases of osteomalacia, seldom or never resorting to the Cæsarian section, except in the worst cases of rickety distortion.

Certain dangers attend the operation of craniotomy. What are these? Certain injuries may be inflicted upon or result to the mother.

1. The perforator has been known to strike the promontory of the sacrum, or to lacerate the cervix uteri.

2. Spicula of cranial bones resulting from perforation may scratch or tear the soft parts.

3. The crotchet may slip, and lacerate the soft parts.

The above, of course, may be avoided with care.

4. But serious evil is likely to result from deferring the operation too long—*i.e.*, until after exhaustion has set in—and under a too protracted operation in an unsuitable case. Long-continued dragging of the head upon a brim which it cannot pass, jamming the soft parts more especially at the two points of greatest projection, the promontory and the symphysis, ends by stopping the circulation in the parts compressed, bruising them, actually grinding through them. In this way, after severe operations, it has been found that a large hole has been made through the posterior cervix uteri. Such injury, added to the shock and exhaustion of the system, may be fatal.

5. If the immediate injury above described do not occur, the long-continued pressure may cause mortification of a limited portion of the neck of the uterus. Thus, in the course of a few days, a slough is formed between the vagina and bladder, resulting in vesico-

vaginal fistula. Dr. F. H. Ramsbotham said* that, "in almost all the cases which he had seen of a fistulous opening into the neck of the bladder consequent on labour, the child had been putrid; and he attributed the slough more to contact with the putrid head than to simple pressure on the part." However this may be, I am very certain that simple pressure, especially if bearing upon a limited spot, is enough to cause the accident; and I have quite recently seen a case of recto-vaginal fistula ensuing on the birth of a living child delivered by the natural powers.

In the possibility of attendant danger, craniotomy differs essentially from the forceps. Whilst under craniotomy mischief or death may ensue, the forceps, if used rightly and in suitable cases, is an innocuous instrument. Statistics, professing to show that the mortality from the use of the forceps is at the rate of 1 in 20, are flagrant examples of the fallacy of arguing "*post hoc, ergo propter hoc.*" Properly speaking, the mortality from the forceps is *nil.* Women die because the instrument is used too late.

* "Medical Times and Gazette," 1862.

LECTURE XIX.

THE CÆSARIAN SECTION — THE INDICATIONS FOR — THE MORAL ASPECT OF THE OPERATION AS BETWEEN MOTHER AND CHILD—THE CONDITIONS THAT RENDER THE OPERATION NECESSARY — QUESTION BETWEEN CÆSARIAN SECTION AND TURNING IN MORIBUND AND DEAD WOMEN TO SAVE THE CHILD — THE TIME TO SELECT FOR THE OPERATION — PREPARATION FOR OPERATING—THE OPERATION—THE DANGERS ATTENDING IT, AND THE PROGNOSIS—SYMPHYSEOTOMY.

THE CÆSARIAN SECTION.

THE Cæsarian section occupies a doubtful place between Conservative and Sacrificial Midwifery. It is conservative in its design, in its ambition; it is too often sacrificial in fact. It is resorted to with a feeling akin to despair for the fate of the mother, which is scarcely tempered by the hope of rescuing the child. It is looked upon by the great majority of obstetricians as the last desperate resource, as the most forcible example of that kind of surgery which John Hunter regarded as the reproach of surgeons, being a confession that their art was baffled. On the other hand, it is regarded by some enthusiastic practitioners, dazzled perhaps by its false brilliancy, as an operation deserving to be raised into competition with turning, craniotomy, or cephalotripsy. At different times and in different countries it has been looked upon with favour, because promising salvation to the child. The child is weighed against its mother; and conscience is silenced with the reflection that the possible or probable rescue of the child may rightly be purchased by subjecting the mother to the most imminent peril. It is held that this double chance, made up of odds in favour of the child and against the mother, is to be preferred to the single chance afforded by an operation which gives up the child for the sake of odds in favour of the mother.

The situation is painful, and may well perplex those who are not steady in their allegiance to those moral laws which ought to rule over all professions, and which certainly recognise no exception here.

None of us, probably, will claim exemption from these laws; but possibly some may interpret them differently. And here, again, we must call to mind a lesser law to which reference has before been made—namely, that the choice between two operations will be influenced by the comparative skill in them which the operator happens to possess. Under this influence the favoured operation will be more and more cultivated, and its competitor more and more neglected. Thus, to apply this law to the present discussion: the man confident in his skill in the extraction of a dead child by the natural passages with safety to the mother will be disposed to assign the narrowest possible limits to the Cæsarian section; and, on the other hand, the man who has not this confidence will be disposed to prefer the Cæsarian section, an easy operation.

The operation must be studied under two aspects—*first*, as one imposed by *necessity*, as the *only* means of effecting delivery; *secondly*, as one of *election*, deliberately chosen as the *best* means of effecting delivery.

The distinction is very important to be borne in mind; for, under that fatal fascination which seems to oppress the reasoning faculty in statisticians, conclusions drawn from figures representing the most dissimilar facts are accepted and put forth as the legitimate deductions from experience. With our present materials, I believe it is the most idle and unprofitable waste of intellect and of time to seek to draw rules for practice from statistical operations. It seems to me impossible in a great number of cases to distinguish or to estimate the relative shares in causing death that arise from causes antecedent to the labour, from causes arising during and in consequence of the labour, and from the operation itself. All cases, therefore, which were not selected—that is, deliberately operated upon, under simple conditions, especially freedom from dangerous disease and from protracted labour—must be put aside or considered apart. How many simple cases of this kind do we possess? A dozen? Twenty? Perhaps a few more; but certainly not enough to deduce a law of mortality to estimate the risk to life from the Cæsarian section. M. Pihan-Dufeillay,* indeed, declared that the operation performed under favourable circumstances, as early as the impossibility of delivery *per vias naturales* is recognised, gives nearly

* "Archives Gén. de Méd.," 1861.

75 per cent. of recoveries. But what assurance have we that an undue proportion of successful cases are not recorded, unsuccessful ones remaining in the dark?

And if this element of the comparison, the risk to life of Cæsarian section, is so defective, vague, and uncertain, it follows that comparison or ponderation in the statistical scales is impossible, even assuming that the other element, the risk to life under craniotomy, were determined. The known may be ponderable; the unknown is certainly imponderable. But do we know more of this second element, the risk to life in cases of exactly similar pelvic deformity under other modes of delivery? I am not aware that any attempt, not vitiated by the most transparent fallacies, has been made to establish what this risk is. Obviously, we cannot recognise fatal cases of craniotomy in extreme deformity, say of conjugate diameter reduced to 2″ or to 1·75″, unless the operation was begun under selected circumstances—that is, before exhaustion had set in—and conducted with due skill, and after the most approved methods. We are fairly called upon to reject all fatal cases in which craniotomy was performed with bad instruments, in which the skull was either not crushed down by the cephalotribe, or the calvarium not removed so as to leave nothing but the base to bring through the brim, edge on, or the head and trunk not reduced by sections, as by my method. The point to determine is, what is the limit of contraction that admits of this proceeding, or some better proceeding, being carried out with a reasonable presumption of safety? When that is determined, it follows, as a logical necessity, that this proceeding ought to be adopted in cases falling within that limit.

To dispute this proposition—and it has been disputed, the disputants not seeing the dilemma prepared for themselves—is to dispute the propriety of performing craniotomy in any case. For why do we perform craniotomy or cephalotripsy in the case of a pelvis contracted to 3″ or 3·50″? Is it not because, possessing means of extracting a dead child through such a pelvis with a reasonable prospect of safety to the mother, we acknowledge it to be our duty to use those means, and to save the mother? Now, assume that we also possess means of extracting a dead child through a pelvis narrowed to 2·00″ or 1·50″ with a reasonable prospect of safety to the mother, who will venture to dispute, whether as a matter of logic or of morals, that the mother shall be denied that prospect?

The case, then, as far as relates to pelvic contraction, stands on the old ground. It may be stated broadly as follows:—Embryotomy stands first, and must be adopted in every case where it can be carried out without injuring the mother. The Cæsarian section comes last, and must be resorted to in those cases where embryotomy is either impracticable, or cannot be carried out without injuring the mother. There is, therefore, no election. The law is defined and clear. The Cæsarian section is the last refuge of stern necessity.

The formula of Dr. Hull may be accepted. It affirms "that *every circumstance which can render delivery per vias naturales impracticable ought to be considered as requiring the Cæsarian operation.*"

Of course, those who advocate the Cæsarian section in cases of contraction below 2·25″, may deny the possibility of extracting a dead child through the natural passages with a reasonable prospect of safety to the mother. But this denial, of course, applies only to the experience of those who deny. I can but refer to the testimony of Pajot and many Continental authors, as to the positive fact as regards cephalotripsy. I refer to that of Kelly, Osborn, Professor D. Davis, Dr. Braxton Hicks, and myself, as regards craniotomy. (*See* Lecture XVIII.) It is right, however, to state that Scanzoni* declares the limit generally adopted in Germany to be 2·50″. I repeat, with all the emphasis that conviction based upon experience dictates, that delivery by the natural passages, either by cephalotripsy, by the craniotomy-forceps, or by my new method of embryotomy, if the conjugate diameter measures 1·50″, is perfectly practicable, and with a presumption of safety to the mother much greater than that attending the Cæsarian section.

INFANT MORTALITY ATTENDING THE CÆSARIAN SECTION.

The probability of saving the child must obviously vary according to circumstances. Sometimes the operation is determined upon without any consideration for the child, simply as a means of delivering the woman. The child may be known to be dead. The most practical question is—What is the probability of saving the child when the operation is performed under the most favourable selected circum-

* "Lehrbuch der Geburtshülfe," 1867.

stances? Statistics, even upon this point, are not conclusive. But it may be assumed that the prospect of the child is less than in the case of ordinary labour. The experience of the Royal Maternity Charity gives rather more than three per cent. of still-born children. Dr. Radford says the risk to the infant is not much greater than that which is contingent on natural labour. Scanzoni* ascertained the fate of 81 children out of 120 operations performed between 1841 and 1853; 53 children, or 60 per cent., were born alive. In all probability several of these died within the first week. The mortality, then, is greater than in ordinary labour.

But assuming the probability of saving the child to be ever so high, are we justified by law, or by religion, the basis of law, in taking the woman's life as it were into our hands, and deliberately subjecting it to the most imminent hazard for the sake of probably saving her child? It has been urged that, the woman having a prospect of a miserable existence for a few months or weeks only, whilst the child is likely to be saved and live to maturity, the child's life is the more valuable, and ought therefore to be preferred. Or, taking the case of a woman who cannot give birth to a living child by the natural passages, calling upon the obstetrician time after time to deliver her by craniotomy, it is asked: ought we not at length to refuse craniotomy and insist upon the Cæsarian section?

Dr. Denman states the question perspicuously thus: —"I cannot,† however, relinquish the subject with-

* "Lehrbuch der Geburtshülfe," 1867.

† "Introduction to Midwifery."

out mentioning another statement of this question, which has often employed my mind, especially when the subject has been actually passing before me. Suppose, for instance, a woman married, who was so unfortunately framed that she could not have a living child. The first time of her being in labour, no reasonable person could hesitate to afford relief at the expense of her child; even a second and a third trial might be justifiable to ascertain the fact of the impossibility. But it might be doubted in morals whether children should be begotten under such circumstances; or whether, after a determination that she cannot bear a living child, a woman be entitled to have a number of children (more than ten have sometimes been sacrificed with this view) destroyed for the purpose of saving her life; or whether, after many trials, she ought not to submit to the Cæsarian operation, as the means of preserving the child at the risk of her own life. *This thing ought to be considered.*"

The question is indeed a trying one. We may easily go astray in the labyrinth of casuistry, unless we hold steadily to the clue laid down by the moral law. I think it will not be disputed that, in law, *he who accelerates death is held responsible for having caused death.* We are not justified in regarding that life as less sacred because we believe that, in the ordinary course of disease, it will not last beyond a very short time. We are, therefore, not justified in preferring the Cæsarian section to craniotomy in a case where the latter operation offers a fair prospect of safe delivery, because the woman is suffering from osteomalacia, which is commonly, but not always, a progressive and fatal disease. We have no right to lessen the woman's chance of life because it is already small. We cannot, even in the case of osteomalacia, be certain of our prognosis. I have known women live for many years with osteomalacia, and even recover.

Then take the case put by Denman. The conduct of the woman is assumed to be culpable, and we are assumed to be in the position of accomplices or abettors in her fault if we repeatedly relieve her by craniotomy. But are we entitled to take upon ourselves the office of the Judge? Are we to make ourselves the ministers of Justice? Vengeance is not ours. When did Medicine ever withhold her merciful hand from the degraded, the sinful, the criminal? Shall we dare to put a mere vegetative life—that of an unborn child—into the scale against that of a being like ourselves accountable to the Almighty. Can we take upon ourselves the awful weight of deciding that the wretched woman was wrong—criminal, in becoming a mother? She is subject to her husband. If punishment is due, must it fall upon her? and are we to inflict it? I cannot, therefore, hesitate in expressing my conviction that we should be traitors to our trust if we were to perform the Cæsarian section, when craniotomy is safer for the woman, because, in our judgment, she was culpable in becoming a mother. Nor is the life of the mother to be estimated alone. It must be weighed also in reference to her value as a wife, and as the mother of existing children whose fate is intimately involved in her own.

THE CONDITIONS THAT RENDER THE CÆSARIAN SECTION NECESSARY.

The most frequent condition is *deformity with contraction of the pelvis.* The operation is justified whenever the contraction is such as to render it impossible to extract a dead child through the natural passages. This may be stated at 1·50″ conjugate diameter and below, at the higher limit of 1·50″ coming into competition with craniotomy. Cases may also occur in which a conjugate diameter of 2·00″ may call for Cæsarian section if the pelvis is much distorted, so that the diagonal and transverse diameters offer insufficient compensation for the narrow conjugate. This is more especially the case when, as in osteomalacia, the outlet of the pelvis is also so contracted as to render the necessary manipulation and the introduction of instruments impossible.

The most frequent form of distortion calling for the Cæsarian section is that which arises from osteomalacia. In malacosteon the sides of the triangle forming the brim of the pelvis are all pressed inwards, and are more or less convex. The result is that the brim is practically divided into two parts, neither of which is available for the passage of the head. Rickets, also, will sometimes produce a pelvis that will leave no alternative. The slipping down of the lumbar vertebræ—spondylolisthesis—into the pelvic cavity, if to any great extent, leaves no other resource.*

The next most frequent causes are tumours of various kinds growing into the pelvis, such as bony or malignant tumours springing from the wall of the

* See a memoir on this subject, "Obstetrical Transactions," 1865, by Robert Barnes.

pelvis; tumours of the ovary descending into the pelvic cavity, and getting fixed there. Dr. Sadler, of Barnsley, has recorded a case * in which the operation became necessary from the pelvis being filled up with an enormous *hydatid cyst* springing from the liver. Other exceptional causes, chiefly remarkable on account of their extreme rarity, have been observed. Atresia of the cervix uteri and vagina may be so extensive and unyielding that the Cæsarian section may be less hazardous than the attempt to open a canal through the cicatricial tissues.

We are sometimes driven to the operation after having exhausted other modes of proceeding — for example, when craniotomy may have failed to deliver. It is a remark of Dr. Radford that most of these cases are progressive in their character.

The Cæsarian section, or rather gastrotomy simple, is indicated in certain cases of rupture of the uterus, when the child cannot be extracted with advantage through the pelvis.

It is resorted to when the mother has died undelivered, in the hope of rescuing the child. In this way several children have been saved. The success, of course, will depend greatly upon opening the uterus very soon after the mother's death. How soon? It is difficult to assign a precise limit beyond which the child's life may be regarded as lost. Harvey said, "Children have been frequently taken out of the womb alive hours after the death of the mother." I am not aware of instances to justify this statement. But Burns says: "The uterus may live longer than the body; and after the mother has been quite dead

* "Medical Times and Gazette," 1864.

the child still continues its functions." The fœtus in lower animals will live some time if the ovum is not opened. It seems not improbable that a modified degree of placental circulation may continue for some little time after the mother's general circulation has stopped; and it is, further, even more probable that the child may survive for some little time in a state of asphyxia capable of restoration on being brought into the air. Certainly children have been extracted alive ten minutes after the mother's death. On the other hand, children have been extracted dead when the operation has been delayed fifteen minutes. This has occurred twice in my own experience. In Lecture IV. I have adduced further evidence bearing on this question. The chance of preserving the child by the Cæsarian section *post mortem* will be much influenced by the circumstances of the mother's death. If she dies by sudden injury, the child may survive a little longer. If she dies from hæmorrhage or rupture of the uterus, the child's death is likely to have preceded that of the mother. If she dies from phthisis or other gradually exhausting disease, the child may survive some minutes.

The Cæsarian section after death comes into competition with forced delivery *per vias naturales*. Sometimes, if the cervix is dilated, the child may be extracted by turning as quickly as by Cæsarian section, and it has this advantage — that, being less likely to shock the friends, it may be practised when the Cæsarian section would be rejected, or resisted until too late.

This view has recently found favour in the Italian schools. Professor Rizzoli proposed to deliver in this manner as soon as the woman was dead. The late Professor Esterlé* related a case, occurring in 1858, in which he successfully executed this proceeding in a woman dying of cerebral apoplexy. One argument he adduced in support was the fact that in similar cases the child survives its mother's death so short a time that to save it delivery must be effected whilst the mother is still living. He says that, in his judgment, this is the proper course to adopt—of course requiring that all possible gentleness shall be practised, and that, if necessary, the plug and vaginal douche be used beforehand.

Cases in support of this practice are recorded by Belluzzi,† Ferratini, and others. Out of three cases Belluzzi saved two children. Ferratini ‡ saved the child of a woman dying of phthisis. The following conclusions are submitted by Ferratini :—

1. The Cæsarian section must be absolutely excluded from common practice, reserving it for cases of organic defect not admitting of operations through the natural passages.

2. Whenever the obstetrist is called to dead pregnant women, he should resort to forced delivery, cases of apparent death being not rare.

3. This operation is absolutely necessary whenever pregnancy has gone to 180 days.

4. It does not aggravate the condition of the mother (?), and gives the best chances to the child.

* "Annali Universali di Medicina," 1861.

† "Nuovi fatti in appoggio dell' estrazione del feto col parto forzato durante l' agonia delle donne incinte, onde salvare più facilmente il feto stesso in sostituzione a tale operazione, o al taglio cesareo post mortem." Bologna, 1867.

‡ "Un nuovo fatto in appoggio," etc. Genova, 1868.

5. It leaves no trace upon the corpse, requires no instrument, and is less likely to excite objection on the part of the friends.*

A condition which has several times led to the Cæsarian section has been *disease, mostly malignant, of the lower segment of the uterus*, preventing its due dilatation. In this case, the opportunity is commonly presented of inducing labour or abortion. The conditions of choice are often perplexing. If labour be induced before seven months, the prospect of a viable child is small, but the uterus may be able to dilate sufficiently without injury. On the other hand, the life of the mother is probably doomed to be of short duration from the progress of her disease. Any injury to the diseased structures, as even by premature labour, is liable to accelerate her death. Is it not then better, both in her interest and in that of the infant, to let things go on to the natural term of gestation? She may live two or three months longer, and the child will undoubtedly have a better chance.

What is the best time to select for the operation?

1. Sometimes, of course, all chance is denied us, or the range of time offered is extremely limited. If called to a woman in labour at term, and under the conditions assumed to require the operation, it ought to be performed without delay. There is, if possible, greater reason than in conditions requiring other less formidable proceedings, to anticipate the exhaustion and local injury that follow upon protracted labour. It is a misfortune, tending fatally to compromise success, to be obliged to operate when the system is

* This subject is also referred to by Dr. Hyernaux, of Brussels, in the second edition of his treatise on Obstetrics.

prostrated; when the structures that have to be wounded are so worn and injured that the power of reaction and repair is seriously reduced; and when the blood is deteriorated by the products of nervous and muscular overwork. It is, then, a clear indication to operate early in labour.

2. If the patient come under observation early in pregnancy, we have the double opportunity of considering the propriety of inducing labour, with the object of avoiding the operation, and of selecting the time for its performance should it be unavoidable. Is it an advantage to operate during labour?—that is, does the process of labour conduce to the success of the Cæsarian section? Amongst Continental practitioners, and indeed generally here, the Cæsarian section being regarded as a mode of delivery, it is held to be a primary indication to respect the laws of parturition, and to enlist the natural powers in our aid as much as possible. It has been, therefore, almost universally considered proper, in cases where the ultimate necessity of resorting to the Cæsarian section is recognised, to postpone the operation until the advent of labour. It is presumed that the epoch which Nature fixes for labour is that when the most favourable conditions for the process and for the recovery are present. The whole organization is better prepared, and the uterine muscles having acquired their highest stage of development, and contraction having actually set in, it seems reasonable to anticipate that the wound made in the uterus will close better, and that the necessary changes attending delivery will be more safely carried out.

Dr. Ludwig Winckel,* whose experience in this

* "Monatsschrift für Geburtskunde," 1863.

operation is the greatest, says the most favourable time is the end of the second stage of labour, when the membranes are ready to burst. He advises not to rupture the membranes. The escape of liquor amnii into the abdomen does no harm, and the extraction of the child is more easy if the membranes are kept entire until the moment of seizing the child.

But admitting spontaneous labour at term to be a favouring condition, may not labour induced artificially before term be an equally favourable condition? The arguments in support of the affirmative deserve attention. Dr. Braxton Hicks brought on labour a fortnight before term as a preparation for the Cæsarian section, influenced by the opinion that by so doing, the uterus, taken at a period prior to the highest degree of degeneration of its muscular fibres, would heal better. I am not disposed to think that the degree of fatty change observed in the mature uterus is any impediment to reparation. There are, at any rate, too many examples of complete repair after section by the knife, and even after rupture, to admit of a doubt that the mature uterus is in a condition not unfavourable for the operation. On the other hand, it cannot be doubted that the uterus at seven or eight months is also capable of complete repair after injury. We may, then, very properly consider whether, assuming things to be equal, *quoad* the uterus, there may not be other circumstances that may rightly turn the scale in favour of premature delivery. Such, I think, do exist. For example, if we wait for the advent of natural labour, we may be called upon to operate in the middle of the night, and surrounded by many difficulties, all concurring to lessen the prospect of success. By selecting our own time, we may have daylight, the assistance of colleagues, and every appliance that may be thought useful.

The best time to select, then, would be as near the natural term of gestation as possible, and this may be determined approximately by taking some day in the estimated last fortnight of gestation.

Then comes the question, Shall we start labour before operating, or proceed to the operation at once without exciting any preparatory action of the uterus? I think the preponderance of reason is in favour of operating upon a uterus already in the act of labour. The *first step* will be to pass up an elastic bougie into the uterus overnight. This will excite some degree of uterine action. Next day, the hour of operating being fixed, say, for 1 p.m., we may in the morning ascertain to what extent labour has proceeded. If the os uteri is not open more than enough to allow a finger to pass, it will be useful to dilate it a little more with the caoutchouc bag No. 2. This will probably induce further contraction of the uterus, and secure one most desirable object—namely, a free outlet for liquor amnii and other discharges by the natural passages.

The labour then being started thus far, we are ready for the operation. The question between general and local anæsthesia arises. In all abdominal operations, the vomiting so liable to attend or follow chloroform is a serious drawback. The violent straining is apt to open the uterine wound, to stretch the abdominal wound, to destroy the "rest" which is such an important condition of repair, and thus to compromise the success of the operation. The ether spray is at least free from this objection; and possibly, further experience will show that it ought to be preferred.

THE INSTRUMENTS AND ASSISTANTS.

The *instruments* required are:—1. A sharp bistoury. 2. A bistoury having a blunt end. 3. A director such as is used in ovariotomy. 4. A large probang armed with sponge. 5. Artery-forceps and ligatures in case of bleeding from the abdominal wound. 6. New sponges. 7. Ice. 8. Two powerful apparatuses for inducing local anæsthesia by congelation, or chloroform. 9. Silver or silk-sutures for uterine and abdominal wounds. 10. Lint. 11. Many-tailed bandage and adhesive plaster.

Assistants.—Skilled assistants should stand one on each side of the patient. Another should be free to hand instruments and assist in sponging, &c. A nurse or two to help will complete *the necessary staff.*

Preparation.—The bowels should be emptied by castor-oil or by enema on the morning of the operation.

THE OPERATION.

Position.—The patient is laid on a table on her back, with the head and shoulders slightly raised. The operator stands on the patient's right or in front; an assistant stands on each side.

The *catheter* is introduced to empty the bladder. If the case be one of osteomalacia, explore for the last time carefully to ascertain if the pelvis can be opened up by dilatation by the hand. In this way, twice at least, the operation has been avoided. It is related that a young surgeon, burning with the ambition to perform the Cæsarian section, invited his *confrères* to assist. Everything was ready, when Osiander requested permission to examine. He opened up the pelvis, turned, and extracted a live child; and thus, says he, "I saved the woman from the Cæsarian section and from death." Dr. Tyler Smith has also related a case in which he was able to dilate the pelvis with his hand, and deliver after craniotomy. I have already adverted to the testimony of Professor Lazzati, the distinguished Director of the Lying-in Hospital of Milan, who informs me that, although osteomalacia is very frequently observed there, he does not often find it necessary to resort to Cæsarian section even in cases of great deformity. The common practice is to turn, and the bones yield to admit the child, either mutilated or not. When the child has passed, the bones collapse again. He has more frequently been called on to perform the Cæsarian section on account of extreme distortion from rickets.

In cases of great distortion, the uterus is not seldom found considerably displaced. Where there is great prominence of the sacral promontory, and squatting of the chest down upon the flanks, the uterus is necessarily thrown much forward, sometimes so as to overhang the symphysis. With this there is occasionally marked lateral obliquity; and, what is less common, but still likely, a twisting of the uterus on its long axis, bringing one of its sides to look more or less forward. This was noticed in Dr. Hicks's case.

It is also desirable to take note, by auscultation, of

the seat of attachment of the placenta. Assistance may also be obtained by laying the hand flat on the uterus, when, if the walls are thin, a peculiar thrill or vibration marks the seat of the placenta, which is confirmed by feeling the part bulging a little, as if a segment of a smaller globe were seated on a large spheroid. This has been pointed out by Dr. Pfeiffer.*

The uterus is then brought into proper relation with the linea alba, so that the two incisions may correspond, unless it be found that the placenta can be avoided by cutting a little on one side of the median line. The abdominal incision is best made in the *linea alba*, extending from below the umbilicus to within about three inches of the symphysis pubis. When nearly through, it is desirable to get a finger through a small opening, and, using this as a director, to cut from within outwards, so as to avoid scratching the uterus. The assistants support the abdominal wall on either side, looking out to prevent the escape of intestine. The *uterine incision* is made in the middle line, sparing the fundus and lower segment as much as possible, as these parts are not well adapted to close by contraction. Circular fibres predominating near the cervix tend to make the wound gape. A manœuvre described by Winckel, and forced upon him by his being often called upon to operate with insufficient assistance, here deserves attention. An assistant hooks the forefinger of each hand in the upper and lower angle of the uterine wound, and, lifting them up, fixes them in contact with the corresponding angles of the abdominal wound. This shuts out the

* "Monatsschrift für Geburtskunde," 1868.

intestine effectually, and tends to prevent the blood from running into the abdominal cavity.

If the placenta is found directly behind the wound, the hand of the operator is insinuated between the placenta and the uterine wall, detaching it until the edge is felt, when the membranes are pierced, and *the child is seized by the feet.* This point obstetricians will pardon me for prominently emphasising, because I have seen a surgeon pull at the arm and fail to extract the child. Sometimes the neck is tightly grasped by the uterine wound. If the constriction does not soon yield, it is better, says Scanzoni, to extend the incision than to drag overmuch, lest the womb be torn.

When the child and placenta are removed, attention is required to watch the bleeding; this generally takes place from the cut sinuses in the uterine walls, less frequently from the inner surface from which the placenta has been separated. Hæmorrhage is best checked by direct compression of the uterus with the hand; if the uterus contract well, the hæmorrhage ceases. Ice may be applied to the wound, and a piece placed in the uterine cavity.

Before closing the wound, *thrust the probang through the os uteri and vagina,* to make sure that the natural passages are clear for the discharges from the uterus.

The Closure of the Wound.—When the bleeding has ceased, and any blood that may have found its way into the abdomen is removed, we have to consider the question of applying sutures to the uterine wound. It is a matter of observation that in many fatal cases the edges of this wound have been found flaccid and gaping. But in the majority of, if not in all, these cases the operation had been performed on women exhausted by protracted labour; and, on the other hand, in women operated upon at a selected time, when the powers are unimpaired, the uterus commonly contracts well. Winckel says he has never lost a case from hæmorrhage, and has not stitched the uterine wound. Mr. Spencer Wells has had a successful case in which he used a long piece of silk as an uninterrupted suture, leaving one end hanging out through the cervix and vagina. By pulling on this end the suture was removed after several days.

Another mode of suture would be to carry the same suture through the uterine and the abdominal walls, so as to secure adhesion between the two parts. A serious source of danger is from vomiting; the straining and relaxation attending this accident tend to promote the opening of both wounds, and to force the discharges through them. This risk would be lessened by uterine sutures. Upon the whole the case may be stated thus:—If the patient is operated upon at a selected time, if the danger of vomiting is lessened by not taking chloroform, and if the uterus contracts well during the operation, the sutures may be dispensed with; but under the opposite circumstances, it would be better to stitch the uterus as was done in Mr. Spencer Wells's case.

Closure of the Abdominal Wound. — The methods adopted in ovariotomy may be followed. I am inclined to prefer the uninterrupted silk suture. Winckel's cases were closed by the more common method of interrupted sutures, with intermediate skin-sutures. The important point is to *close the wound completely.*

After-treatment.—A full dose of opium should be given

immediately, either in form of pill or suppository. Light nourishment and perfect repose are the things to be observed. The dressings should not be removed for five or six days. To obviate foulness, sprinkling with Condy's fluid or weak carbolic acid may be resorted to. The sutures may be removed on the seventh or eighth day. The bowels may be relieved by enema on the fourth or fifth day.

What is the risk to life attending the Cæsarian section, numerically expressed? I have already made some remarks upon this, the statistical aspect of the question. I doubt if any satisfactory answer can be given. Can any quasi-analogical deduction be drawn from the mortality attending ovariotomy? Take this to be one death to two or three recoveries; may we expect a similar result from the Cæsarian section—I mean, of course, when performed at a selected time? It would be rash to expect an equal success; and there is a consideration which, I think, is generally neglected by statisticians. It is this: we are not justified in treating the particular patient whose case is before us as an abstract entity, a mere arithmetical unit. Her fate is not to be decided by what are called statistical laws, but which are in reality too often nothing better than the accidental issues of blind gropings. We must study case by case; compare them; analyse clinically. "Non numerandæ, sed perpendendæ sunt observationes."

THE DANGERS OF THE OPERATION; THE PROGNOSIS.

The principal risks run are as follows:—

1. If the operation is performed as the last resource

after protracted attempts to deliver by other means, the woman is liable to sink from shock and exhaustion within a few hours; or if she survive beyond a few hours, there is the risk of hæmorrhage, of peritonitis, and of puerperal fever. It may be said that the prospect of recovery, when the operation is performed under these circumstances, is very small.

2. If the operation is performed at a selected time, the woman escapes the shock attendant upon the protracted labour, and encounters the shock of the operation with unimpaired strength. Still, the shock is very great, and is not seldom fatal *per se*. This is the first and most pressing danger. Could it be in any way modified or controlled, the Cæsarian section might be undertaken with more confidence. But shock necessarily attends all severe abdominal injury. It affects different persons in different degrees. Nor can we readily predicate of any given person that she will bear shock well or badly. It is an uncertain element, and must probably ever perplex all calculation as to the result of the Cæsarian section in any particular case. I do not think that chloroform materially lessens the shock; and it adds the danger of vomiting.

3. The next danger is *hæmorrhage*, and as hæmorrhage is often associated with prostration as cause and as effect, the danger is serious. This may come on within a few hours. It might be expected that hæmorrhage would be liable to come from the inner uterine surface, as after ordinary labour, but the more common source is probably from the sinuses divided in the uterine wound. The quantity lost may be enough to cause a fatal anæmia. But the more

common evil is from the irritation caused by the blood collecting in the abdominal cavity giving rise to

4. *Secondary shock* and *peritonitis.* That secondary shock precedes peritonitis I have no doubt. Intense pain, even tenderness on pressure, rapid small pulse, accelerated and impeded breathing, suggest the diagnosis of peritonitis; but if at this stage the patient die and be examined, probably no trace of peritonitis, as revealed by redness or effusion, is discovered. Peritonitis may come on the day following the operation. It may be met by fomentations to the abdomen, by opiate suppositories; and the prostration soon ensuing must be combated with wine, brandy, beef-tea, chicken-broth. Salines are often useful, especially at first.

5. If the patient escape the preceding dangers, there is still the risk of septic infection, of *septicæmic puerperal fever.* The source of this is the absorption of septic matter from the cavity of the womb or from the edges of the wound; or it may arise from general blood-dyscrasia resulting from the accumulation in the circulation of effete matters which the excreting organs are unable to dispose of.

6. In addition to the dangers incident to the operation and to the puerperal state, there is the danger inherent to the disease which rendered the operation necessary, liable in some cases, as in cancer, to be aggravated by the operation, which may accelerate the fatal issue.

Winckel says that osteomalacia is much more unfavourable than rickets in connection with Cæsarian section. Still, osteomalacic patients bear wounds well, and the power of repair is often great.

The uterus often contracts adhesions with the abdominal wall during repair. These adhesions do not appear to entail any serious inconvenience; and should pregnancy again occur, and the Cæsarian section be again necessary, they render the operation less dangerous. The peritoneal cavity is shut off; the incision through the abdominal wall leads directly through the adhesions to the uterus. Thus the dangers of hæmorrhage, of effusions into the abdomen, are eliminated, and it is even probable that the shock is less. On the other hand, no adhesions may be found, and the wound in the uterus heals so completely that years afterwards no trace of cicatrix is found (Radford).

Several cases are now known in which the Cæsarian section has been performed twice, thrice, and even four times on the same woman.

These cases of repeated success would seem to indicate a special tolerance of severe injury in the subjects, and cannot wisely be taken as evidence, absolute or cumulative and statistical, in reduction of the danger of the operation. This is further illustrated in the following history:—Dr. Freericks* performed the section on account of contracted pelvis. Mother and child recovered. When again pregnant, premature labour was induced about the eighth month. When labour had begun, collapse set in; the uterus had ruptured. The child was removed from the abdomen by gastrotomy. Vomiting caused extrusion of the intestines. To effect reposition, numerous pricks were made in them to let gas escape, without effect until an incision was made with a bistoury,

* "Nederl. Tijdschr. v. Geneeskunde," 1858.

and much thin pappy matter was evacuated. The intestines were then replaced, and the wound was closed. She recovered completely. How many women would be as tolerant?

LECTURE XX.

THE INDUCTION OF PREMATURE LABOUR—THE MORAL BEARING OF THE OPERATION—THE FITNESS OF THE SYSTEM AND OF THE GENITAL ORGANS FOR PREMATURE LABOUR—THE INSUFFICIENCY OF SIMPLY PROVOCATIVE MEANS—TWO STAGES OF PREMATURE LABOUR ARTIFICIALLY INDUCED—THE PROVOCATIVE; THE ACCELERATIVE—DISCUSSION OF THE VARIOUS PROVOCATIVE AGENTS—DANGER OF THE DOUCHE—ACTION OF THE VARIOUS DILATORS—THE MODE OF PROCEEDING—PROVOCATION OVERNIGHT—ACCELERATION AND CONCLUSION OF LABOUR NEXT DAY—DESCRIPTION OF CASES DEMANDING INDUCTION OF LABOUR—MODE OF DETERMINING EPOCH OF GESTATION—PROCEEDING IN CONTRACTED PELVIS OR OTHER MECHANICAL OBSTRUCTIONS—IN CASES OF URGENT DISTRESS OF MOTHER.

We now come to an operation which carries us fairly back within the domains of Conservative Midwifery. The induction of premature labour is designed to save the mother and child, or at least the mother, from those perils which one or both would have to encounter at or before the natural term of gestation. In many cases those perils increase in an accelerated ratio with the advance of gestation. By anticipating the ordinary epoch of delivery, by selecting a time when these perils have either not yet arisen or are still comparatively small, we may make the labour auspicious, indeed natural in everything but in the moment of its occurrence. In many other cases we may obtain an equally auspicious result by commanding the entire course of the labour, overcoming certain difficulties by appropriate proceedings. I shall show that it has been too much the custom to consider that our resources are limited to the first order of cases—to leave too much to accident—and that, by taking the whole conduct of labour into our own hands, we may greatly extend the application of this most beneficent operation, save much suffering, and greatly add to the probability of saving the lives of mother and child.

We have thus three great conservative operations. Two of these — the forceps and the induction of premature labour — have been contributed by the London school.

Denman tells us that he learned from Dr. C. Kelly, that "about the year 1756 there was a consultation of the most eminent men at that time in London to consider the moral rectitude of, and advantages which might be expected from, this practice." It met with their general approbation; and under this sanction the operation was resorted to with success in many instances.

It is happily no longer necessary to prove the moral rectitude of the practice. Its justification rests upon the same basis as that from which the whole art of Medicine derives its authority. Its design and its general effect are to save life, in many cases the lives of both mother and child; and in the rest, where the child cannot be rescued, to increase at least the chances of safety to the mother. The moral aspect of the question is now reversed; the accuser and defendant have changed places; it rests with those who neglect that which will rescue a woman and her offspring from impending danger, who suffer one or both to drift to destruction, to justify their neglect.

We may therefore proceed at once to discuss what are the advantages to be derived from the practice.

A preliminary question arises, the solution of which is necessary to the right appreciation of what can be gained by inducing labour prematurely, and of the means of accomplishing this purpose. This question, one which has been almost wholly neglected hitherto, is—What is the condition of the uterus and the system generally in reference to its fitness to assume the work of parturition prematurely?

First, as to the fitness of the general system to enter upon labour and the puerperal state. Upon this point little need be said, since we must be content to accept the conditions as they exist at the time of our selection for the operation. We cannot materially modify them. Experience, moreover, has amply shown that the system is fairly competent to assume the duties cast upon it at any time after the end of seven months. To apply the physiological formula of the Genesial Cycle so beautifully described by Tyler Smith, we observe that when gestation is brought to a term, the breasts enter upon their office, milk is secreted for the nourishment of the infant, and the uterus, thrown out of work, undergoes involution. These processes are usually carried on with scarcely less efficiency when labour occurs prematurely.

Almost all the consequences of labour at term may follow premature labour and even abortion—inflammation and abscess of the breast; peritonitis, including pelvic peritonitis and pelvic cellulitis; thrombosis, including phlegmasia dolens; and all the forms of puerperal fever. But experience does not seem to indicate that these complications are in any sensible degree more likely to attend premature labour.

Secondly, as to the fitness of the parturient organs. Here the case is widely different; for, any lack of efficiency in these must be made up by art. *When labour comes prematurely, the uterus is overtaken in an imperfect state of development.* It is taken by surprise. This implies imperfection in the contractile power of the body of the uterus, and greater resistance in the cervix. It is true that the child, the body to be expelled, is smaller, and that in this way the balance between power and resistance is to some extent restored. But this is certainly not always so. It frequently, nay commonly, happens that the uterus is slow to respond to the unexpected call made upon it. It is but reasonable to anticipate that help will often be useful. And help can be given both to facilitate the dilatation of the cervix, and to supplement the contractile energy, if this cannot be aroused.

Now, the expediency of giving this help, and the means of doing it, have been almost entirely overlooked. Action has been limited to attempts at provoking the uterus to expel its contents, leaving the rest very much to chance. The consequence too frequently has been that the child has been born at some unforseen, inopportune time, before aid could be procured, and has perished from one of those accidents, such as preternatural presentation or descent of the cord, which are so likely to occur in premature labour. Thus, supposing it was determined to bring on labour at the eighth month by detaching the membranes, by puncturing them, or by inserting a bougie in the uterus. This done, it has been considered that there was nothing to do but to wait patiently until active labour should set in, when the medical attendant should be sent for. Now this may come to pass in twelve hours, in twenty-four hours, or in two, three, four days, or even later. There is no certainty about it. When labour comes the child is expelled with little warning, almost suddenly, and before the medical attendant can be fetched. And it has to run the gauntlet of all those perils which especially surround premature labour unaided.

Does it not follow that it is desirable to keep a control over the whole course of labour—to take care that nothing adverse to mother or child shall happen in our absence—to substitute, in short, skill and foresight for accident? Few, perhaps, will hesitate to answer this question in the affirmative. But another question must follow. Can we so regulate a provoked labour throughout as to limit and define the time expended, and to conduct the delivery so as to give more security to the child and to the mother? This also I am prepared to answer in the affirmative. Repeated experience justifies the declaration made by me in 1862, "That it is just as feasible to make an appointment at any distance from home to carry out at one sitting the induction of labour, as it is to cut for the stone." The operation may be brought entirely within the control of the operator. Instead of being the slave of circumstances, waiting anxiously for the response of Nature to his provocations, he should be master of the position.

Assuming, then, that it is both desirable and possible to control and regulate the entire course of a labour prematurely induced, let me describe the method after which the proceeding should be conducted.

The act of artificial labour may be divided into two stages.

The first stage is provocative and preparatory. This includes some amount of dilatation of the cervix uteri, and implies a certain amount of uterine action, and lubrication of the cervix and vagina.

The second stage is the accelerative or concluding stage. It consists in the expulsion or extraction of the fœtus and placenta.

The ordinary modes of conducting an induced labour almost ignore the last stage, or the means of accelerating delivery.

We have divided the agents at our command for effecting delivery at will into the *provocative* and the *accelerative.* Let us first examine *the means we possess of provoking labour.* These are numerous. In a course of lectures designed to be practical rather than historical, it is not desirable to discuss them in detail. I have endeavoured to do this in a memoir "On the Indications and Operations for the Induction of Premature Labour, and for the Acceleration of Labour."* It may be stated, as a general fact, that all the means employed act by stimulating the spinal centre to exert itself in causing contraction of the uterus. Some of these agents act directly upon the spinal marrow, being carried thither in the blood. Such are ergot of rye, borax, cinnamon, and other drugs. Some evoke the energies of the diastaltic system by stimulating various peripheral nerves. Such are rectal injections, the vaginal douche, the colpeurynter, the carbonic-acid douche, probably the irritation of the breasts by sinapism and the air-pump, the cervical plug, the separation of the membranes, the placing a flexible bougie in the uterus, the intra-uterine injection, the evacuation of the liquor amnii, and galvanism.

The artificial dilatation of the cervix, the evacuation of the liquor amnii, and the intra-uterine injection act in a more complicated manner, and not simply through the diastaltic system. Some of the above agents are altogether uncertain and untrustworthy; some are in the highest degree dangerous; and some are both efficient and safe. Ergot, borax, cinnamon, and all other drugs may be dismissed on account of their uselessness or uncertainty. Ergot is not only uncertain, but when it acts it is liable to prove fatal to the child. Rectal injections may be harmless, but cannot be relied upon. Irritation of the breasts often fails, and it is liable to be followed by inflammation and abscess. The vaginal douche (Kiwisch's plan), which consists in playing a stream of water against the cervix uteri, is often tedious, and is not free from

* "Obstetrical Transactions," 1862.

danger. It requires to be repeated at intervals during one, two, or more days. It is liable to cause congestion of the lower segment of the uterus. Serious shock, metritis, and death have followed.

The *intra-uterine douche*, sometimes described as Kiwisch's plan, was in reality introduced by Schweighäuser in 1825, and by Cohen in 1846.* It is known in Germany as Cohen's method. It was recommended by Schweighäuser as a better means of detaching the membranes than the use of the finger or sound adopted by Hamilton. Cohen thought the injected fluid acted, not by detaching the membranes, but through its being absorbed by the surface of the uterus. Professor Simpson† says that he at first used the vaginal douche of Kiwisch, but "he soon found it a simpler and more direct plan to introduce the end of the syringe through the uterine orifice;" he became convinced that the douche was liable to fail, unless the injected fluid accumulated in, and distended the vagina, so as to expand that canal and enter the os uteri; and that its efficiency was great in proportion to the extent to which it separated the membranes.

The *intra-uterine douche*, although more certain, is even more dangerous than the vaginal douche. Lazzati relates two fatal cases. Taurin saw, in January, 1860, in Dubois' clinique, such grave symptoms follow, that death was apprehended. Salmon, of Chartres, related to the Académie de Médecine (July, 1862) a fatal case. Depaul communicated to the Parisian Surgical Society (1860) a case of death occurring suddenly from the uterine douche. Blot

* "Neue Zeitschrift für Geburtskunde," Band xxi.

† "Obstetric Memoirs and Contributions," vol. i., 1855.

had to deplore a similar accident in the Clinique d'Accouchements. Tarnier relates two similar cases. Esterlé relates a case* in which serious obstruction to the cardiac circulation, ending in death, occurred. Two eminent physicians have informed me of a fatal case which occurred within their knowledge in the neighbourhood of London. It may be asked, How is it that the injection of a stream of water into the vagina or uterus can prove fatal? The cases cited, and they are by no means all that are known, leave no doubt as to the fact. It seems to me that danger results in three ways. The first is by *shock*. If water is injected into the gravid uterus, it can only find room by stretching the tissues of the uterus. This sudden tension is the cause of shock. It has been supposed that some of the fluid finds its way through the Fallopian tubes into the peritoneum, causing shock. And the following case, related by Ulrich,† suggests another solution:—

"H. W., aged twenty-nine, was, at the end of her second pregnancy, carrying twins. Three vaginal douches were used to accelerate labour, the last one by a midwife. The temperature of the water was 30° R. The 'clysopompe' was used. Eight hours after the injections had been going on, the patient got up in bed, and instantly fell down senseless, and died in a minute at most, with convulsive respiration-movements and distortion of the face. Five minutes afterwards, crepitation was felt on touching the body. Venesection was tried in the median vein. Only a few drops of blood came. On section, the cerebral

* "Annali Universali di Medicina," March, 1858.

† "Monatsschrift für Geburtskunde," 1858.

sinuses were found full of dark fluid blood; the membranes not very hyperæmic; the brain normal. The heart was lying quite transversely, the left ventricle strongly contracted, the right ventricle quite flaccid; the coronary vessels contained a quantity of air-bubbles. The left heart contained scarcely any blood; the right had a little; it was quite frothy."

It is probable, then, that air may get into the uterine sinuses.

Dr. Simpson relates the following*:—"He had been greatly alarmed by seeing a patient faint under an injection, probably from some of the fluid getting into the circulation. And he had seen two more alarming cases still, where both the patients died. In both, only a few ounces of water were injected; and yet rupture of the uterus took place. The occurrence of the rupture was to be explained by the fact that the uterus, being already fully distended, could not admit the few ounces of fluid without being stretched and fissured to some extent; and during labour these slight fissures might easily be converted into fatal ruptures. In one case the patient died before labour was completed; in the other, in twelve hours after its termination."

Another objection urged by Dr. Simpson is, that in injecting water we have no control over the direction it will take in the uterine cavity, and that the placenta may be detached. Cohen's cases show that this accident may happen.

It is also apt to displace the head, and cause transverse presentation.

Of course no degree of efficiency could justify the use

* "Edinburgh Medical Journal," 1862.

of a method fraught with such terrible danger. But the douche does not possess even the merit of certainty. It has been repeated many times during several days before labour ensued. Lazzati, having tried it in thirty-six cases, found that the number of injections required ranged from 1 to 12; the quantity used was about forty pints; the duration of the injections was from ten to fifteen minutes at a time; the temperature of the water 28° to 30° R. The time expended from the first injection to labour varied from one to fourteen days, the average being four days.

It has also been found that a large proportion of the children were lost.

The douche, therefore, whether vaginal or intra-uterine, ought to be absolutely condemned as a means of inducing labour. I think it necessary to repeat this emphatically, because, notwithstanding the warnings conveyed by many fatal catastrophes, I find that the use of the method is still taught and practised.

Mr. James, formerly surgeon to the City of London Lying-in Hospital, described * a plan of intra-uterine injection which he had practised since 1848. He passes an elastic male catheter to the extent of four or five inches through the os, between the uterine wall and the membranes, and then injects about eight ounces of cold water. Of eight children, only two were still-born.

Recently,† Professor Lazarewitch, of Charkoff, has explained, modified, and given more precision to this method. He proves, by observations and experiments, that the nearer to the fundus of the uterus the

* "Lancet," 1861.

† See "Obstetrical Transactions," 1868.

irritation acts the more sure and speedy is the result, and *vice versâ*. He contends that the frequent failure of the douche was due to the stream not being carried much beyond the os. He found that when the stream was carried up to the fundus, one injection was commonly enough. He therefore introduces a tube as near to the fundus as possible, and then injects several ounces of water. The cases he relates (twelve in number) sufficiently establish his proposition that this method is more sure than other modes of applying the douche; but they are too few to prove that it is more safe. I feel very sure that, if it be at all frequently adopted, fatal catastrophes will ensue.

It may, moreover, be doubted whether, in cases managed according to the principle of James and Lazarewitch, the injection of water was not really superfluous. The passage of a catheter five or six inches into the uterus detaches the membranes along its course, and this, it has been seen, is usually quite enough to provoke labour. Why not, then, rest satisfied with that portion of the proceeding which is efficient and safe, and discard that which is superfluous and dangerous?

It is instructive to compare the histories of some cases of intra-uterine injection with those of accidental or intra-uterine hæmorrhage depending upon detachment of the placenta. Sudden severe pain in the abdomen at the seat of effusion, shivering, vomiting, collapse, are all observed in both cases. In the case of hæmorrhage, these are certainly not in proportion to, or due alone to, the loss of blood. They seem to be the direct effect of injury to the uterus from sudden distension of fibre. The uterus will *grow* to keep pace with developmental stimulus of a body contained in it; but *it will not stretch* to accommodate several cubic inches of fluid suddenly thrust into it. Yet this is what it is called upon to do when water is injected. If the water escapes as fast as it enters, the shock may be avoided, but the operation is also liable to fail in inducing labour.

The injection of carbonic acid gas or even common air is more dangerous still than the injection of fluids. Scanzoni has related two fatal cases from the injection of carbonic acid, and Professor Simpson relates one where the patient died in a few minutes after the injection of common air.

Another agent is *galvanism.* Herder suggested this as a direct stimulant, to cause the uterus to expel its contents, in 1803. In 1844, Hörninger and Jacoby brought on labour by this agent. Dr. Radford showed the value of galvanism in labour and in controlling hæmorrhage. In 1853, I published* a memoir on this subject. I succeeded, in three cases, in inducing labour by it. But the method is tedious, and sometimes distressing to the patient. I have, therefore, abandoned it.

Another exciting or provoking agent consists in the insertion of some form of *plug or expanding body in the os or cervix uteri.* A great variety of contrivances for this purpose have been proposed and tried. It is unnecessary to describe the greater part of them. Those most in use are the sponge-tent, the laminaria-tent, and the elastic air or water dilator. There is no doubt labour can be induced by these agents. But it appears to me that their use to provoke labour is not

* "Lancet" and "L'Union Médicale."

based on a rational view of the physiological or clinical history of the process. I agree with Lazarewitch, that irritants applied to the cervix are slow and uncertain. And I believe that in most cases some further means, such as rupturing the membranes, will be necessary. The laminaria-tent is, however, extremely useful in expediting the dilatation and evacuation of the uterus in some cases of abortion.

The method known as Professor Hamilton's, which consists in *detaching*, by means of the finger or sound, *the membranes* of the ovum from the lower segment of the uterus, has the recommendation of safety; but it is uncertain in its operation.

The success that commonly attends the plan of *introducing a bougie into the uterus* between the ovum and the uterine wall is perhaps evidence of the truth of Lazarewitch's proposition, that irritation should be applied to the fundus. I find that the bougie should be passed at least six or seven inches through the os uteri in order to insure action. Probably, in many cases where it has failed, the bougie has only penetrated a short way. By passing the bougie gently, letting it worm its own way, as it were, it will naturally run between the membranes and the uterus where there is least resistance, turning round the edge of the placenta. Dr. Simpson says we may always avoid the placenta by ascertaining its position by the stethoscope.

Some use an elastic catheter supported in its stilet, and withdraw the stilet when the catheter is passed. The stilet converts the catheter into a rigid instrument, which is objectionable. An elastic bougie answers perfectly.

If a rigid instrument be used, there is great likelihood of rupturing the membranes; and, although this may happen at some distance from the os uteri, premature escape of the liquor amnii may follow. The bougie owes part of its efficacy, no doubt, to the necessary detachment of the membranes from the uterus, but not all, since it is found that labour more surely supervenes if the bougie be left *in situ* for several hours.

I believe this method is now the one most generally adopted. No other method combines safety and certainty in an equal degree.

Puncturing the membranes as a provocative of labour is practised in two ways. The direct puncture at the point opposite the os uteri is probably the oldest method of inducing labour. It is one of the surest. The immediate effect of draining off the liquor amnii is to cause concentric collapse of the uterine walls, diminishing its cavity in adaptation to the diminished bulk of its contents. This involves some disturbance, probably in the utero-placental circulation. The parts of the fœtus come into contact with the uterine wall. Hence uterine contraction is promoted both by diastaltic excitation and by the impulse given by the concentric collapse.

In certain cases, the puncture of the membranes is the most convenient, as where the object is to lessen the bulk of the uterus, and ensure labour quickly. But it is open to the following objection:—It is an inversion of the natural order of parturient events. Some uterine action, lubrication, and expansion of the cervix ought to precede the evacuation of liquor amnii. If this order be not observed, the child is apt to be driven down upon the unyielding cervix, and the uterus, still contracting concentrically, compresses the child and kills it. And this is all the more likely to happen in premature labour from the greater liability to shoulder-presentation and descent of the funis.

This objection is to some extent obviated by a modification of this method. Hopkins * recommended to pass the sound some distance between the ovum and the uterine walls, and then to tap the amniotic sac at a point remote from the os. By this mode it was sought to provide for the *gradual* escape of the liquor amnii. This operation may be regarded as a compromise between the direct evacuation of the liquor amnii and Hamilton's method of detaching the membranes. It is an important improvement, and is still successfully adopted in this country and in Germany.

Vaginal Dilatation.—In 1842,† Dr. Hüter described a method for exciting labour by placing a calf's bladder, smeared with oil of hyoscyamus, in the vagina, and distending it with warm water. This proceeding he repeated every day until labour set in, which usually happened in from three to seven days. Professor Braun ‡ substituted a caoutchouc bladder, to which, from the purpose to which it was devoted, he gave the name of *colpeurynter*. Von Siebold, Von Ritgen, Germann, Birnbaum, and others, adopted this modification. Another form of vaginal dilator is the air-pessary of Gariel. The earlier trials with

* "Accoucheur's Vade Mecum." Fourth edition. London, 1826.

† "Neue Zeitschrift für Geburtskunde," 1843.

‡ "Zeitscrift für Wiener Aerzte," 1851.

this instrument seem to have been especially unfortunate, since six mothers died out of fourteen; and Breit saw inflammation of the genitals and death caused by it. I do not think these dangers are inherent in the method, if carefully pursued; but the principle of vaginal dilatation and excitation is certainly untrustworthy.

Direct Cervical Dilatation.—For the last fifty years various contrivances for mechanically dilating the cervix have been tried. The idea of dilating the cervix by sponge-tents was announced by Brünninghausen in 1820. This was again advocated in 1841 by Scholler. It has since been in constant employment at home and abroad. From personal observation, I am in a position to affirm that this method is very uncertain as to time. Symptoms like those of pyæmia have ensued from the absorption of the foul discharges caused. This accident may possibly be obviated by the use of tents charged with antiseptic agents.

Osiander, Von Busch, Krause, Jobert, Dr. Graham Weir, Rigby, invented other forms of dilatatorium more or less resembling the urethral dilators which have lately come into use. These numerous contrivances attest the strength and prevalence of the opinion that it was desirable to possess a power of dilating the os and cervix uteri at will. The subject attracted the attention of Dr. Keiller, in Edinburgh, early in 1859, and in March of that year he, Dr. Graham Weir assisting, accelerated a labour which had been provoked by other means, by introducing within the os uteri the simple caoutchouc bag, and gently distending it.

The case of Mr. Jardine Murray * is the first published case I am acquainted with in which fluid pressure was used to dilate the uterus to accelerate labour. It was a case of placenta prævia. Mr. Murray first detached the placenta from the cervical zone after my method, then introduced a flattened air-pessary between the wall of the uterus and the presenting surface of the placenta, and inflated by means of a syringe.

Dr. Storer published a case in 1859,† in which he introduced "the uterine dilator" within the cavity of the uterus. He especially insisted that the dilatation "was *from above downwards.*" I saw inconveniences in the use of elastic bags expanding inside the uterus, even more serious than those attending the vaginal dilator or colpeurynter of Braun. The cervix it was that required dilating, and a bag expanding below it in the vagina, or above it in the uterus, could only act upon it indirectly, imperfectly, and uncertainly. Besides, the uterine dilator seemed unsafe; during dilatation it must distend, stretch the uterine walls at the risk of injury and shock, and it was very likely to displace the head from the os uteri.

I had long felt the desirability of bringing the further progress of labour with placenta prævia, after having arrested the hæmorrhage by detaching the placenta from the cervical zone, under more complete control. I had always strongly insisted upon the danger of forcibly dilating the cervix with the hand, and when I read Mr. Murray's case I was engaged in devising an elastic dilator capable of

* "Medical Times and Gazette," 1859.

† "American Journal of Medical Science," July, 1859.

expanding the cervix with safety. The first form I devised was an elastic bag, with a long tube mounted on a permanent flexible metal tube, having apertures at the end inside the bag. The metal tube served as a stem to introduce the bag inside the cervix, to keep it there, and to carry the water for distension. This form was modified and adopted by Tarnier, of Paris, and others, when I had abandoned it for the fiddle-shaped bags, which are now in general use. The constriction in the middle is seized by the cervix, whilst the two ends expanding serve to prevent the instrument from slipping up or down. This instrument imitates very closely the natural action of the bag of membranes. By its aid it is very possible, in many cases, to expand the cervix sufficiently to admit of delivery within an hour, although generally it is desirable to expend more time. I have completed delivery in five hours, in four hours, and even in one hour from the commencement of any proceedings. In many cases of placenta prævia where there was scarcely any cervical dilatation, I have effected full dilatation in half an hour.

In the paper in the *Edinburgh Medical Journal* (1862) I proposed that the first step in the induction of labour should be the full dilatation of the cervix uteri, and after that to proceed to further provocation and acceleration. I related cases in which I began with dilating the cervix, afterwards rupturing the membranes, further dilating, and turning. I am now convinced that, although this rapid method is very feasible, and is even proper under some circumstances where prompt delivery is urgently indicated, it is desirable, under ordinary conditions, to prepare the uterus by some preliminary excitation.

The proceeding recommended.—Having discussed the various methods of provoking labour which have been practised, we are now in a position to select the most safe, convenient, and efficient. It has been already said that no method is so much in harmony with the principle of acting *cito, tuto, et jucunde*, as the introduction of the elastic bougie into the uterus. The plan I have successfully practised for some years is the following:—First, overnight pass an elastic bougie six or seven inches into the uterus, and coil up the remainder of the instrument in the vagina; this will keep it *in situ*. Next morning some uterine action will have set in. In the afternoon, at an appointed time, proceed to *accelerative* measures.

Before rupturing the membranes, adapt a binder to the abdomen, and let this be tightened, so as to keep the head in close apposition to the cervix. This will often prevent the cord from being washed down by the rush of liquor amnii. Dilate the cervix by the medium or large bag, until the cervix will admit three or four fingers. Then rupture the membranes, and, before all the liquor amnii has escaped, introduce the dilator again, and expand until the uterus is open for the passage of the child. If the presentation is natural, if there is room, and if there are pains, leave the rest to Nature, watching the progress of the labour. If these conditions are not present, and one or other is very likely to be wanting, proceed with accelerative methods,—that is, to the forceps or turning, or, in cases where the passage of a live child is hopeless, to craniotomy. By pursuing this method we may predicate with great accuracy the term of the labour. Twenty-four hours in all—counting from the insertion of the bougie—should see the completion of the labour. The personal attendance of the physician during two hours is generally enough. The mode of proceeding must vary according to the conditions of the case.*

What are the conditions that call for the induction of labour?

Gestation may be divided arbitrarily into two parts. During the first part, terminating at 6½ or 7 months, or at the end of 180 or 200 days, it is scarcely probable that a viable fœtus will be expelled. To induce labour within this period is really to bring about abortion. It is, therefore, only done under the pressure of conditions that preclude waiting till the child is viable, and out of regard solely to the safety of the mother. Between 200 and 230 days is a stage of very doubtful viability, and the physician will still endeavour to postpone interference until after the latter date, or the second part of gestation, when, the child being viable, the operation may be undertaken in the interest of both mother and offspring.

In a large proportion of cases we are able to select our time within certain limits. For example, where there is moderate pelvic contraction, admitting of the safe passage of a child a little below the full size, we may be justified in waiting until the end of eight months—say, 250 days. The difficulty is to determine the starting-point of the pregnancy. There is a very probable range of error of at least 15 days.

* For a series of cases illustrative of this practice, see "St George's Hospital Reports," 1868.

If we count 15 days too many, we reduce the duration of pregnancy to 235 days—that is, we run the risk of falling within the first part, when the child is of doubtful viability. If, on the other hand, we count 15 days too few, we run the contrary risk of approaching too near the natural term of gestation, and of having a child too large to pass the narrow pelvis alive.

The best way, perhaps, of avoiding these two rocks is to reckon the pregnancy from the day after the cessation of the last menstrual period, the most probable time of conception. Count 230 days from that epoch, and add 20 days for a margin of safety. This will leave a full month, or 30 days, to complete the development of the child. The cases are few, if all the resources in the acceleration of labour are turned to account, in which a child of 250 days may not be delivered alive. But if we fall upon a child of 215 days or less, the chances of its surviving are very slight. I regard the error of procrastination as being generally of less moment than the error of anticipation. Of course, if the pelvic contraction is great—say to 2·50″—it will be prudent not to calculate beyond 240 days, but rather to incur the risk of bringing a non-viable child.

It will be convenient to enumerate first those conditions which, in the interest of the mother, and disregarding the child, demand the interruption of gestation during the first part.

These are, A. Certain cases of extreme contraction in the bony or soft parts—*e.g.*, distortion and narrowing of the pelvis below 2·00″; the encroachment of considerable tumours, especially if they are unyielding, upon the pelvic canal; some cases of advancing and extensive cystic disease of the ovary; great contraction from cicatrices of the os uteri and vagina, not admitting of free dilatation; retroversion or retroflexion of the uterus not admitting of reduction; some cases of carcinoma of the uterus or vagina; some of tumours of the uterus.

B. Certain cases of urgent disease of the mother, depending upon and complicating pregnancy—*e. g.*, obstinate vomiting, with progressive emaciation, and a pulse persistent for some days above 120; some cases of advancing jaundice, with diarrhœa; some cases of albuminuria, convulsions being present or apprehended; some cases of chorea; hæmorrhages producing marked anæmia, especially if depending upon commencing abortion or placenta prævia; some cases of disease of the heart and lungs, attended with extreme dyspnœa; such are aneurism, great hypertrophy, valvular disease, œdema of the lungs, pleurisy.

If, in the presence of any of the foregoing complications, we have been fortunate enough to carry the patient over the first part of pregnancy, reaching the period when the child is viable, we may still be compelled to induce labour. The indications from disease beginning in the first part, as hæmorrhage, convulsions, cardiac distress, vomiting, jaundice, may grow more urgent, or they may arise during the second part.

My experience leads me to conclude that in cases of urgent disease there is more frequent occasion to regret having delayed the operation too long, than having had recourse to it too soon. When through obstinate vomiting, for example, nutrition has long been arrested, the starved tissues craving for supplies, and falling into disintegration, feed the blood with degraded and noxious materials; the system feeds upon itself, and poisons itself; the poisoned blood irritates the nervous centres, and these centres, wrought to a state of extreme morbid irritability, respond to the slightest peripheral, uterine, or emotional excitation. All nervous energy is thus diverted from its natural destination, and exhausted in destructive morbid action. Irritative fever ensues, the pulse rises to 140 or more, no organ in the body is capable of discharging its functions, for the pabulum of life is cut off at the very source. At this point labour, whether it occur spontaneously, as it often does, or be induced artificially, comes too late. The tissues are altered, the powers are impaired beyond recovery, and death soon follows delivery.

The most generally recognised indication is the presence of such a degree of pelvic contraction as to forbid the birth of a live child at term.

No one, I believe, disputes that, where we have the choice, induction of labour should be performed where the ultimate alternative is the Cæsarian section; and this rule should hold whether the proceeding hold out a hope of saving the child or not. It should also be resorted to for the sake of avoiding craniotomy.

In the great majority of cases, we are led to determine upon the expediency of inducing labour by the history of antecedent labours. Where craniotomy has been performed on account of contracted pelvis clearly recognised, there can be little ground for doubt. But why should one or more children be sacrificed in order to teach the physician that the pelvis is too small? Is there no other gauge of the capacity of the pelvis than a child's head? Of course it will be admitted that a woman pregnant for the first time is equally entitled to the benefit of the premature induction of labour, if it be known that her pelvis is too small. The difficulty is to know this. In this country, and generally in private practice, the opportunity of making an obstetric estimate of the pelvis before labour is very rarely afforded. The first labour at term is, therefore, the common practical test of a woman's aptitude for child-bearing. But on the Continent, where a very large proportion of women are delivered in hospitals, where they are received one or two months before the end of pregnancy, examination of the pelvis is made on admission, and thus they and their children come within the benefit of this proceeding.

The modifications proper to be adopted in different cases are as follows:—

1. In the case of pelvic deformity not admitting the birth of a live child at term.

There are three degrees of contraction to be considered. The *first* or least degree, say, giving a conjugate diameter of 3·50 in. In such a case a child of seven or eight months' development will probably pass without difficulty. Here it may be enough to provoke the labour, and watch its course, as in ordinary labour.

The *second* degree, giving, say, a conjugate of 3·00″. In such a case, unless the child prove very small or timely aid be given, its head may be delayed so long in the brim that it will be lost. Here it will be proper to provoke the labour by inserting the elastic bougie overnight; to accelerate the labour by dilating the cervix, rupturing the membranes, applying the forceps, or turning.

The *third* degree, giving, say, a conjugate below 3·00″. Here, also, it may be necessary to accelerate the labour by turning, or possibly by craniotomy.

A double advantage is gained by bringing on labour prematurely when the pelvis is greatly contracted. We not only secure a fœtal head that is smaller, but also one that is more compressible. During the last month of gestation, ossification of the cranial bones proceeds rapidly. Taking two heads, the one at eight months and the other at the full term of gestation, of equal size, the head of eight months' gestation will, on account of its less perfect bony development, come through the same contracted pelvis with more ease, or may even come through alive when the equal-sized head at term would have to be perforated. This is especially seen in those cases where turning is resorted to as an accelerative proceeding.

The course to pursue is as follows:—If the uterus act with sufficient power, and the pelvic contraction be not so great as to impede the passage of the child's head, and the cord do not fall through, watch and let Nature do her work. But if the head be delayed, or the cord fall through, we must intervene. There are two alternatives. We may first try the forceps. But if the conjugate is reduced to 3·00″ or below, turning is the true accelerative means. If I may trust my own experience, I should, without hesitation, say the prospect of a child being born alive under the conditions postulated is much better than under any other mode of delivery, and even better than is the prospect under turning in ordinary circumstances at the full period of gestation. The explanation is this:—The smaller and more plastic head is caught at the smaller or bi-temporal diameter between the projecting promontory and the symphysis pubis; the jutting promontory leaves abundant room on either side in the sacro-iliac region of the brim for the cord to lie protected from pressure; and, if care be taken that the cervix uteri be adequately expanded, the head comes through so quickly that the danger of asphyxia is not great. The mode of turning deserves attention. The object being to secure a quick delivery, the soft passages must be well prepared. We might turn by the bi-polar method, without passing more than two fingers through the os uteri. But I have found that, although it is always well to avail ourselves more or less of the bi-polar principle, it is desirable, in this case, to pass the greater part of the hand through the cervix to grasp the further knee. The reason is this:—The cervix that will admit the hand will, in all probability, permit the ready transit of the child. We thus secure adequate dilatation.

When the turning is completed, extraction must follow. It should be performed gently, drawing upon the one leg until the breech has passed the outlet; the extraction of the trunk should be slow; and a loop of cord should be drawn down to take off tension. When the arms are liberated, the neck of the child is in danger of being nipped in the circle of the cervix. This is the moment for acceleration. The two legs are held at the ankles by the left hand, whilst the right-hand fingers are crutched over the back of the neck. The head is sure to enter the contracted brim in the transverse diameter. It has then to describe the circle round the point of the jutting promontory, which I have described (*see* Lecture XVII.) as the "curve of the false promontory." Traction must, therefore, be at first carefully exerted in the direction of this curve or orbit—that is, well backwards—so as to bring the head round and *under* the promontory. When it has cleared the strait, and is in the pelvis, the occiput commonly comes forward, and traction is changed to the direction of Carus' curve to bring the head through the outlet. Unless rigorous attention be paid to the above rule for bringing the head through the brim, so much time may be lost as to imperil the success of the operation.

In cases of extreme deformity, in which it is difficult or impossible to perforate or to seize a leg, if we have induced labour at six months, the fœtus may still pass, if we give time. After making a reasonable attempt to snare a foot, by extreme manipulation and the wire-écraseur, if we leave the uterus to act for twelve or twenty-four hours, the child having perished and become moulded, some part of it, a foot or shoulder, will come within reach. This can be drawn down; the head can be perforated, and then traction will deliver. The placenta should also, if not following readily, be left for two or three hours to be expelled. By adopting this method, I have lately delivered, in St. Thomas's Hospital, a six months' fœtus, in a case of great osteomalacic deformity, in which it was impossible to get two fingers through the brim. By thus calling Nature to our aid, and practising a little "masterly inaction," we avoided the Cæsarian section, and saved the woman. This course applies especially to cases of osteomalacia, in which some amount of yielding or unfolding of the pelvic bones may be generally obtained.

In dealing with cases in which the induction of labour is indicated by urgent distress in the mother, we must again be governed by a careful estimate of the circumstances. There is no common rule.

Take first the case of *convulsions.* It has been seen over and over again that the convulsions have ceased soon after the uterus has been emptied. Everything conspires to prove that the convulsions are due to conditions arising out of the pregnancy. What, then, more logical than to terminate the pregnancy as soon as possible? Yet experience suggests caution as to the mode of acting. In not a few cases the completion of labour has failed to put an end to the convulsions. In not a few cases death has followed labour, whether this have occurred spontaneously or have been induced. Is the unfortunate issue the consequence of procrastination in inducing labour, or of over-haste, of want of precaution in the mode of proceeding? I believe it is due sometimes to one cause and sometimes to the other.

The question of inducing labour before the actual outbreak of convulsions—that is, during the conditions that lead up to convulsions—rarely comes practically before us. We have, therefore, mainly to do with the question how best to carry out a labour the indication for inducing which is clear. Is it to be done *citissime?* Is it to be done slowly and deliberately? I believe the latter principle is the more judicious. The proceedings should involve the least possible manual or other operative interference. The detachment of the membranes or the insertion of a bougie is too slow in results. It is better to puncture the membranes. This at once lessens the bulk of the uterus, and diminishes the pressure upon the abdominal vessels. If the convulsions remit, we may leave the labour to Nature. If urgent symptoms persist, we may dilate the cervix carefully by the cervical dilators, and accelerate by forceps, by turning, or even by craniotomy, according to the special indications.

A similar rule applies in almost all cases where the induction of labour is indicated by urgent distress of the mother, such as heart disease, or chorea. In the case of dangerous vomiting in the early months, it will be useful, as a preliminary measure, to insert a laminaria-bougie as far as it will readily pass into the uterus. This will answer the double purpose of detaching the ovum and dilating the cervix.

In retroversion of the uterus irreducible, and with urgent symptoms, the puncture of the membranes is the proper course. We immediately gain relief by permitting the concentric diminution of the volume of the uterus.

Lastly, there is a series of cases in which the indication is simply or primarily to save the child. There are certain conditions which tend to destroy fœtal life before the term of gestation. If we can bring the child into the world before the anticipated period of its death *in utero*, we may hope, by bringing it under fresh influences, to save it. Denman gives the case of a woman who lost her children about the eighth month, a rigor preceding. He suggested the induction of labour. There are various diseases which are known to endanger the child as they advance. Such are hydrocephalus, syphilis of the child; fatty degeneration, hypertrophy, dropsy of the placenta. In cases where there has been no sufficient opportunity of treating the mothers before or during pregnancy, and where there is a history of labours ending in the birth of dead children, the induction of labour is indicated.

There are cases in which the wisest medical and ethical judgment is required. A woman pregnant about six months is dying of phthisis. Would it prolong her life or improve her condition if labour were induced? and should we be justified in sacrificing the child with that object?

A woman pregnant about seven months is dying of phthisis. The child is assumed to be viable; its life hangs upon the fragile thread of its mother's life, which may break before the natural term of gestation is accomplished. Are we justified in inducing labour to rescue the child, disregarding the mother? or is such a course likely to prolong her life or to accelerate her death?

The decision in such cases is both perplexing and painful.

Observation of the course of things when pregnancy is complicated with phthisis, lends material help in arriving at a solution. It was long thought, and I believe some people still think, that pregnancy is antagonistic to the advance of phthisis. If this were true, the decision would be obvious. Let the pregnancy alone. But experience, I think, is adverse to the opinion that pregnancy exerts any beneficial influence upon phthisis. I am sure I have seen in numerous instances phthisis advance with accelerated speed towards a fatal issue when complicated with pregnancy, the sufferer either dying before the term, or sinking rapidly after labour.

It is an idea founded more, I am afraid, on imagination than on facts, that Nature, in her solicitude to perpetuate species, will struggle with unwonted energy to sustain the life of the expectant parent until the embryo is matured. Faith in this hypothesis would lead us to procrastinate.

Putting aside poetry not supported by facts, there are two considerations that offer material aid in arriving at a decision. First, pregnancy is commonly less trying to a phthisical patient than labour and childbed. The puerperal state especially throws such an increase of work upon the circulation, that the system often breaks down at this period. It is therefore desirable in the interest of the mother to postpone labour as long as possible.

Secondly, the prognosis in phthisis, even in cases apparently the most desperate, is often open to grave fallacy. Who has not seen patients whose days, whose hours almost were counted, survive for months and years? In the interest of mother and child, then, it is not wise to take precipitately the irrevocable step.

I will conclude this subject by recalling attention to the rule urged by Denman and his contemporaries, namely, that the artificial induction of labour should only be undertaken after deliberate consultation. When we consider the many and weighty medical, social, and moral questions involved in the arbitrary interruption of pregnancy, we shall see abundant reasons for seeking assistance in avoiding possible clinical error, and for sharing serious professional and social responsibility. I have never performed the operation without the knowledge or sanction of some professional brother. If this rule be universally acknowledged, and religiously followed, by the medical profession, then shall we strike with authoritative condemnation those miscreants who use the trusted garb of medicine to cover deeds that shun the light.

LECTURE XXI.

UTERINE HÆMORRHAGE—VARIETIES: CLASSIFICATION OF—FROM ABORTION—CAUSES OF ABORTION: MATERNAL, OVULINE, AND FŒTAL—COURSE AND SYMPTOMS OF ABORTION: TREATMENT—THE PLACENTAL POLYPUS—THE BLOOD POLYPUS—THE PLUG—PERCHLORIDE OF IRON—PROPHYLAXIS—HYDATIGINOUS DEGENERATION OF THE OVUM.

THE management of flooding involves so much operative treatment, that a course of lectures on obstetric operations cannot be complete without a history of this subject. I therefore consider it desirable to give a condensed description of uterine hæmorrhage. Hæmorrhage is certainly one of the most frequent and most perilous of all the accidents that threaten the pregnant woman. Often occurring without warning, in impetuous torrents, quick death from shock and exhaustion may ensue. If not immediately fatal, the drained and enfeebled system, ill adapted to resist injurious influences, may sink in a few days from some form of puerperal fever, from thrombosis, or other complication. And, if these secondary dangers be escaped from, there are still to be encountered the remote effects, sapping the strength, perverting nutrition, and predisposing the patient to various protracted diseases. I have seen almost complete blindness, deafness, hemiplegia, and other forms of paralysis persist as consequences of hæmorrhage.

The action of the physician must obviously, in the first place, be fixed on the endeavour to save life from the imminent peril. But his duty is far from ending here. He has to take measures to prevent the return of the flooding; to rally the patient from the depression already present; and to secure her against the secondary and remote consequences. He will draw the most valuable indications in practice from a careful clinical study of the various conditions upon which uterine hæmorrhage depends. He may rescue the patient from instant death, but he must not, therefore, think his work complete. He must act on the aphorism that every ounce, every drop of blood that he can save, will be useful in averting subsequent evil. I insist upon this, because I have observed that many men, perhaps most, are more afraid lest the energetic use of the most efficient agent in stopping bleeding may be more hazardous than the continuance of the bleeding itself. This dread, I believe, I know, is unfounded; and I earnestly beg those who feel it, and whose hands are thereby paralyzed, to reflect that the dangers of hæmorrhage are great and certain, that the time for action is brief, that lost opportunity is irretrievable, whilst the dangers apprehended from the prompt use of the great hæmostatic agent, perchloride of iron, are at most problematical. Whilst we are hesitating, the woman dies. The emergency justifies some risk. Even those who distrust the remedy should remember the Celsian maxim: "Anceps remedium melius quàm nullum." It is eminently applicable here. We should not choose to founder against Scylla even because Charybdis lay beyond. We should try and clear Scylla at all hazards, trusting to skill and Providence to clear Charybdis too. If, therefore, means be pointed out of almost infallible efficacy in stopping hæmorrhage, thus avoiding the first rock, is it rational to withhold those means for fear of shock, air in the veins, or thrombosis—rocks ahead indeed, but not necessarily in our course?

This study will supply the rationale of treatment, pointing out the ends that have to be accomplished. The means of accomplishing those ends are, to a great extent, empirical. We have, then, physiological and empirical agencies at our command in seeking to arrest hæmorrhage. For clinical purposes, hæmorrhages may be most conveniently divided into those which occur during pregnancy, and those which occur during and after labour.

The first order of cases includes these three principal forms:—

1. The hæmorrhages of abortion.
2. The hæmorrhages from placenta prævia.
3. The hæmorrhages, so-called "accidental," from premature detachment of the placenta.

The hæmorrhages which occur during and after labour—that is, after the birth of the child—will be discussed in the next lecture.

1. *The Hæmorrhage of Abortion.*

Hæmorrhage occurring during early pregnancy is both a cause and a symptom of abortion.

Under normal conditions, the relations between the ovum and the uterus, both in structure and function, are so harmoniously balanced, that the abundant supply of blood attracted to the uterine vessels by the developmental stimulus of the embryo finds its natural employment. The demand keeps place with the supply, and the structures being healthy, there is no extravasation; but if this healthy correlation be in any way disturbed, hæmorrhage is likely to occur. Whatever produces a state of hyperæmia predisposes to hæmorrhage. In pregnancy this hyperæmia exists in a high degree. The developmental nisus acts as a *vis à fronte*, attracting blood powerfully to the vessels of the uterus. The congestion is so great that a flaw anywhere will almost certainly lead to an escape of blood. The following defect is the most frequently observed:—*A morbid condition of the mucous membrane*, combined or not, with a morbid condition of the muscular wall of the uterus, and frequently attended by *abrasion or loss of epithelium.* The part thus affected is commonly the vaginal portion of the cervix uteri. The morbid action here is aggravated by the physiological developmental stimulus; and at the menstrual epochs the additional stimulus of ovulation, attracting still more blood, the vessels, ill-protected by the diseased tissues, break down, and hæmorrhage results. Simple emotion, or physical shock, will often act in like manner, by directing a sudden excess of blood to the uterus. So long as the blood only comes from the os and cervix uteri, the embryo may be safe, but the hæmorrhage itself may be injurious, and a remedy should be applied. In such cases, lightly touching the abraded congested surface with nitrate of silver a few times at intervals of a week will commonly effect a cure. It may be feared that such local treatment will provoke abortion instead of preventing it. Experience, however, corroborates what physiological reasoning suggests, namely, that the cure of a diseased action in the organs of parturition will contribute to the security of the pregnancy. It is not desirable to let disease go on because a woman is pregnant.

An analogous form of hæmorrhage is that which arises from *the decidual cavity* during the first six weeks of pregnancy, *i.e.*, before the decidua reflexa and decidua vera have coalesced. The mucous membrane of the uterine cavity, being unhealthy, pours out blood which escapes externally. This involves far more danger to the embryo, because the extravasation is not likely to be strictly limited to the free surface of the decidua; it will probably spread to parts in connection with the chorion and young placenta. The embryo then perishes; and blood being retained in the substance of the placenta, and some insinuating itself between the decidua and the uterine wall, the uterine fibre, put suddenly on the stretch, is irritated, and spasmodic contractions, tending to expel the contents, are set up. This spasmodic action, itself a new source of local hyperæmia, adds to and keeps up the hæmorrhage.

There are, of course, many causes of abortion; but whatever the primary or predisposing cause, the immediate or efficient cause of abortion is extravasation of blood into the decidua, and between the decidua and uterine wall, leading to partial separation of the ovum. When things have gone so far that the integrity of the ovum is impaired by extravasation of blood into the structure of the decidua and placenta, abortion is commonly inevitable; the indication then is to accelerate the entire removal of the ovum.

I cannot in this place discuss minutely the numerous causes, maternal and ovuline, which lead to abortion.

In dealing with threatening or present abortion, whatsoever the cause, the course of action is very much the same. But, as a guide to prophylaxis, a knowledge of these causes is important; it is well to have a clear idea of them. They may be classified as follows:—

A. Maternal Causes of Abortion.

I. Poisons circulating in the mother's blood.	α. Introduced from without: as fevers, syphilis, various gases, lead, copper, &c. β. Products of morbid action: as jaundice, albuminuria, carbonic acid from asphyxia, and in the moribund, &c.
II. Diseases degrading the mother's blood.	Anæmia, obstinate vomiting, over-lactation.
III. Diseases disturbing the circulation dynamically: as liver-disease, heart-disease, lung-disease.	
IV. Causes acting through the nervous system.	α. Some nervous diseases. β. Mental shock. γ. Diversion or exhaustion of nerve-force: as from obstinate vomiting.
V. Local disease.	α. Uterine disease: as fibroid tumours, inflammation, hypertrophy, &c., of uterine mucous membrane. β. Mechanical anomalies: as retroversion, pressure of tumours external to uterus, &c.
VI. Climacteric abortion.	
VII. Abortion artificially induced.	

B. The Fœtal Causes of Abortion.

I. Diseases of the membranes of the ovum, primary or secondary upon diseases of the maternal structures or blood, as—
- Fatty degeneration of the chorion or placenta.
- Hydatidiform degeneration of " "
- Inflammation, congestion of " "
- Apoplexy of " "
- Fibrous deposits of " "

II. Diseases of the embryo :—
- α. Malformation.
- β. Inflammation of serous membranes.
- γ. Diseases of nervous system.
- δ. " of kidney, liver, &c.
- ε. Mechanical, as from torsion of the cord.

In short, anything causing the death of the embryo. Often the causes are complicated, arising partly from the maternal, partly from the fœtal side; and often it is difficult to unravel these, or to discover the efficient cause.

The careful study of the etiology of abortion has a use beyond that of guiding prophylaxis. In a given case of threatening abortion, it will often furnish important indications. For instance, in some cases it is desirable to endeavour to avert the abortion, to arrest the process; but where certain diseases, as fever, severe blood-change, or conditions causing dynamical disturbance of the circulation exist, abortion is a means by which Nature seeks relief from the double burthen.

This double burthen may overwhelm the powers of the system. By eliminating one burthen, the pregnancy, the case is reduced to its most simple expression, and the patient may successfully struggle through the uncomplicated disease. In such a case, the indication is to second Nature by accelerating the accomplishment of the abortion.

A similar indication arises in those cases in which there is local disease rendering the continuance of pregnancy hazardous or impossible, as retroversion, tumours, &c.

Also in certain cases of embryonic disease, where the previous history, or our knowledge of the state of the uterus or its contents, establishes a strong presumption that the embryo is either dead or is not likely to survive.

Bearing these things in mind, let us trace the *ordinary course and symptoms of abortion.* These will vary somewhat, according to the cause and condition of the ovum and uterus. The two great symptoms are pain and hæmorrhage. There is reason to apprehend abortion if, after a woman has missed one or more monthly periods, she complains, about a menstrual epoch, of unusual heavy aching lumbar pains, followed by spasmodic or colicky pains in the lower abdomen and pelvis; and still more if a discharge of blood takes place. But these symptoms may pass away under rest, and the pregnancy go on. When this occurs, we may infer that there is no disease of the fœtus or its membranes incompatible with its life; and that the symptoms are due to the recurrence of the ovarian stimulus which is the physiological cause of menstruation.

The congestion attending the menstrual nisus may be the cause of extravasations of blood into the decidua and between the villi. There, coagulating and hardening, through the absorption of the watery part of the blood, it may help to form a solid mass, which may retain connection with the uterus for some time longer, and be eventually expelled as a *fleshy mole.* No trace of an embryo may be discovered; but a careful microscopical examination will reveal the elementary constituents of the ovum, especially chorion-villi.

If things are going on to abortion, the hæmorrhage will continue, perhaps in great abundance; pains of a forcing or expulsive character persist, until amongst the discharges, the embryo, enclosed or not in the amnion and chorion and decidua, is found. In early abortions, occurring at from six weeks to two months, the embryo will often pass enveloped in the chorion. The decidua, or mucous membrane of the uterus, may persist for some time longer; and so long as it remains, the hæmorrhage and pain, and the danger of the patient last.

In abortions occurring at three or four months, the embryo is sometimes expelled alone in the first instance, the ovum being burst under the contractions of the uterus. Then the membranes, amnion, chorion and decidua, and placenta follow. But it may happen that the ovum is not burst, and then the whole ovum will come away in one mass. This is more likely to be the case when the embryo has perished some time before its expulsion, and when the process of retrogression of the media between uterus and placenta has made some progress. Some diseased ova are the most liable to be thrown off in this entire form. But if the abortion be the result of hydatiginous degeneration of the chorion, it is extremely likely that only a part of the diseased structure will come at a time. Then hæmorrhage will still continue. You must carefully examine what has passed, to see if there is any appearance of parts being torn off, giving reason to infer that more remains behind. You must also, in every case, examine *per vaginam*, to ascertain if there is anything in the uterus.

The first question of a practical nature is, *Can the abortion be averted?* Can gestation continue?

If you have found any portion of the embryo or membranes in the matters expelled, you must give up the hope of this.

If you find very active expulsive pains, attended with free hæmorrhage, affecting the patient's strength, and if the os uteri is opening so as to admit the finger freely, you may, even if no portion of the ovum has been expelled, give up the hope of averting abortion.

The sooner the ovum, or the remains of it, are voided, the sooner the patient will be out of danger. This, then, is the first indication: *Empty the uterus.* How? Sometimes you will feel the ovum projecting partly through the os. It feels like a polypus. In this case it is probably detached, and under proper manipulation it will come away. The method of proceeding is as follows:—The patient laying on her back or on her side, with the thighs flexed so as to relax the abdominal walls, you press with the palm of one hand above the symphysis upon the fundus of the uterus, so as to depress the organ well into the pelvic cavity. By this manœuvre you carry the os lower down, making it more accessible to the finger passed internally, and you support the organ by providing counter pressure. This singularly facilitates the penetration of the cavity by the finger, which must generally pass to the fundus, in order to get hold of the ovum and scoop it out.

Generally the uterus, in abortion, is low in the

pelvis; and the vagina being relaxed, it is not difficult to reach the os; but not seldom, although the os can be reached, it may be necessary, in order to command the uterus and its contents, to pass the hand into the vagina. This is a very painful proceeding, and calls for chloroform. This is more especially the case if the ovum adheres; then the finger must be swept well round the cavity of the uterus—even at the fundus. The decidua may thus be broken up, and only pieces will come at a time.

Various ovum-forceps have been devised for the purpose of seizing the ovum and bringing it away. Levret and Hohl, amongst others, invented forceps of this kind, and Stark contrived a form of spoon. I have used several; but after trial of them I prefer the hand. The hand gives you information of what you are doing as you go on; and it acts by insinuating the finger between the ovum and the uterus, peeling one from the other, not by avulsion, as the forceps must do. Besides, you know if you use your fingers exactly what has been accomplished, if any part of the ovum has been left behind, and you avoid the risk of injuring the uterus.

I have said that there is no security against return of hæmorrhage whilst the ovum or a portion remains behind. But there is an exception to this rule. I have frequently observed in cases where it was not possible to bring away the whole ovum, that the remains, if much broken up, did no harm, and that the hæmorrhage ceased. The *débris* gradually become disintegrated; and involution goes on unimpeded.

Sometimes a portion of decidua and placenta

adheres so intimately that it cannot be removed by the fingers. Projecting into the uterine cavity, it forms the *placental polypus*. If hæmorrhage continue, this may be removed by the wire-écraseur, which will shave it off smoothly from the surface of the uterus. I believe this is the best and safest way of dealing with these masses. There is another condition to be considered here. Sometimes, blood collecting in the cavity of the uterus gets compressed, so that the fluid portion being squeezed out, the fibrin forms a firm body, which may adhere to the wall of the uterus. This is the *blood-polypus*. Just like an ovum or bit of decidua, it is a cause of hæmorrhage by irritating the uterus, inducing spasm or colic. It can commonly be scooped out by the finger.

When the embryo has been expelled, the uterus will often close upon the membranes, the cervix contracts, imprisoning them, and the uterus in its imperfectly-developed state having little expulsive power, there is *retention of the membranes*. Retention is often complicatied with *adhesion*. Then I have often found fibrinous deposits in the placenta. These two conditions sometimes require time to overcome. The consulting practitioner here occasionally reaps credit which is scarcely his due. He is called in, perhaps, on the third day, or later, when the adhesion of the decidua to the uterus is breaking down. He passes in his fingers and extracts at once. But had he tried the day before he might have failed, like the medical attendant in charge.

The first difficulty to overcome is the narrowing of the cervix. Whilst using means to dilate the cervix, you may use the ordinary means to control the hæmorrhage. There are certain medicines, as ergot, cinchona, strychnine, turpentine, which are to a certain extent useful. The most frequent resource is cold, either in the form of cold water applied to the abdomen and vulva, or in the form of ice in the vagina.

The most efficient internal remedy is turpentine; but, like all others, it is apt to fail. It is better not to rely upon these agents, but to proceed at once,

2. *To Plug.*—The usual plan is to plug the vagina; but this is really unscientific and illusory. In a short time, the vagina contracting squeezes the plug, compressing it to a calibre that no longer fits the canal; blood flows freely past it or collects about it. The proper way to plug was pointed out to me some years ago by Dr. Henry Bennet. You must plug the cervix uteri itself. This may be done through a speculum, pushing small portions of lint or sponge into the cavity by means of a sound. It is well to tie a bit of string round each bit of plug, to facilitate removal. This is a true plug; it arrests hæmorrhage and favours the dilatation of the cervix.

But in the sea-tangle, or laminaria digitata, introduced by Dr. Sloan, of Ayr, we possess an agent that surpasses all others for the purpose.

The tangle may be used either perforated as a tube two or three inches long, or solid. For ordinary purposes, especially in the non-pregnant state, where the os and cervix are very small, the tubular form is most convenient. To introduce a plug of this kind, it may be mounted on a wire attached to a stem provided with a canula. Thus mounted, the plug is carried into the uterus like a sound; and when in, the wire is withdrawn, and leaves the plug *in situ.* The instrument-makers sell an instrument of this kind designed by me. But a flexible catheter, cut down at the end so as to leave about two inches of the wire-stilet bare, upon which the tangle-tube can be mounted, answers very well. If desirable, three or four tubes can be introduced together, forming a faggot. By this means more rapid and complete dilatation of the cervix can be obtained. In cases of abortion, however, I have found it more convenient to take a solid, smooth piece of laminaria about four inches long, the calibre of a No. 8 or No. 9 bougie, and to give it a slight curvature at one end. This curve very much facilitates introduction, and the length of the plug enables one to pass it as easily as a sound. It may be passed as far as it will go without obstruction, that is, generally about three inches within the os uteri, leaving an inch projecting into the vagina. Thus it is maintained *in situ.* In a few hours it will have swollen, so that, if passed overnight, next morning, on removing it, the finger will be admitted with ease. During its expansion, the hæmorrhage will have stopped; and on its removal you will generally find the ovum fit to be detached. During the stretching of the cervix by the laminaria, vomiting occasionally occurs. For this it is well to be prepared. Vomiting is commonly excited whenever uterine tissue is put at all suddenly on the stretch.

I have heard that my caoutchouc water-dilators are sometimes used for this purpose; but they are hardly adapted for the early stages of pregnancy. The laminaria is far more convenient.

When the cervix has been well dilated, and the

uterus has been emptied of its contents, there is rarely any more hæmorrhage; the patient is then generally safe. But if there should be any return, I strongly urge the application of perchloride of iron to the inner surface of the uterus. This acts as an immediate styptic, and may be relied upon to check any further bleeding. The mode I prefer in these cases, where the cervix is widely open, is to soak a piece of sponge, fixed on a whalebone stem, in a mixture of one part of the strong liquor ferri perchloridi of the British Pharmacopœia with three of water, and to pass this into the cavity of the uterus as a swab. A difficulty sometimes experienced is, that the moment the charged swab touches the cervix, the constriction caused arrests its further progress. Where this is the case, it is necessary to pass a small tube like a catheter, and to inject *very slowly* a small quantity, say half-an-ounce, of the solution through it.

I must caution you not to rely upon the topical use of perchloride of iron *before* the uterus is cleared. There is no substitute for, no evasion of, this primary duty. It must be remembered that a patient who has suffered abortion is liable to all the puerperal affections that assail women who have been delivered at term: for example, hæmorrhage, phlegmasia dolens, septicæmia, and other forms of puerperal fever, and pelvic peritonitis. To most of these affections loss of blood strongly predisposes; hence the importance of stopping the hæmorrhage as quickly as we can. The after-treatment of abortion is:—1. *Immediate*, directed to obviate or mitigate the concomitants or consequences of abortion, that is, anæmia, the risk of inflammation, and other puerperal accidents.

The first kind of treatment is *restorative*, and the great condition of restoration is *rest*. The woman who has suffered abortion should be kept for ten days in bed, and made to conform to all the rules that govern the lying-in chamber. The neglect of this precaution is the source of much immediate danger, and of many ulterior affections that sap the health, and lead to protracted misery.

Women are apt to think lightly of a miscarriage; and many medical practitioners who have not seen the more severe cases countenance this error. They can scarcely be persuaded that abortion may cause death. It is a common belief that the hæmorrhage, however profuse, will stop in time, and that the patient is sure to rally; but this is not the experience of those who are largely consulted in difficult cases. I have known not a few deaths from primary hæmorrhage and shock, not a few from septicæmia, some from inflammation; and I have seen many women who have, indeed, escaped with their lives, but only to suffer for years afterwards from anæmia and other disastrous consequences.

2. *Curative and Prophylactic*, that is, aimed at the causes which led to the abortion. This is much neglected in practice. In every case there is an efficient cause. The hope of preventing recurrence of abortions depends upon our tracing it out. Do not sit down disarmed and passive under the influence of that ignorant dogma which asserts that women have a habit of aborting, that they are labouring under an abortive diathesis. This is no more than saying that women abort because they abort, which is not very instructive.

The management of hæmorrhage from hydatidiform degeneration of the ovum requires a little special consideration. As I have already said, hæmorrhage may recur in repeated attacks, alternating with watery discharges, and only portions of the ovum may come away. The discharges, if carefully examined, will show shreds of membranes having the characteristic cystic enlargements. Sometimes these can scarcely be recognised without a magnifying glass. Whenever they are recognised, the indication is clear to proceed to empty the uterus. There is little or no prospect of a live fœtus when cystic disease has once begun. In the majority of cases, no embryo at all is formed. But it is an error to suppose, with Mikschik and Graily Hewitt, that this disease is always secondary upon the death of the embryo. Many instances of living fœtuses with a cystic placenta have been recorded, for example, by Perfect, Martin of Berlin, Villers, and Krieger. There is every reason to believe that, in many cases at least, it is the advancing cystic degeneration of the chorion that destroys the embryo. In many cases the cause of the chorion-change must be sought in a primary disease of the decidua. Virchow and others describe the decidua as hypertrophied. There is hyperplasia with intimate adhesion of the mucous membrane to the uterine wall. In this state, the chorion-villi shoot into the substance of the morbid decidua; and not only that, but may actually penetrate deeply into the wall of the uterus itself. When this happens, there is such continuity of structure that complete detachment of the ovum is impossible. Volkmann relates (Virchow's Archiv. 1868) a case of this kind. In consultation with Dr. Hassall, of Richmond, I saw a remarkable example. These extreme cases are, of course, exceptional; but they illustrate the common condition, that there is disease, hypertrophy, and more or less intimate adhesion of the decidua. The lesson in practice is this: although we ought always to endeavour to detach the whole ovum by passing the fingers, or the hand, if necessary, into the uterus, we must not persevere too pertinaciously in this attempt, lest we inflict injury upon the uterus. Take away all that will come readily, and then, if hæmorrhage return, apply the perchloride of iron. Sometimes the diseased ovum will come away in one mass, moulded to the shape of the cavity of the uterus. Still, in these cases, the decidua will show evidence of being hypertrophied. The surface will be covered with thick shreds; and decidua may be traced into the substance of the diseased mass.

In every case of abortion, I think it is important to apply a firm bandage to the abdomen afterwards. This, by depressing the fundus of the uterus and promoting contraction, obviates the danger of sucking in air, and consequent septicæmia. It is sometimes desirable, also, when the discharges are at all offensive, to wash out the vagina with Condy's fluid.

But I would submit it as a rule, that whensoever the discharges are offensive, the presumption is that there is something in the uterus that ought to be removed. To clear up this point, never fail to make a thorough exploration.

LECTURE XXII.

PLACENTA PRÆVA—HISTORICAL REFERENCES: MAURICEAU, PORTAL, LEVRET, AND RIGBY—MODERN DOCTRINES EXPRESSED BY DENMAN, INGLEBY, AND CHURCHILL—THE OLD AND THE AUTHOR'S THEORIES OF THE CAUSE OF HÆMORRHAGE IN PLACENTA PRÆVIA—THE PRACTICE OF FORCED DELIVERY, DANGERS OF—THEORY OF PLACENTAL SOURCE OF HÆMORRHAGE, AND METHOD OF TOTALLY DETACHING THE PLACENTA—THE COURSE, SYMPTOMS, AND PROGNOSIS OF PLACENTA PRÆVIA—THE AUTHOR'S THEORY OF PLACENTA PRÆVIA, AND TREATMENT—THE PLUG—PUNCTURE OF THE MEMBRANES—DETACHMENT OF PLACENTA FROM CERVICAL ZONE—DILATATION OF CERVIX BY WATER-PRESSURE — DELIVERY — SERIES OF PHYSIOLOGICAL PROPOSITIONS IN REFERENCE TO PLACENTA PRÆVIA —SERIES OF THERAPEUTICAL PROPOSITIONS—THE SO-CALLED "ACCIDENTAL HÆMORRHAGE."

PLACENTA PRÆVIA.

Greatly-extended personal experience, and the testimony of many excellent observers in this country and abroad, confirm the soundness of the theory of placenta prævia which I first enunciated in the *Lancet* in 1847, more fully unfolded in the "Lettsomian Lectures" of 1857, and illustrated in various subsequent publications. From this theory, as from preceding theories, distinct rules of practice logically follow. To obtain a clear appreciation of the principles which should govern us in the difficult task of managing a case of placenta prævia, it is therefore desirable to begin with a statement of the physiology of the subject. This I will make as brief as possible, bearing chiefly in mind the practical purpose before us.

Before the time of Levret and Rigby, there could scarcely be said to be a clearly-reasoned-out theory of placenta prævia. The treatment was purely empirical, using the word in the Celsian sense. Although it was known to obstetricians, especially to Giffard and Portal, that the placenta might be implanted upon the cervix uteri, the distinction between hæmorrhages depending upon this, and upon other conditions was not accurately defined. In all hæmorrhages of a severe character occurring during pregnancy, the indiscriminate practice inculcated by Guillemeau, Mauriceau, and their successors, of emptying the uterus, was blindly followed. This practice, rightly called the *accouchement forcé*, consists in forcing the hand, if necessary, through the cervix, in seizing and extracting the child, and in removing the secundines as quickly as possible. The guiding theory was very simple: it was a rough deduction from observations showing that the hæmorrhage generally stopped when delivery was complete, and that unless delivery were effected, the hæmorrhage went on.

Levret and Rigby, observing Nature more closely, grasped with more precision the special bearing of the implantation of the placenta over the cervix as a cause of hæmorrhage. They believed that so long as the labour continued the hæmorrhage would go on, even increasing. The logical conclusion was, that it was necessary to empty the uterus as quickly as possible. Rigby at one time drew a broad distinction between the management of hæmorrhage depending on placenta prævia, which he called "unavoidable," and hæmorrhage arising on detachment of placenta growing at its normal site, which he called "accidental." In the first, he says, "Manual extraction of the fœtus by the feet is absolutely necessary to save the life of the mother; in the second species such practice is never required." Later, however, he admitted that "assistance might be required in accidental hæmorrhages."

Almost all subsequent authorities concur in accepting the doctrines, and in adopting the practice of Levret and Rigby.

This is Denman's opinion: "It is a practice established by high and multiplied authority, and sanctioned by success, to deliver women by art, in all cases of dangerous hæmorrhage, without confiding in the resources of the constitution. This practice is no longer a matter of partial opinion, on the propriety of which we may think ourselves *at liberty* to debate; it has for near two centuries met the consent and approbation of every practitioner of judgment and reputation in this and many other countries."

The following quotation from Ingleby may be taken, as expressing the theory almost universally recognised until recently:—"And thus the placenta will undergo a continuous separation corresponding to the successive expansion of the neck, *until nearly the whole of the surface* is dissevered from its uterine connection. From this it is evident that when the placenta is affixed either to the cervix or os uteri, *whether wholly or partially*, the vessels will become exposed on each successive detachment, and the ultimate safety of the patient will depend upon delivery by turning the child, excepting, perhaps, in two peculiar states, in which rupture of membranes is the only treatment offered to us in one case, and the safest, and therefore the most eligible, in the other. *Pain*, efficacious as it is in the accidental form of hæmorrhage, unless adequate to the expulsion of the child, *is neither to be expected nor to be desired*, to any material extent, in the unavoidable form, as it only renders the effusion more abundant; for though a certain degree of relaxation is necessary, it must be remembered that in exact ratio as the cervix uteri is successively developed, and the os internum progressively dilated, will an additional mass of placenta be detached from its connecting medium, and hæmorrhage necessarily be renewed."*

Churchill† speaks no less distinctly: "The flooding is the necessary consequence of the dilatation of the os uteri, by which the connection between the placenta and uterus is separated, and the more the labour advances, the greater the disruption and the more excessive the hæmorrhage. From this very circumstance it follows that the danger is much greater than in the former cases (of accidental hæmorrhage), and also that what in them was the natural mode of relief is here an aggravation of the evil, and cannot be employed as a remedy."

What theory can be more hopeless? Pain is the

* "Uterine Hæmorrhage," 1832.

† "Theory and Practice of Midwifery," 1855.

thing wanted, but if it comes it brings danger with it! Expansion of the cervix uteri is a necessary condition of labour, but the cervix cannot expand without causing more hæmorrhage! Nature is utterly at fault. She is condemned without appeal. Art must take her place.

Those, therefore, who accepted the theory of Levret and Rigby, were confirmed in the old practice of delivering at once at all hazards. The *accouchement forcé* received the sanction of modern science. But absolute laws seldom endure long without question or mitigation in practice. Mauriceau had long ago discovered that in some cases of partial placental presentation, rupture of the membranes and evacuation of the liquor amnii was enough to arrest the hæmorrhage. Subsequently, Puzos defined with more precision the cases in which rupture of the membranes might be practised. Wigand and D'Outrepont avoided the *accouchement forcé* as much as possible, trusting to the plug and rupture of the membranes. Dr. Robert Lee* says: "The operation of turning, which is required in all cases of complete presentation, is not necessary in the greater number of cases in which the edge of the placenta, passing into the membranes, can be distinctly felt through the os uteri. If the os uteri is not much dilated or dilatable, the best practice is to rupture the membranes, to excite the uterus to contract vigorously, by the binder, ergot, and all other means, and to leave the case to Nature."

Since the time of Levret, says Caseaux, the insertion of the placenta upon the neck of the uterus has been considered as an inevitable cause of hæmorrhage

* "Lectures on the Theory and Practice of Midwifery," 1844.

during the last three months of pregnancy and during labour. The loss is then, says Gardien, of the very essence of pregnancy, and especially of labour. The current opinions apply to the separate conditions of pregnancy and of labour. The explanation of hæmorrhage during pregnancy is as follows:—Until the fifth month the body only of the uterus undergoes any considerable change; but from that date the neck participates. The shortening it undergoes is attended by enlargement of its base, at the head of the os internum. The placenta, fixed and immoveable on the spot where it is planted, cannot follow the expansion of the upper part of the neck; hence the connections between placenta and uterus are torn, and hæmorrhage necessarily results.

This doctrine, passed on traditionally, and accepted almost without question, is undoubtedly founded on an anatomical and physiological error. Stoltz, Rœderer, Weitbrecht, and Matthews Duncan,* clearly show that the cervix proper contributes in no way to the reception of the ovum. I have repeatedly felt the cervical canal entire closed above by the narrow os uteri internum at the end of pregnancy. I have also ascertained the truth of this by dissection of women dying after labour. And if the explanation referred to were true, hæmorrhage would be far more frequent; for the rapid expansion of the cervix, if physiological, should be a constant condition, and the uterus must infallibly always grow away from the placenta.

Then, as to the hæmorrhage occurring at the time of labour, which is attributed to the active expansion

* "Researches in Obstetrics," 1868.

D D 2

of the cervix, casting off placenta. Is this consistent with clinical observation? If you reflect upon what is observed, you will see that it can be only partly true. It is an indisputable fact that hæmorrhage, frequently at least, and I believe commonly, begins before there is any expansion of the cervix at all. The true explanation is, I submit, the very reverse of that generally accepted. What is the part endowed with the most active growth? Is it not the ovum, the placenta? The growth of the cervix is secondary; it is the result of the stimulus of the ovum. The first detachment of placenta, then, arises from an excess in rate of growth of the placenta over that of the cervix, a structure which was not designed for placental attachment, and which is not fitted to keep pace with the placenta. Hence loss of relation; hence placenta shoots beyond its site, and hæmorrhage results. Hæmorrhage is most common at the menstrual epochs, and has not necessarily anything to do with labour. At the menstrual nisus there is an increased determination of blood to the uterus and to the placenta. This over-filling of the placenta makes it too big to fit the area of its attachment; it breaks away at the margin of the os, and blood escapes. Under the irritation of this partial detachment, the infiltration of some blood into the substance of the placenta, which increases the bulk of the organ, and the insinuation of some blood, perhaps clotting, between the placenta and the uterine wall, active contraction of the uterus may be excited. Then the retracting cervical portion may detach more placenta. But this is secondary. And what I shall presently set forth will abundantly prove the error of the theory that "the placenta will undergo separation corresponding to the successive expansion of the neck, until nearly the whole of the surface is dissevered from its uterine connection." The severance will never go beyond the cervical zone.

The view just enunciated, that the first loss of relation between the placenta and uterus is due to the excess in rate of growth and periodical hyperæmia of the placenta, is strengthened by the analogous case of Fallopian gestation. Like tubal gestation, gestation in the lower segment of the uterus is an example of *error loci;* both are alike in being instances of ectopic gestation. In both cases the ovum grows to a structure not adapted to harbour it. This want of adaptation consists in unfitness to grow with the advancing growth of the ovum. Hence, in the case of tubal gestation, there comes a time when the growth of the ovum is so rapid that the tubal sac, not able to keep pace with it, bursts. This catastrophe, too, commonly happens near a menstrual period, when increased growth is aggravated by increased afflux of blood. I have elsewhere drawn attention to the fact, that before the rupture of a Fallopian sac, a discharge of blood by vagina often takes place. This is the evidence of the ovum outstripping its habitat, and getting partially detached. It is exactly what happens in placenta prævia.

To return to our history. Still the prevalent practice was to deliver at once; and this was, and is, often done without much regard to the fitness of the parts to undergo this severe operation. So imperious is the dogma of "unavoidable," persistent hæmorrhage, that the difficulty presented by an undilated os uteri is overcome by a special hypothesis, which assumes that in these cases of flooding, the os uteri is, by the flooding, always rendered easily dilatable. Unfortunately, this is not true. Proofs of laceration, of fatal traumatic hæmorrhage from the injured cervix, as the penalty for forcing the hand through the presumed dilatable cervix, abound; but the error is so prevalent, that I think it useful to adduce some evidence upon the subject. The late Professor D. Davis says he had met with many examples of even fatal hæmorrhage unaccompanied by any amount of dilatation of the orifice of the womb. He relates a case where very profuse hæmorrhage had occurred, yet the orifice of the uterus was but slightly dilated, and as rigid as if no hæmorrhage had been sustained. Labour was induced, taking four or five hours to expand the os. Living twins were delivered. On the fifth day after the labour, profuse flooding set in, and caused death. No rupture was found; but the long-continued bearing incident to the introduction of the hand had produced contusion, inflammation, and suppuration of the os uteri, and a portion of its tissue, about the diameter of a sixpence, had sloughed off, and left behind it a deepish ulcer; several branches of arteries were found in the depth of it, and thus was rendered evident the cause of the fatal hæmorrhage.

Dr. Edward Rigby says:* "Cases have occurred where the os uteri has been artificially dilated, where the child was turned and delivered with perfect safety, and the uterus contracted into a hard ball; a continued dribbling of blood has remained after labour; the patient has gradually become exhausted, and at last

* "System of Midwifery," 1844.

died. On examination after death, Professor Naegele has *invariably* found the os uteri more or less torn."

Collins and others relate examples in point. The truth is, that so far from the os uteri in placenta prævia being in a state favourable to dilatation, the conditions are precisely the reverse. The implantation of the placenta involves an enormous increase of vascularity of the parts, and this, added to the imperfect muscular development attained when labour comes on prematurely, renders the dilatation of the cervix especially difficult and dangerous. Hæmorrhage from laceration is not the only danger. One, a little more remote, but scarcely less formidable, is that of inflammation, of pyæmia. Some of the worst cases of puerperal fever I have seen, have, in my opinion, been the direct consequences of the bruising of the vascular tissues of the cervix, caused by forcible delivery for placenta prævia. I will show presently how this injury to the uterus can almost always be averted.

Next there came a new theory, based upon a presumed physiological basis. Levret believed that the placenta supplied a portion of the blood in this form of uterine hæmorrhage. Rawlins, of Oxford,* says: "The blood proceeds more from the vessels of the detached portion of the placenta than from the denuded vessels of the uterus." The late Professor Hamilton entertained the same opinion. Kinder Wood, Radford, and Simpson adopted this opinion; but they did not agree in their reasoning from it. Simpson drew from it the practical conclusion, that to stop the hæmorrhage we should completely detach

* "Dissertation on the Obstetric Forceps," 1793.

the placenta. Radford, who had previously adopted this practice, was led to it mainly by the clinical observation that the hæmorrhage had ceased on the spontaneous separation and expulsion of the placenta. Another resource, then, was added to our means of dealing with placenta prævia. We might choose between forced delivery and artificial total detachment of the placenta.

How is the total detachment of the placenta accomplished? The method prescribed is to pass the hand, if necessary, into the vagina, then to pass two fingers through the cervix uteri, and with them to separate the placenta.

The fatal objection I urged against this proceeding was that it was impracticable. In by far the greater number of cases the placenta extends higher than the meridian of the uterus, often reaching the fundus. The fingers are not long enough to reach even half-way towards the further margin of the placenta. The diameter of the placenta is nine or ten inches; the fingers will barely reach three inches. In the greater number of cases, therefore, in which the directions prescribed have been followed, the placenta has not been wholly detached, and the result, when successful, cannot be attributed to an operation which was not performed. This is further proved by the history of some of the cases narrated as examples of this practice. The child was born alive. Is it consistent with our knowledge of the conditions upon which the child's life depends, to suppose that the child will survive if the whole placenta be detached, unless the birth follow the detachment immediately?

And this condition, under the postulates of the hypothesis, is wanting.

The entire detachment of the placenta has been urged, on the ground that it can be executed at a stage when the dilatation of the cervix is insufficient to admit of turning. But if it cannot be executed without passing the whole hand into the uterus, that is, without forcing the hand through the undilated cervix, in what respect is the operation less severe than that of delivery by turning? Is it not reasonable to conclude that, since the forcible entry has been effected, the seizure and extraction of the child, as well as the detachment of the placenta, had better, to give the child a chance, be completed at the same time?

If so, the special character of the proceeding vanishes; it is even more severe than turning, which does not require the hand to be passed through the cervix.

But it is contended that clinical observations prove that hæmorrhage has stopped on the total detachment of the placenta. These observations are partly true, partly fallacious. The true observations are those in which the placenta has been *spontaneously* cast off and expelled before the birth of the child. These cases are not numerous. They do not justify the conclusion drawn, that the *artificial* total detachment of the placenta will be equally followed by arrest of hæmorrhage. There is a fundamental physiological distinction between the two cases. When the placenta is cast off spontaneously, it is because the uterus contracts powerfully. This contraction stops the bleeding. When the placenta is detached artificially, there may be, and probably is, defective uterine contraction. The bleeding will be likely to continue. There is no independent virtue in the mere detachment of the placenta, as *post-mortem* hæmorrhage abundantly proves.

Here, then, we have true observations erroneously interpreted. The fallacious observations are of the following nature. The physician, seeking to imitate the spontaneous detachment, obeys the precept to pass two fingers into the uterus, and concludes that he has wholly detached the placenta. In this he has, in all probability, deceived himself. The fingers can only detach as far as they will reach, that is, for an area of two or three inches from the os uteri. The operator has unconsciously failed in doing what he tried to do; he has unwittingly done very nearly what he ought to do. The hæmorrhage stops; he sees in his success a proof of the truth of the theory that total detachment of the placenta is the security against hæmorrhage. But he has not detached the placenta, as he imagined; he is simply giving proof of the truth of a very different theory.

I will now endeavour to give a succinct view of my theory of placenta prævia, and of the principles that should guide us in the treatment of this complication.

The chief points in the physiology of placental attachment are expressed in the annexed diagram:—

The inner surface of the uterus may be divided into three zones or regions, by two latitudinal circles. The upper circle may be called the upper polar circle. Above this is the fundus of the uterus. This is the seat of fundal placenta, the most natural position.

It is the zone or region of safe attachment. The lower circle is the lower polar circle. It divides the cervical zone or region from the meridional zone. The meridional space comprised between the two circles is the region of lateral placenta. This placenta is not liable to previous detachment. Attachment here may, however, cause obliquity of the uterus, oblique position of the child, lingering labour, and dispose to retention of the placenta and post-partum hæmorrhage.

Fig. 91.

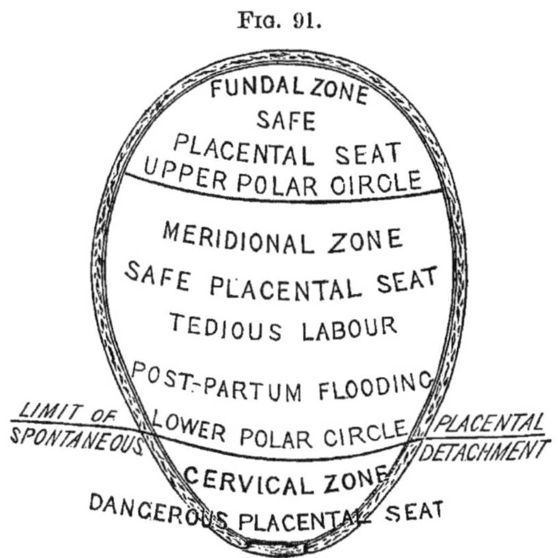

Below the lower circle is the cervical zone, the region of dangerous placental attachment. All placenta fixed here, whether it consist in a flap encroaching downwards from the meridional zone, or whether it be the entire placenta, is liable to previous detachment. The mouth of the womb *must* open to give passage to the child. This opening, which implies retraction or shortening of the cervical zone, is incompatible with the preservation of the adhesion of the placenta within its scope. In every other part of the womb there is an easy relation between the contractile limits of the muscular structure and that of the cohering placenta. Within the cervical region this relation is lost. The diminution in surface of the uterine tissue is in excess.

The lower polar circle is, then, the physiological line of demarcation between prævial and lateral placenta. It is the boundary-line *below* which you have spontaneous placental detachment and hæmorrhage; *above* which spontaneous placental detachment and hæmorrhage cease.

When the dilatation of the cervix has reached the stage at which the head can pass, and all that part of the placenta which had been originally adherent within the cervical zone is detached, and if, as is the constant tendency of Nature to effect, the intermitting active uterine contractions arrest the hæmorrhage, a stage is reached when the labour is freed from all prævial placental complication; the lateral portion of placenta retains its connection, supporting the child's life; the labour is, in all respects, a natural labour.

This is the course which Nature strives to accomplish. I have verified it frequently at the bedside. Many cases are recorded by old and recent authors in which this course was successfully accomplished, although the narrators failed to interpret the phenomena correctly. If observations in point are not more abundant, it is simply because men, acting servilely under the thraldom of the "unavoidable hæmorrhage" theory, fear to let Nature have a chance of vindicating her powers. The instant resort to the *accouchement forcé* interrupts the physiological process; we can see no more.

Mercier seems to have been so struck with the occasional absence of hæmorrhage, that he wrote an essay under the title, "Les accouchements où le placenta se trouve opposé sur le cot de la matrice sont-ils constamment accompagnés de l'hémorrhagie."* He thought the exhalant vessels were in a state of constriction which opposed the passage of blood. Moreau and Simpson thought the previous death of the fœtus, and the consequent diversion of blood from the uterus, explained the absence of hæmorrhage. This to a certain extent may be true; but I am in a position to affirm, from repeated observation, two things—first, that hæmorrhage occurs when the child has long been dead; secondly, that there may be no hæmorrhage although it is born alive.

Caseaux † also says, "The hæmorrhage which has been generally regarded as inevitable under these circumstances may nevertheless not occur, not even during labour; and the dilatation of the neck may be effected without the escape of a drop of blood."

In an appendix to my work on "Placenta Prævia," I have related the histories of several cases, some quoted from various authors and some occurring under my own observation, illustrating this point. Indeed, it was the reflection upon a case of this kind, which

* "Journal de Médecine," vol. lv.

† "Traité théorique et pratique de l'Art des Accouchements." Sixième edition, 1862.

I saw in 1845, which led me to the working out of my theory. The cases in this appendix referred to, prove the fallacy of the explanations before offered as to the cause of the cessation or absence of hæmorrhage. They show that it is not accounted for by the death of the child; nor by the pressure of the head or other part of the child upon the bleeding surface. They afford strong presumption, at least, that the hæmorrhage ceases when the physiological limit I have described has been reached, tonic or active contraction of the uterus concurring. I am far from affirming that Nature is generally equal to the carrying out of this process. Not seldom she is utterly unequal to the task. It is important to know what are the causes of her break-down; knowing these, we obtain valuable indications in practice.

There are two conditions commonly present in flooding from placenta prævia. The *first is an immature uterus.* Flooding frequently occurs before the term of gestation is complete. The uterus is therefore taken by surprise, before its tissue is developed, before it has acquired its normal contractile power. Besides this, the tissue is ill-adapted to expand. The *second is the loss of blood* itself, impairing the vital power, causing shock and prostration. The several or joint action of these conditions is a powerless labour, the absence of contraction. Hence the continuance of hæmorrhage. We feel we cannot depend upon contractile force when all force is ebbing away with the blood. We are compelled to act, to assist Nature in her extremity.

Let us now endeavour to determine the exact position of the lower polar circle or boundary-line between hæmorrhage and safety. This can be done with considerable accuracy.

The lower segment of the womb must open to an extent corresponding to the circumference of the child's head in order to permit its extrusion. By noting, therefore, the amount of necessary recession

FIG. 92.

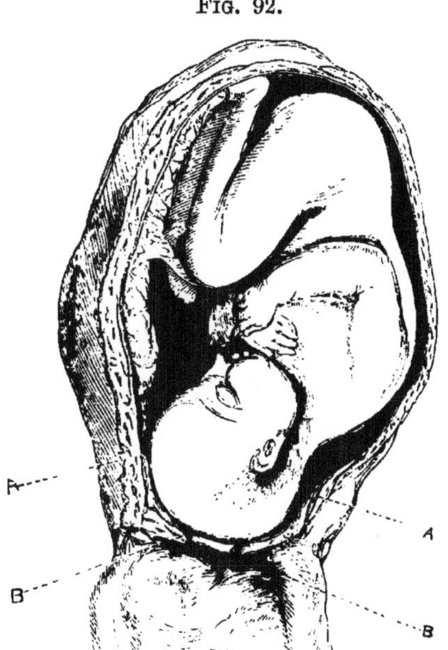

SHOWING A LATERAL PLACENTA DESCENDING TO A A, THE BOUNDARY-LINE BETWEEN THE MERIDIONAL AND CERVICAL ZONES.

The placenta descends to the very fullest expansion of the os, and therefore remains just within the limit of safe attachment. The space between A A and B B is the range of orificial expansion necessary to permit the passage of the head.

or shortening of the lower segment of the womb to reach this extent of expansion, we shall obtain the exact measure of the original depth of the cervical zone, the region of prævial placental attachment. Fig. 92 will serve to illustrate this position.

This point may be further demonstrated by the following simple proceeding:—Take a fœtal skull, and marking the left parietal protuberance for a centre, stretch a vulcanized india-rubber ring over the circle of greatest circumference of the skull, preserving it at equal distance from the centre. This ring will represent exactly the os uteri at the utmost stage of expansion necessary for the passage of the head. To this extent the os uteri *must* expand; beyond this it *need not, and will not* expand. It therefore marks the limit between the orificial and lateral portions of the uterus.

If we now measure the distance between the presenting parietal protuberance of the fœtal head and any part of the line of greatest circumference, we shall have the utmost extent of the cervical zone. In a full-sized fœtal head this is about three inches. If we now describe a circle within the womb at three inches' distance from the undilated os, we shall have drawn the lower polar circle. This is nearly exactly what the fingers introduced inside the womb can do.

There is another demonstration of the extent of the orificial zone, derived from examination of the state of the placenta after its delivery. The part which came nearest to the os uteri, and which may have been felt during labour, is exactly indicated by the rent in the membranes being close to it, it being obvious, as was pointed out by Levret, Von Ritgen,* and Hugh Carmichael,† that, as the membranes must burst or be ruptured at the os uteri when the child passes, so, by measuring the distance of this rent from the lower

* "Monatsschr. f. Geburtsk," 1855.

† "Dublin Quarterly Journal," 1839-40.

edge of the placenta, we shall obtain exactly the elevation of the placenta from the os uteri.

This part of the placenta, which had been detached during labour, is found infiltrated with extravasated blood, making a thick, firm, and black flap, quite distinct in appearance from the rest of the placenta, which, growing above the lower polar circle, had preserved its adhesion in the ordinary way until after the expulsion of the child. The area of this ecchymosed portion corresponds to the area of the orificial zone.

I believe, however, that the boundary-line of safety is often practically reached before the expansion of the mouth of the womb has reached the full diameter of the child's head. I have observed that the hæmorrhage has completely stopped when the os uteri had opened to the size of the rim of a wine-glass, or even less.

We may now consider

The *Course, Symptoms, and Diagnosis of Placenta Prævia.* When the placenta grows wholly or in part within the cervical zone, its relation to the uterine wall at this part is always liable to be disturbed. At any time during the course of pregnancy, hæmorrhage from this cause is apt to take place. It is most probable that some cases of presumed ordinary abortion at the third or fourth month are in reality due to this cause. At least, in aborted ova of this period, I have frequently verified the fact that there was placental structure within the lower zone; but, usually, placenta prævia is not recognised as such earlier than the end of the fifth month. At and after this time, the woman may be overtaken without warning by a smart flooding of fresh florid blood. This often occurs when the patient is at rest, even in the night; sometimes when she is out of doors, or away from home, so little is she prepared for any accident; sometimes it comes after unusual exertion. These attacks of hæmorrhage are usually quite independent of labour or of uterine contraction; they occur most frequently at the menstrual epochs. It is not uncommon to observe a recurrence of these hæmorrhages at intervals of about a month. It cannot be doubted that the disturbance of relation between placenta and uterus is brought about by the increased afflux of blood brought to the uterus and placenta at those periods. Sometimes the bleeding subsides, and the patient is reprieved for a time. It is even possible that she may go the full term of pregnancy, after having suffered several attacks of hæmorrhage, and be delivered naturally with little loss. But most frequently, premature labour will be excited either at the first or second attack. The seventh and eighth months are frequent critical epochs. The bleeding having begun, some blood is extravasated between the uterine and placental surfaces, the cervical part of the placenta gets thickened and hardened with extravasated blood, and thus the uterus is excited to contraction. When once the uterus is set in action, the termination in labour is highly probable.

The Diagnosis.—The os uteri is often found scarcely, if at all, dilated more than is usual in pluriparæ—and it is in pluriparæ that placenta prævia most frequently occurs — if examination is made early during the flooding; but the cervix or vaginal portion is commonly thicker than ordinary. The finger passed into the os will miss the head or other presenting part of the child, especially if the case be one of placenta centralis. In this case, also, ballottement will be difficult or not made out; but, instead, you may feel the quaggy, spongy placenta or a blood-clot. The cervix is generally more tender to the touch, and pain is often felt during gestation at the lower segment and on the side to which the placenta grows. Levret says, the uterus, instead of being rounded or pointed, is flattened, as if divided into two parts as in twin-pregnancy, but the division is more on one side, causing oblique irregularity of form; and in the first months the patient has been conscious of a swelling, with pain and hardness, in one side. Gendrin says, a pulsation, not synchronous with the mother's pulse, may be felt at the os uteri.

The stethoscope will, as Hardy and McClintock * have shown, often determine the seat of the placenta at once.

It is usual to teach, that in accidental hæmorrhage the bleeding is arrested during a pain, whilst in placenta prævia the hæmorrhage, although continuing during the intervals, is greatly increased during a pain. Nothing, I believe, can be more illusory than trusting to this distinction. Certainly, at the onset, there is often no pain in placenta prævia; and, as the case proceeds, active pains will often stop the hæmorrhage.

The Prognosis applies to three principal questions: 1. What is the immediate danger of the patient? 2. What is the ultimate danger? 3. Will the case

* "Practical Observations."

go on at once to the completion of delivery, or will the symptoms subside?

1. The immediate danger to life from shock and loss of blood is serious, if the hæmorrhage is profuse, if the cervix remain unexpanded, and if delivery and contraction of the uterus be not secured within a short time. Whenever the loss is rapid and great, and the patient is clearly affected by the loss of blood, the indication is strong to abandon at once the prospect of postponing labour, and to proceed immediately to accelerate delivery.

2. The ultimate dangers, supposing immediate sinking from hæmorrhage is cleared, arise from anæmia. The secondary effects of hæmorrhage are: malnutrition, nervous disorders; the local injury to the cervix uteri during labour, the contusion, laceration, dispose to inflammation, and blood-infection from the necrosis of tissue about the mouths of the utero-placental vessels; phlegmasia dolens is not uncommon, and is sometimes of very severe, even fatal type, being complicated with more than the usual degree of blood-infection; all the other forms of puerperal fever are more common after placenta prævia; secondary hæmorrhage from laceration of the cervix; and lastly, there is the prospect of imperfect involution of the uterus, and of chronic endocervicitis.

In 1864, I had occasion to review my experience of the terminations in 69 cases of placenta prævia. The deaths were 6, *i.e.*, 1 in 11½, a proportion much smaller than that usually given in statistical tables. But upon this point I think it idle to dwell, for general statistical tables, drawn from miscellaneous sources, are utterly untrustworthy in this matter.

Of the 6 fatal cases, 1 died, three weeks after labour, of pyæmia; 1 died in a few days of pyæmia following forced delivery, performed by a practitioner who prided himself on his promptitude in the treatment of these cases; 2 were moribund when first seen; 1 died of exhaustion (she had had eleven children); 1 died of puerperal fever, aggravated by ill-usage.

I feel very sure that if we could always see these cases at the earliest stage of hæmorrhage, and if they were treated on the principles I have laid down, the mortality would be brought very much below anything hitherto known. In the cases above referred to, there occurred a series of twenty-nine successful cases uninterrupted by a single death.

3. Will the case go on to delivery? If the hæmorrhage is moderate, if the os does not dilate, if there is little or no sign of uterine action, there is the probability of the utero-placental relations being so little disturbed that the pregnancy may go on. But this question is often practically settled by the physician, who, governed by his estimate of the strength of the patient, the stage of pregnancy, and the urgency, absolute and relative, of the symptoms, may resolve to accelerate the labour. If the pregnancy have advanced beyond the seventh month, it will, as a general rule, I think, be wise to proceed to delivery, for the next hæmorrhage may be fatal; we cannot foretell the time or the extent of its occurrence, and when it occurs, all, perhaps, that we shall have the opportunity of doing will be to regret that we did not act when we had the chance.

The Prognosis as to the Child.—This will depend very much upon the conduct of the case. But under any method of treatment the risk to the child is great, so great, indeed, that Simpson and Churchill have expressed the opinion, that the hope of saving the child ought scarcely to influence the treatment. In this view I cannot concur. It is true that in many cases the child is dead before there is any opportunity of treatment. The child dies of asphyxia, the result of the mother's loss of blood; this blood, which is the means of aëration of the child's blood, comes in too small quantity and too feebly to effect the necessary change. Again, the child is frequently premature, sometimes not even viable; frequently there is transverse presentation, frequently the cord is prolapsed, and then the child has to run the gauntlet of artificial modes of delivery. Exposed to these perils, it is not surprising that the child will often perish. But still the broad fact remains, that a considerable proportion of children are born alive; and nothing can be more certain than that some of the sources of peril to the child may be lessened or altogether avoided, and that not only without adding to the danger of the mother, but even increasing her security. Out of sixty-two cases of placenta prævia which had come under my observation in 1864, but which had not all been treated after my method, or under favourable circumstances, twenty-three children were born alive. I am confident that a better result than this may be obtained, if the proper principles of treatment are applied, beginning at the onset of the flooding.

The total detachment of the placenta before the birth of the child is almost necessarily fatal to it. The precipitate forcible delivery of the child is scarcely less hazardous. The method I recommend is calculated to give the child every possible chance, and at the same time conduces in the highest degree to the safety of the mother.

TREATMENT OF PLACENTA PRÆVIA.

The treatment must vary according to the nature of the case, and cases of placenta prævia vary greatly. I have already adverted to the distinction drawn by many authors between partial and complete placental presentation, and to the fact that a different rule of treatment applies in the two cases. It is generally taught that in partial presentation there is less danger, and that less active treatment is called for. As a general proposition, this is true. But I have frequently seen the most severe flooding attend partial presentation, so as to call for action as energetic as any flooding from central placenta. I think a more practical division of cases is into,

A. Cases of placenta prævia, whether partial or complete, in which there is active contractile power in the uterus, with spontaneous dilatation.

B. Cases in which the contractile power in the uterus is absent, with or without dilatation.

The first question, we have seen, is to decide whether the pregnancy can be allowed to go on. If the pregnancy is only of five or six months, the os not dilated, all pain absent, and the hæmorrhage very moderate, we may temporise, watching, however, most vigilantly. But if the hæmorrhage be at all profuse, and there be any sign of uterine action, no matter what the stage of pregnancy, act at once;

accelerate the labour. Above all, do not trust to the weak conventional means of keeping the patient in the recumbent posture, in a cool room, with cold cloths to the vulva, mineral acids, acetate of lead. This is but trifling in the presence of a great emergency. Commonly, they do no good whatever; always, they lose precious time.

The great hæmostatic agent is contraction of the uterine fibre. To obtain contraction is, therefore, the end to be sought. Although the powers of the system may still be good, the uterus will not always act well, especially in premature labour, whilst it is fully distended. To evoke contractile energy, it is often enough to puncture the membranes. This done, some liquor amnii runs off; the uterus collapsing, is excited to contract, and being diminished in bulk, it acts at advantage. Labour being active, the cervix expands promptly, the placenta gets detached from the orificial zone, the bared uterine vessels get closed by the retracting tissue and by the pressure of the advancing head; the hæmorrhage ceases spontaneously.

1. *The puncture of the membranes* is the first thing to be done in all cases of flooding sufficient to cause anxiety before labour. *It is the most generally efficacious remedy, and it can always be applied.* The patient lying on her side, a finger is passed up into the os uteri, guiding a stilet, quill-pen, or porcupine's quill to the membranes, whilst the uterus is supported externally.

2. At the same time *apply a firm binder over the uterus.* This further promotes contraction; and by propelling the child towards the os uteri, it accelerates the expansion of the os and moderates the hæmorrhage.

3. Puncture of the membranes is in many cases enough of itself. But if the hæmorrhage continues, and especially if the patient shows signs of exhaustion, the os uteri being undilated, the *plug* may be tried. The best plan is to plug the os uteri itself with laminaria-tents, as described at pp. 392, 393. This will expand the os, and prepare for further proceedings. The various forms of vaginal-plugs, such as the silk-handkerchief, tow, lint, even Braun's colpeurynter, are treacherous aids, requiring the most vigilant watching. Do not go to sleep in the false security that, because you have tightly packed the vagina, the bleeding will cease. The plug you have introduced with so much pain to the patient soon becomes compressed, blood flows past it or accumulates around and above it, and the tide of life ebbs away unsuspected. Never leave a patient more than an hour, trusting to the plug. Feel her pulse frequently; watch her face attentively; examine to see if any blood is oozing. Remove the plug in an hour, and feel if the os uteri is dilating. If it be dilating, and the hæmorrhage has stopped, you may trust Nature a little further, watching her closely. The labour may now go on spontaneously, probably issuing in the birth of a living child.

4. You must not, however, be surprised to find that the hæmorrhage continues, that the os uteri does not expand, and that there is no active labour. Expectancy has its limits; the time has come for more decided measures. Will you force your hand in and turn? Remember what I have said of the dangers of this proceeding. There are two means of accomplishing the end in view, without violence, with more

certainty and with more safety to the patient and her child.

Consider, what are the ends to be attained?

We want to check the hæmorrhage at once;

We want security against the renewal of hæmorrhage.

To attain these ends, the uterus must be placed in a condition to contract. The essential steps towards this consummation are—first, the free dilatation of the cervix; secondly, the completion of the labour.

The first difficulty is to *effect the dilatation of the cervix.* Under any process, this must take a little time. Can anything be done in the meantime to moderate the bleeding?

Something very effectual may be done. If the physiology of placenta prævia, which I have laid down, be correct, it may be expected that by accelerating the completion of the necessary process of *separating all the placenta which adheres within the orificial zone*, we shall, at any rate, get over the period of danger more quickly. If we can do this, we shall gain other advantages: we remove an obstacle to the dilatation of the cervix, for the adherent placenta acts as a mechanical impediment; and we lessen the risk of laceration of the placenta, an accident very likely to happen in the ordinary course and under turning, and which, by rupturing the fœtal vessels, adds to the peril of the child.

The operation is this: Pass one or two fingers as far as they will go through the os uteri, the hand being passed into the vagina, if necessary; feeling the placenta, insinuate the finger between it and the uterine wall; sweep the finger round in a circle, so as to separate the placenta as far as the finger can reach; if you feel the edge of the placenta where the membranes begin, tear open the membranes freely, especially if these have not been previously ruptured; ascertain, if you can, what is the presentation of the child before withdrawing your hand. Commonly, some amount of retraction of the cervix takes place after this operation, and *often the hæmorrhage ceases.* You have gained time. You have given the patient the precious opportunity of rallying from the shock of previous loss, and of gathering up strength for further proceedings.

If, the cervix being now liberated, under the pressure of a firm binder, ergot, or stimulants, uterine action return so as to drive down the head, it is pretty certain there will be no more hæmorrhage; you may leave Nature to expand the cervix and to complete the delivery. The labour, freed from the placental complication, has become natural.

5. If, on the other hand, the uterus continue inert, the hæmorrhage may not stop; and you must proceed to the next step, *the artificial dilatation of the cervix.* This is accomplished by the use of my hydrostatic dilators. Insert the middle or largest-sized bag into the os; distend with water gently and gradually, feeling the effect of the eccentric strain upon the edge of the os. When the bag is fully distended, keep it *in situ* for half-an-hour or an hour if necessary. During this time, the hæmorrhage is commonly suspended; probably the intra-uterine portion of the bag presses upon the mouths of the bared vessels; certainly retraction of the cervix goes on, which is the direct means of closing these vessels; and under the combined effect of pressure from below by the

dilator, and from above by the binder, the contents of the uterus are kept in close contact with its inner surface, thus keeping pressure on the vessels of the cervix, and stimulating the entire organ to contract. When this object is accomplished, and you find the cervix freely open, you may withdraw the bag.

Again, you may wait and observe if Nature is able to carry on the work. If contraction persists, if the head presents, the labour is henceforth essentially normal, and may be allowed to go on spontaneously. You must nevertheless watch attentively.

If contraction is inefficient, if hæmorrhage goes on, if another part than the head is presenting—a condition very frequent in placenta prævia—we must carry our help further. We must do what Nature cannot do: we must deliver. This is done by seizing the child's leg and extracting. Now, this can almost always be accomplished without passing the hand into the uterus. The bi-polar method of turning here finds one of its most valuable applications. It avoids the danger and difficulty of forcing your hand through an imperfectly-expanded os, through imperfectly-developed and abnormally-vascular structures. Having seized a leg, it must be drawn down gently, so as to bring the half-breech into the cervix. Traction must be so regulated as to bring the trunk through with the least amount of force necessary for the purpose. Whilst delivery is going on, the hæmorrhage is generally arrested. Rapid extraction involves a certain amount of violence and shock. Gentle extraction, giving the cervix time to dilate gradually, avoids this mischief.

As soon as the child is born, re-adjust the pressure

upon the uterus; and if there is no hæmorrhage, allow three or four minutes for the system to rally, before attempting to remove the placenta. If hæmorrhage occurs, and the placenta does not come on moderate traction, you must pass the hand into the uterus to detach it. The portion growing to the meridional zone is not always readily cast. Examine the placenta carefully on its uterine surface to see if it is entire.

In every labour, the cervix, having to suffer great distension and contusion under the passage of the child, and possessing less contractile elements in its structure than the rest of the uterus, is liable to paralysis for a time. This state is more likely to occur in labour with placenta prævia, and it is doubly dangerous because the cervix is the placental site. Here is another reason for sparing the cervix to the utmost.

The chief facts in relation to placenta prævia may be summed up in the following series of physiological and pathological propositions. In the "Obstetrical Transactions," and elsewhere, I have published the histories of cases in illustration and proof.

I.—*Series of Physiological Propositions.*

1. That in the progress of a labour with placenta prævia, there is a period or stage when the flooding becomes spontaneously arrested.

2. That this hæmostatic process does not depend upon total detachment of the placenta; upon death of the child; upon syncope in the mother; nor upon

pressure on the cervix bared of placenta, by the presenting part of the child, or distended membranes.

3. That the one constant condition of this physiological arrest of the flooding is contraction, active or tonic, of the muscular structure of the uterus.

4. That this physiological arrest of flooding is neither permanent nor secure until the whole of that portion of the placenta which had adhered within the lower zone of the uterus is detached—that being the portion which is liable to be separated during the opening of the lower segment of the uterus to the extent necessary to give passage to the child.

5. That when this stage of detachment has been reached, there is no physiological reason why any further detachment or flooding should take place until after the expulsion of the child, when, and not till then, the remainder of the placenta, which adheres to the middle and fundal zones of the uterus, is cast off, as in normal labour.

6. That the rent made in the sac of the amnior to give passage to the child, being necessarily at that part opposite to the os uteri, marks the spot at which the placenta was attached to the uterus; thus, in cases of partial placental presentation, the rent is commonly found close to the edge of the placenta.

7. That attachment of the placenta to the lower segment of the uterus at the posterior part is a frequent cause of transverse presentations.

8. That in the case of partial placental presentation, where an edge of the placenta dips down to near the edge of the os uteri internum, the umbilical cord commonly springs from that edge; and that thus

is explained the liability to prolapsus of the funis in cases of placenta prævia.

9. That adhesion of the placenta to the os uteri internum impedes the regular dilatation of the part, and, consequently, whilst such adhesion lasts, the proper course of labour is hindered.

10. That inflammation of the uterine structures, particularly of the cervix, is especially likely to supervene upon delivery attended by placental presentation; and that the danger of this complication is increased by the forcible dilatation and contusion of the vascular cervix, caused by the introduction of the hand, or by the extraction of the child. That one of the purposes intended by Nature in fixing the seat of the placenta at the fundal and middle zones of the uterus, is the preservation of the parts rendered highly vascular by connection with the placenta, from the distension, pressure, and contusion attending the passage of the child.

II.—*Series of Therapeutical Propositions.*

1. That the greatest amount of flooding generally takes place at the *commencement* of the labour, when the os is beginning to expand. That the os is always, from its being near the seat of the placental attachment, highly vascular, and is frequently, at this stage, very rigid. That any attempt to force the hand through this structure, at this stage, either for the purpose of wholly detaching the placenta, or of turning the child, must be made at the risk of injuring the womb; and that the dragging the child through the

os when in this condition—even when it has not been necessary to pass the hand into the uterus—is a proceeding affording slender chance of life to the child, and dangerous to the mother.

2. That the entire detachment of the placenta is not necessary to ensure the arrest of the hæmorrhage; and that, therefore, it is not necessary either to wholly detach the placenta before the birth of the child, or, in cases uncomplicated with cross-presentation, to proceed to forced delivery with a view to wholly detach the placenta after the birth of the child.

3. That, since the dilatation of the cervical portion of the womb must take place in order to give passage to the child, and since, during the earlier stages of this necessary dilatation, hæmorrhage is liable to occur, it is desirable to expedite this stage of labour as much as possible.

4. That in cases where labour appears imminent, with considerable flooding, whilst the os internum uteri is still closed, the arrest of the flooding and the expansion of the os may be favoured by plugging the vagina, and especially the cervix, and by the use of ergot.

5. That, since a cross-presentation, or other unfavourable position of the child at the os internum uteri is apt to impede or destroy the regular contractions of the uterus which are necessary to the arrest of flooding, it is mostly desirable, in cases complicated with unfavourable positions of the child, to deliver as soon as the condition of the os uteri will permit.

6. That, in some cases, the simple use of means to excite contraction of the uterus, such as ergot, rupturing the membranes, the administration of a

purgative, or the employment of galvanism, will suffice to arrest the hæmorrhage.

7. That, in some cases where it is observed that the os uteri has moderately expanded—namely, to the size of a crown-piece, or less—the placenta being felt to be detached from the cervical zone, and the hæmorrhage having ceased, it is not necessary to interfere with the course of the labour, now become normal.

8. That, at the critical period, when the total detachment of the placenta or forcible delivery are dangerous or impracticable operations, the introduction of the index finger through the os, and the artificial separation of that portion of the placenta which lies within the lower or cervical zone of the uterus, is a safe and feasible operation.

9. That the artificial detachment of all that portion of the placenta which adheres within the cervical zone of the uterus will at once liberate the os internum uteri from those adhesions which impede its equable dilatation; and, by facilitating the regular contraction of this segment of the uterus, favour the arrest of hæmorrhage, and convert a labour complicated with placental presentation into a natural labour.

10. That the immature uterus, partly paralyzed by loss of blood, cannot be trusted to assume the vigorous action necessary to effect delivery; that it is, therefore, often desirable to aid by dilating the cervix artificially; that this can be done safely and quickly by the caoutchouc water-dilators.

The management of the case after the removal of the placenta falls within the subject of post-partum hæmorrhage, and will be discussed in the next lecture.

III. *The so-called "Accidental Hæmorrhage."*

We have now to consider the *hæmorrhage, so-called "accidental,"* from premature detachment of the placenta.

Pathology, Symptoms, and Prognosis.—In this case, towards the end of gestation, the placenta, which had grown normally to the fundal or lateral regions of the uterus, becomes detached wholly or in part; and the uterus, remaining distended by the ovum, bleeds freely from the exposed placental site. No form of flooding is more formidable than this. Three circumstances especially concur in making it dangerous: First, it most commonly occurs in women who have borne many children, whose constitutions are worn by sickness or poverty, and whose tissues, therefore, are badly nourished, wanting in tone, tending to atrophy or degeneration. Secondly, the hæmorrhage is often wholly or partially "concealed." The detachment taking place at a distance from the cervix, the blood accumulates in the cavity of the uterus; we lose, therefore, the common evidence of flooding. Thirdly, this sudden irruption of blood into the cavity of the uterus stretches the uterine fibre, producing shock, perhaps collapse. The peritoneal coat has even been known to be torn by this sudden stretching. In reference to this form of laceration of the uterus, I must recall attention to the fact, observed by Professor Simpson, that forcible injection of water or air into the uterus, for the purpose of inducing labour, has caused similar laceration. Fatal mischief is often done before the physician is called in, or has the opportunity of acting.

We must seek for other symptoms than hæmorrhage to guide our action. The most characteristic are—first, *acute pain* in some part, generally the fundus, of the uterus; second, *collapse;* third, *great distension of the fundus of the uterus;* it protrudes more than naturally into the epigastric region, and communicates a doughy feel, the form of the fœtus being lost at this part. These three signs are the result of the stretching of the uterine fibre by the accumulation of blood in a circumscribed part of the uterus. The histories of cases* show that the detachment almost always begins in the middle of the placenta, and proceeds towards the margin under the pressure of blood accumulating. A cavity is formed for the reception of the blood, partly by inward compression of the placenta, which tends to be separated, and partly by the bulging outwards of the uterine wall. The placenta examined after expulsion, instead of being convex on its maternal surface, is cup-shaped. Dr. Oldham described a remarkable case, in which the placenta retained its adhesion all round the margin only, a large mass of blood being imprisoned in the hollow formed between uterus and placenta. The preparation is in Guy's Museum. A fourth result of the sudden shock and uterine injury is commonly *absence of all true labour-pains.*

There are indeed the general signs of loss of blood: fainting, blanching, agitation, perhaps deafness or blindness. The skin is cold and clammy, the pulse feeble, dicrotous, or almost extinguished, the features

* See Oldham, "Guy's Reports, 1856," and Braxton Hicks, "Obstetrical Transactions," vol. ii.

are pinched; the whole aspect indicates suffering and depression.

These symptoms come on suddenly. The causes are, I believe, primarily, lax, feeble tissue, and a circulation easily disturbed, fatty degeneration, calcareous deposits, fibrinous effusion of the placenta. In one fatal case, I found fatty degeneration of the muscular fibres of the heart. The immediate causes are, sudden emotion, over-exertion, vomiting, direct injury to the uterus, previous death of the child.

In some presumed cases there is a fallacy often overlooked. The placenta dips down into the lower or orificial zone, and although on examination the edge of the placenta may not be felt, the case is really one of placenta prævia. This is afterwards proved by observing that the rent in the membranes is within an inch or two of the edge of the placenta.

As a main cause of hæmorrhage during pregnancy, Gendrin insists upon irregular contractions of the uterine walls. The muscular structure is disposed in two layers, an external and an internal. The relations of these two layers with the vascular layer account for the influence they exercise in the production of hæmorrhage. Moral or physical impressions, or the movements of the fœtus, excite spasmodic contractions; the intra-uterine vascular plexus being pressed irregularly by these muscular contractions, blood must flow in some points of the placental disc; hence a partial congestion, which may cause a rupture of the weak venous ramifications. These contractions, by causing circumscribed puckerings on segments of the uterine globe, necessarily drag upon the placental connections, and may cause their rupture.

In the perfectly healthy uterus and placenta I do not think these contractions, which, if moderate, must be considered normal, exert any unfavourable influences; but where the structures are diseased or weakened, where the blood is degraded, where the emotional and other causes excite unusually active contractions, there can be no doubt that things will happen as Gendrin says.

Treatment.—The first thing to be done is to *rupture the membranes.* This, by letting off the liquor amnii, takes off the strain upon the uterine fibre, allows the walls to resume their natural condition, and provokes labour. This done, it is desirable to watch and give opportunity to rally. To proceed quickly to forcible delivery, might prove fatal by adding to the shock. Ergot, I think, is not very useful. If there is great depression, the drug is either not absorbed, and is therefore inert; or, if absorbed, it is injurious by adding to the prostration. *Stimulants* internally, *warmth* to the extremities, and *friction*, are useful in promoting reaction. This effected, the uterus may be able to contract, and labour may go on spontaneously. If not, the next thing is to *dilate the cervix gradually* by means of the water-dilators. When there is sufficient dilatation, you may deliver by the forceps if the head present, by bi-polar turning if any other part present. The ruling principle should be, to proceed with as little precipitation and force as possible.

When the child is delivered, the placenta comes away with a mass of clotted and fluid blood; and often the prostration is increased, death sometimes quickly following. As soon as the placenta is extracted, I strongly advise to *inject the uterus with perchloride of iron*, instead of trusting to cold or kneading. The great depression contra-indicates the resort to any agents that depend for their efficacy upon a reserve of power in the system. The paralyzed, injured uterine fibre predisposes to fresh hæmorrhage. It is all-important to secure the patient against further loss by the most prompt and trustworthy means.

In these cases the child is almost always born dead. It perishes of asphyxia, arising from the mother's loss of blood and collapse, and the partial or complete detachment of the placenta.

LECTURE XXIII.

HÆMORRHAGE AFTER THE BIRTH OF THE CHILD—CASES IN WHICH THE PLACENTA IS RETAINED—CASES AFTER THE REMOVAL OF THE PLACENTA—SECONDARY HÆMORRHAGE—CAUSES OF RETENTION OF PLACENTA—MODE OF CONDUCTING THIRD STAGE OF LABOUR—CONSEQUENCES OF RETENTION OF PLACENTA—ENCYSTED OR INCARCERATED PLACENTA—HOUR-GLASS CONTRACTION—MEANS OF EFFECTING DETACHMENT AND EXPULSION OF PLACENTA—ADHESION OF THE PLACENTA—CAUSES—TREATMENT—PLACENTA SUCCENTURIATA—PLACENTA DUPLEX—PLACENTA VELAMENTOSA—HÆMORRHAGE FROM FIBROID TUMOUR AND POLYPUS—INVERSION OF THE UTERUS, RECENT AND CHRONIC—HOW PRODUCED—TREATMENT—DIAGNOSIS.

We have now to discuss the nature and management of the *hæmorrhages which occur during and after labour*—that is, after the birth of the child. These may be usefully divided into

a. Cases in which the placenta is retained.

b. Cases in which bleeding persists, or occurs after the removal of the placenta.

c. Cases in which bleeding persists, or occurs some days after labour: the so-called "secondary puerperal hæmorrhage."

a. Hæmorrhage in which the Placenta is Retained.

In the most healthy labour, the supreme effort of expelling the child is commonly attended by so much pain and expenditure of nervous force, that a period of rest, the result of temporary exhaustion, follows. Probably the placenta is in great measure detached during the final contractions which expel the child. Soon sufficient nerve-force is reproduced, the uterus contracts again, and completes the detachment of the placenta. If the placenta has been adherent to the fundal or meridional zones of the uterus, its connection is preserved entire until these final expulsive contractions take place, when it is cast *en masse.* In the case of fundal attachment, the separation begins at the centre of the placenta and extends to the circumference.

But, if a portion of the placenta has dipped down within the orificial zone, this part may have been detached during the expulsion of the child, and hæmorrhage will continue afterwards; or, which is as frequent, the part of the placenta which had grown to the upper part of the orificial zone had not been detached during the second stage of labour, and thus, when the uterus contracts, it is only the fundal and meridional regions of the organ which contract so uniformly as to throw off the placenta; the orificial region not contracting to the same extent, the placenta corresponding to it remains adherent, and again hæmorrhage results. In either of these cases, separation begins at the margin.

If, from any other cause than the foregoing, the nervous force necessary for complete uterine contraction be wanting, so that the uterus is affected unequally, even if the placenta were normally attached, this unequal contraction will cause partial detachment of the placenta and hæmorrhage.

It is not to be forgotten that weakness during expulsion of the child commonly continues into the stage of placental expulsion. This disturbance is the greater, the greater the degree of weakness of the pains, the longer the labour, and the more rapid the artificial delivery. Hence the axiom: in delivery by forceps proceed deliberately, so as to give the uterus opportunity to act and help. Help the uterus, do not supplant it.

Or, again, if after the placenta has been wholly detached, it remains in the uterus, and inertia come on, hæmorrhage will also follow.

Not only does the presence of the placenta excite irregular spasmodic action, but, so long as it remains, the full, equable contraction necessary to completely close the uterine sinuses cannot take place. It follows, as a corollary, that the removal of the placenta is the first great end to be attained as a security against hæmorrhage.

How is this best effected? There is error both in precipitation and in delay. If, immediately after the birth of the child, you begin to pull upon the cord, you irritate the uterus at the moment of temporary exhaustion, you make that temporary exhaustion persistent, you induce irregular spasmodic contraction. This will rarely have any other effect than to cause partial detachment, prolonged retention of the placenta, and hæmorrhage.

Hunter and Denman encouraged the practice of leaving the extrusion of the placenta to the natural efforts, even for several hours. This practice had already been tried in Holland under the authority of Ruysch, and abandoned. Hæmorrhage and puerperal fever so frequently followed, that it was abandoned here also. In France, the rule is still to wait for an hour before interfering. But this sort of moral hand-cuffing for an arbitrary time, can only be applicable to persons who cannot be trusted to observe and interpret accurately the condition of the patient. The uterus will often, indeed, detach the placenta and extrude it into the vagina; but there it will lie for an indefinite time, for the vagina has rarely the power to expel it. What useful purpose can it answer to leave it there? The prevalent practice in this country seems to me the most reasonable. It is to watch for contraction of the uterus, which is ascertained by the hand, and by the consciousness of expulsive pain by the patient. If the uterus is felt, hard, and of the size of a child's head behind the symphysis, and if, on using the cord for a clue, you can feel the insertion of the cord without passing your hand into the uterus, you know the placenta is cast and partially extruded into the vagina. Once beyond the action of the uterus, the spongy mass of the placenta will fill the vagina, adapting itself to the cavity of the pelvis. The vagina, recently distended, has little contractile power, and here the placenta will remain for an indefinite time. The indication clearly is to complete what the uterus and vagina cannot do, by gently removing it. This you may do sometimes in five minutes; and it is rarely desirable to wait more than ten or fifteen minutes.

The consequences of retention of the placenta are these:—

1. Commonly hæmorrhage and spasmodic pain.

2. Sometimes no hæmorrhage, but expulsion after an indefinite number of hours or days.

3. Decomposition of blood in the cavity of the uterus, and imprisonment of the products, by the placenta blocking up the orifice, constituting physometra. In this condition the uterus sometimes enlarges, becoming tympanitic; the most horribly offensive discharges escape.

4. Septicæmia, from the absorption of the foul products.

5. Inflammation of the uterus and peritoneum, possibly from escape of foul products by the Fallopian tubes into the peritoneal cavity.

6. Disappearance of the placenta by disintegration, liquefaction, or absorption. I very much doubt whether absorption of the placenta can be established on good evidence. I feel disposed to regard it in the same light as Velpeau regarded "Vagitus uterinus." Since men of credit affirm that they have seen it, I believe it; but if I had seen it myself, I should doubt.

This may be laid down as an axiom in obstetrics: *by the proper management of labour, including the delivery of the placenta, you greatly secure the patient against hæmorrhage and many other dangers.*

In what does this management consist?

The rule laid down by the late Dr. Joseph Clarke, of Dublin, and endorsed by Collins and Beatty, is the one I recommend for your adoption:—When the head and trunk of the child are expelled, follow down the fundus uteri in its contraction by your hand on the abdomen, until the fœtus be entirely expelled, and continue this pressure for some time afterwards, to keep the uterus in a contracted state. When you are satisfied of this, apply the binder. I always apply the binder during the labour; the support thus given to the uterus is invaluable; it tends to preserve the due relation between the axis of the uterus and that of the brim, thus obviating the most serious objection made to the English obstetric position on the left side.

This plan of causing the uterus to contract and expel the placenta by manual compression has within the last few years been introduced into Germany, as a discovery, by Dr. Credé, without a suspicion, apparently, that it has long been the familiar practice in this country; a singular instance of the want of information amongst our Continental brethren as to the state of British midwifery, the comparative small mortality of which might well challenge the most earnest inquiry on their part.

The great point is, to let or make the uterus cast the placenta by its own efforts; not to pull the placenta away.

A retained placenta may become "*encystea*," or "*incarcerated*," that is, locked up in the fundus of the uterus by contraction of the part below. A form of this retention is familiarly spoken of as "*hour-glass contraction*." It is described and figured as a ring-like contraction of the middle part of the uterus, dividing it into two spaces. By others, more accurately, it is described as due to contraction of the os internum uteri. It is agreed by most authors of experience, that hour-glass contraction is very rare. The varieties of irregular contraction of the uterus might be deduced *à priori* from consideration of the arrangement of the muscular fibres. If all the muscular bundles of the uterus contract harmoniously together, there will be the much-desired normal uniform contraction, closing the cavity, and necessarily expelling anything that may happen to be in it. But it is a matter of observation, that occasionally parts of the uterus contract, whilst others remain passive. Now, which are the parts most disposed to contract, and which to remain passive? Naturally, those parts in which muscle is most abundantly provided will have the greater power of contracting. Two parts are especially rich in this respect, that is, each corner around the entrance of the Fallopian tubes. At the lower part of the body of the uterus, again, the bundles of fibres from either side assume a transverse or circular direction.

Now, there are two parts which are specially exposed to conditions that weaken their contractile power, producing even temporary paralysis. These are, the placental site and the cervix. The attachment of the placenta involves a great development of vascular structure; and however it be explained, the fact is certain, that the placental site is very liable to inertia. The paralysis of the cervix is accounted for by the great distension and bruising to which it is exposed during the passage of the child. After delivery it is constantly felt to be quite flaccid, as if it had lost all power of recovering its former shape and tone. Hence it is obvious that the placental site and the cervix are the two sources of hæmorrhage.

If the placental site be at the fundus, extending into the area of Ruysch's muscle on either side, the fundus generally will be liable to paralysis; and the part below being excited to action, there may arise hour-glass contractions, *i.e.*, contraction of the lower part of the body of the uterus.

If the placental site be in one angle of the uterus, occupying the area of one of Ruysch's muscles, the central part of that area will be liable to paralysis, and the circular bundles on the margin of that area, being excited to action, will close in upon the placenta, forming a sac. Clinical observation confirms this. If you feel through the relaxed abdominal wall a spasmodically-acting uterus, you will often find it of irregular shape, and you will perceive a marked prominence at one side of the fundus, caused by the contraction of one circular muscle. Again, if you pass your hand into a uterus contracting irregularly, you will come to the constriction, the seat of which you will be able exactly to determine, between the two hands, to be where I have described.

It is generally admitted that these forms of irregular contraction are most frequently induced by injudicious meddling, by precipitate artificial delivery, by making too early attempts to bring away the placenta, by pulling on the cord, thus irritating, teasing the uterus. You must remember the precept: Make the uterus cast out the placenta; do not drag it out.

A method of exciting the uterus to contract and throw off the placenta is *the injection of cold water into* *the vein of the umbilical cord.* This was proposed by Mojon.* By this means cold is applied directly to the placental site, so that the very part of the uterus with which the placenta is connected is stimulated to contract. I have no personal experience of this method; but Scanzoni speaks highly of it. It is carried out in the following manner:—Cut the cord across, squeeze out the blood from the cord towards the cut surface, insert the point of the largest syringe the vein will carry, tie the vein on the nozzle, and inject cold water carefully, lest you burst the vein and fail.

If the uterus cannot be made to contract, it will be necessary to *pass the hand into the uterus to remove the placenta.* For this purpose the patient may lie either on her back or left side. Support the uterus firmly, pressing the fundus backwards and downwards towards the pelvis with one hand; at the same time pass your other hand, guided by the cord, into the cavity of the uterus. Feel for the lower edge of the placenta, and, with the fingers flattened out between the placenta and uterine wall, insinuate them by a light waving movement upwards, so as to peel off the placenta from the uterine surface; all the while be careful to support the uterus by the hand outside; you gain from its consentaneous action more command over the internal manœuvre, and wonderfully facilitate the accomplishment of your object. When the placenta is entirely separate, seize it well, endeavouring, by outward pressure, to induce the uterus to expel the placenta and hand together. When the placenta is extruded, proceed in like manner to clear the uterine cavity

* "Annali Universali di Medicina," 1826.

perfectly of clots. Apply the binder firmly, aided by a compress, if necessary.

The whole organ is sometimes affected, assuming the character of tetanus, the contractions are so rigid and persistent. This is especially apt to occur after ergot. To overcome them, you must depend upon steady continuous pressure with the hand or the water-bag; opium or chloroform will often assist. Pass the hand in a conical shape into the constriction, carefully pushing the fundus uteri down upon the inside hand by the other hand on the abdomen. You must trust to time to tire out the spasm, not to force. Your hand may very likely be cramped, but you must persevere, or you will only have irritated the uterus, and be obliged to begin again. When you have succeeded in passing the constriction, grasp the placenta, remove it, and keep up steady pressure upon the uterus externally. When the placenta is removed, the harmony of action of the uterus will be restored.

The object is to restore the due relation of contractile energy in the different parts of the uterus. The greater contractile energy should be exerted by the fundus. We must then seek—1, means to relax the crampy constriction of the lower segment; and, 2, means to evoke the supremacy of contraction of the fundus and body.

The cases we have been considering comprise those in which the *placenta is retained simply from want of uterine energy* to detach and expel it. These are far the most common cases. But *the placenta may be retained by morbid adhesion to the uterus.* Cases of this kind are comparatively rare; they are more troublesome and dangerous than the first, and require more decided treatment.

True adhesion of the placenta commonly depends upon a diseased condition of the decidual element. The most frequent is some form of inflammation with thickening, hyperplasia. This, in all probability, began in the mucous membrane before pregnancy. It is liable to aggravation when the mucous membrane is developed into decidua. Sometimes there is distinct fibrinous deposit on the uterine surface of the placenta; sometimes the decidua is studded with calcareous patches. The maternal origin of the forms of diseased placenta leading to adhesion is proved often by the history of previous endometritis or other disease, and by the frequent recurrence of adherent placenta in successive pregnancies.

I have adverted to this subject under "Abortion." Further information will be found in memoirs by Hegar,* Fromont,† and Hüter.‡

Diagnosis of Placental Adhesion.

You may suspect morbid adhesion, if there have been unusual difficulty in removing the placenta in previous labours; if, during the third stage, the uterus contracts at intervals firmly, each contraction being accompanied by blood, and yet, on following up the cord, you feel the placenta still *in utero;* if, on pulling on the cord, two fingers being pressed into the placenta at the root, you feel the placenta and

* "Die Pathol. und Therap. der Placentarretention." Berlin, 1862.

† "Mém. sur la Rétention du Placenta." Bruges, 1857.

‡ "Die Mutterküchenreste, Monatsschr. f. Geburtsk," 1857.

uterus descend in one mass, a sense of dragging pain being elicited; if, during a pain the uterine tumour do not present a globular form, but be more prominent than usual at the place of placental attachment.

The removal of a morbidly adherent placenta must be effected in the same manner as that just described for retained placenta; but you must be prepared to encounter more difficulty. The peeling process must be effected very gently and steadily, keeping carefully in the same place during your progress, being very careful to avoid digging your finger-nails into the substance of the uterus. In some cases the structures of the uterus and of the placenta are so intimately connected, seeming, in fact, to be continuous parts of one organization, that you cannot tell where placenta ends and uterus begins. In endeavouring to detach the placenta, portions tear away, leaving irregular portions projecting on the surface of the uterus. In trying to take away these adherent portions you must proceed with the utmost caution. The connected uterine tissue is, perhaps, morbidly soft and lacerable. It is very easy to push a finger into it, to the extent of producing fatal mischief. A very serious practical question now arises. To what extent must you persevere in trying to pick off the firmly-adherent portions of placenta? If you leave any portions, hæmorrhage, immediate or secondary, is very likely to follow; in decomposing and breaking-up, septicæmia is likely to follow; and there is, besides, the liability to metritis. If a fatal result ensue, and a portion of placenta be found after death in the uterus, it is but too probable that blame will be cast upon the medical practitioner. The nurse and all the anility of the neighbourhood will be sure to cry out, "Mrs. A. died because Dr. Z. did not take away all the afterbirth." The position is a very painful one. The true rule to observe is, simply to do your best; make reasonable effort to remove what adheres; it is safer for the woman to do too little than too much. You cannot repair grave injury to the uterus. To save your own reputation, you must fully explain the nature of the case at the time. You may lessen the risk of hæmorrhage and septicæmia by injecting perchloride of iron and permanganate of potash. In a few days the process of disintegration may loosen the attachment of the placental masses, and they may come away easily. The safest way, if it can be done, of removing these "placental polypi," is to pass a wire-écraseur over them. As this instrument can only shave the uterine surface smoothly, you are secure against injuring the uterus.

As a warning against attempting too much, and as ammunition to repel an unjust charge of having done too little, remember the following passage from Dr. Ramsbotham, the truth of which I can attest from my own experience:—"Instances are sometimes met with in which a portion of the placenta is so closely cemented to the uterine surface that it cannot by any means be detached; nay, I have opened more than one body where a part was left adherent to the uterus, and where, on making a longitudinal section of the organs and examining the cut edges, I could not determine the boundary-line between the uterus and the placenta, so intimate a union had taken place between them." Morgagni, Portal, Simpson, Capuron, relate similar instances; and there is an instructive case reported by

Dr. R. T. Corbett, in the *Edinburgh Monthly Journal*, 1850.

A very *soft placenta*, especially if it be *thin*, of *large superficies*, so as to be diffused over a considerable portion of the surface of the uterus, is a frequent cause of adhesion. The contracting uterus does not easily throw off such a placenta, at least completely. Perhaps a large part may be expelled or withdrawn, and appear to be all; but a portion of a cotyledon remains behind; bleeding and irregular action of the uterus are kept up, until the hand is introduced, and the offending substance removed.

A very rare—but on that account the more likely to be overlooked—event is the leaving behind a lobe of a *placenta succenturiata*. I have seen some singular examples of this. At a distance from the main body of the placenta, perhaps three or four inches or more from the margin, a mass of chorion-villi will be developed into this placental structure, and connected with the main body only by a few vessels. It resembles a lobe or cotyledon which has grown far away from the rest by itself. Such an accidental or supernumerary placenta may very easily remain attached after the main body, which is complete in itself, has been removed. These placentæ succenturiatæ rarely exceed in size that of a single cotyledon, *i.e.*, they measure about two or three inches in diameter. But I was once called by a midwife of the Royal Maternity Charity to a case of a different kind, at first very puzzling. The child was born, the cord tied, and the placenta, apparently entire, removed, when there followed a second placenta. Both were of nearly similar size and form. The first and natural conclusion of the midwife was that there was a second child still *in utero;* but she could not feel it, so sent for me. I passed my hand into the uterus, ascertained that it was empty, and made it contract. The second placental mass was developed on the same chorion as the chief placenta; vessels ran from it across the intervening bald space of chorion to join the single umbilical cord which sprang from the chief placenta. Dr. Hall Davis exhibited a similar double placenta to the Obstetrical Society.

Another form of placenta associated with hæmorrhage is the *velamentous placenta.* In this case, the umbilical vessels, instead of meeting on the surface of the placenta to form the cord, run for some distance along the membranes, uniting perhaps several inches beyond the margin to form the cord. This part of the membranes containing the umbilical vessels spread out, may be placed over the cervix uteri; and these must be torn during the passage of the child. The hæmorrhage thus resulting comes from the placenta, and of course endangers the child. Dr. V. Hüter describes this formation of the placenta fully.* Two cases of this kind are related by Caseaux. Joerg, quoted by Hegar, describes a case in which bundles of vessels were found over the greater part of the chorion, but no proper parenchymatous placenta; thus resembling the *diffused placenta* of the pig.

Hæmorrhage post-partum sometimes depends upon a *fibroid tumour* embedded in the wall of the uterus, or a *fibroid polypus* projecting into the cavity. No complication can well be more serious. The tumour disposes to hæmorrhage in two ways. In the first

* "Monatssch. für Geburtskunde," 1866.

place, by its density and form, it destroys the uniformity of thickness and density of the uterine wall. This impairs the power of the uterus to contract equally, and to maintain its contraction. Secondly, by its independent vitality, it attracts blood in abnormal quantity to the uterus, and besides acting as a foreign body, it irritates the uterus, exciting to irregular spasmodic action. The ordinary means of stimulating contraction are apt either to act badly or to fail. Kneading is especially dangerous, from the liability to bruise and injure the tumour, and even to lacerate the uterine tissue in which it is imbedded. Some amount of injury of this kind will probably have been endured under the force of labour. These cases, then, require the most gentle manipulation. The safest and most effectual plan is at once to apply the perchloride of iron. This, acting to a great extent independently of muscular contraction, will arrest the flooding, even where it is difficult to induce the uterus to act.

The fibroid polypus also requires special management. This form of tumour, too, is commonly influenced by the developmental stimulus in the pregnant uterus; it enlarges considerably; when the child and placenta are expelled, it is liable to be extruded from the uterus, and if of large size, it may even be projected outside the vulva. Hæmorrhage almost constantly follows, partly, perhaps, from uterine paralysis, partly from the attraction of blood. Another pressing danger, besides hæmorrhage, calls for decisive action. The polypus, which has recently undergone accelerated growth, is copiously infiltrated with fluid and new tissue; it has suffered contusion probably during the passage of the child; it is fatally disposed to a low form of inflammation, tending to necrosis; this is a cause of constitutional infection replete with danger. The double question is before us, how to suppress the hæmorrhage, and how to deal with the polypus, so as to avoid the too probable ulterior mischief. The most immediate difficulty is the hæmorrhage. Excite contraction by cold, by friction on the uterus, by ergot, if you can. But lose not much time upon these uncertain remedies. Inject perchloride of iron before much blood has been lost. The polypus itself should, I think, be removed by the wire-écraseur without delay. Cases of the kind are rare, and therefore it is impossible to deduce a rule of action from sufficient clinical observation. But we know the risk of renewed hæmorrhage, of necrosis of the tumour, of systemic infection, is very great if the tumour be left; and there is not much evidence of mischief from the immediate removal of the tumour. If hæmorrhage arise from the pedicle, it can be stopped by compression and perchloride of iron. A case of this kind occurred in the Maternity of St. Thomas' last year, and has been described by Dr. Gervis in the "Obstetrical Transactions" (1869.) The polypoid mass was very large; it projected externally, and underwent the necrotic process. The patient died.

It can scarcely be doubted that the necessity of supporting a large parasitic growth of this kind must, as the least evil, impede the healthy involution of the uterus.

A formidable cause of hæmorrhage is *inversion of the uterus*. I will not now stop to discuss at length the modes in which this accident is produced. I will remind you that *the uterus may be turned inside out by pulling upon the cord;* but *the uterus may*, as Crosse, Hunter, Radford, Hohl, and Dr. Tyler Smith have shown, *turn itself inside out.* This accident is immediately associated with inertia, or *paralysis of the placental site*, which is itself a great cause of hæmorrhage. Upon this we need not dwell, since it is the essential condition of the form of post-partum flooding we have already considered. But we must remember the general fact, that the placental site is more liable than other parts of the uterus to temporary paralysis.

When this occurs, the placental site bulges into the cavity of the uterus. The placental site is thicker than the rest of the walls of the organ, and naturally projects inwards.

Whether the placenta be still adherent or detached, the first condition for inversion is the falling inwards of the placental site. *This simple falling in or depression of the placental site is the first degree of inversion.*

The *second degree is that of introversion or introsusception;* so great a part of the fundus falls in that it comes within the grasp of the portion of the uterus into which it is received. In the extreme form, the fundus reaches to the os uteri, through which it may be felt like an intra-uterine polypus.

The *third degree is that of complete inversion;* the fundus passes through the os uteri. There are degrees of this. In the extreme form, even the cervix and os are inverted.

The first degree passes into the second and third in this way. If the placenta adhere, and be dragged upon by the cord from below; or, if the diaphragm and abdominal walls act, as in a bearing-down effort, the part already disposed to fall inwards is forced further down into the cavity. When things have gone thus far, further pressure or dragging brings the fundus down upon the cervix and os. If this part be contracted, it may prevent the fundus from coming through; or, the pressure continuing long enough, the os may yield and allow it to slip through. On the other hand, the advancing fundus may find the cervix relaxed and offering no opposition. Accordingly, it has been observed that some cases have occurred gradually; others suddenly. Lazzati insists that the necessary condition is general inertia of the uterus. The histories of many cases show that total paralysis existed. But in some cases it has been observed that uterine action was present, and was the real force that produced the inversion. Hunter was, I believe, the first to describe (*see* "Pathological Catalogue and Preparations," Royal College of Surgeons) *spontaneous active self-inversion.* "The contained or inverted part," he says, "becomes an adventitious or extraneous body to the containing, and it continues its action to get rid of the inverted part, similar to an introsusception of an intestine." Denman and Crosse describe the process in a similar manner.*

In some cases the resistance of the cervix is destroyed by laceration; and, next to the placental site, the cervix is the part most liable to temporary paralysis.

The symptoms of inversion are chiefly those of shock, indicating sudden severe injury.

* For further information on Inversion, I urge reference to Crosse's "Essay," perhaps the most complete and masterly monograph on a difficult obstetric subject ever published. The subject is also treated in the article "Uterus," by myself, in the new edition of Cooper's "Surgical Dictionary," now in the press.

They vary with the degree and progress of the inversion. Thus, the first degree may be unattended by pain, and be indicated solely by hæmorrhage and a corresponding depression of the vital powers. The hæmorrhage comes from the relaxed introcedent part. The depression at the fundus may be felt through the abdominal walls as a cup-like hollow. As the descent proceeds, and becomes *introversion*, urgent symptoms arise: a sense of fulness, weight, as of something to be expelled, is felt; expulsive efforts, both uterine and abdominal, sometimes very violent, follow. Hæmorrhage is not constant. It seems that when the inverted portion is firmly compressed, the hæmorrhage is arrested, and that bleeding is a mark of inertia. When the inversion is complete, the uterus is felt in the vagina, or may be seen as well outside the vulva. Then pain and collapse are aggravated; clammy sweats, cold extremities, vomiting, alarming distress, restlessness, extinction of the pulse, occur. During the expulsion the woman has often exclaimed that her intestines were passing from her; a tumour appears in the vagina, or externally, generally covered by the placenta. The shock, either with or without hæmorrhage, is sometimes so severe as to quickly extinguish life; and the severity of the shock is not the simple exponent of the quantity of blood lost.

When we find inversion, the indication is clear to reduce it, and that as soon as possible. The presence of the uterus grasped in its own neck excites contraction, the part gets strangled, swollen, and it becomes more and more difficult to return it. In most cases, probably, the mass might be pressed back with comparative ease, if the attempt were made immediately after the occurrence of the accident, whilst the cervix was still paralyzed. But this opportunity is rarely found. The first question that arises is as to detaching the placenta first or not. To detach the placenta is to lose a little time, and to run the risk of irritating the uterus to contract; if we leave it, there is the greater bulk to pass back through the os uteri. If you have the good fortune to recognise the accident at the moment, you may be able to take advantage of the flaccidity of the cervix, and return uterus and placenta at once; but if this favourable moment is lost, it will be better to detach the placenta first. Look for the margin of the placenta; insinuate one or two fingers between it and the globe of the uterus; supporting this organ by the other hand, continue to peel off the placenta by sweeping the fingers along. When it is wholly detached, proceed to reduction. The mode of manipulation must vary according to circumstances. If the uterus is large, flabby, and the cervix dilated, it may be quickly replaced by depressing the fundus with the fingers gathered into a cone, and carrying the hand onwards through the os. Lazzati says it is better to apply the closed fist to the fundus; this acts better, and we avoid the risk—by no means a slight one—of perforating the soft structure of the uterus. In executing this, two things must on no account be omitted: one is to support the uterus by the other hand, pressing firmly down upon it from above the symphysis pubis, lest we lacerate the vagina; the other is to observe the course of the pelvic axes, and the form of the pelvic brim. Pressure will first be made a little backwards towards the hollow of the sacrum; then the direction must be towards the brim, and at the same time *to one side, so as to avoid the sacral promontory*. As in attempts to reduce a retroverted gravid uterus, failure has often ensued from not understanding this latter point. It was first, I believe, insisted upon by Dr. Skinner, of Liverpool. I can testify to the value of the rule from personal experience. By attention to it mainly, I was enabled to reduce an uterus in fifteen minutes, which had been inverted for ten days, defying repeated efforts by other practitioners. When reduction has been completed, the hand following the receding fundus will occupy the cavity of the uterus, and the organ will be grasped between the hand inside and the hand supporting outside. The opportunity should be taken to induce contraction by pressure externally and by excitation internally. But I would not withdraw the hand from the cavity, lest re-inversion take place, until I had taken the following further security. Pass up along the palm of the hand an uterine tube connected with a Higginson's injecting-syringe; throw up by means of this six or eight ounces of a mixture composed of equal parts of the strong solution of perchloride of iron and water, so as to bathe the whole inner surface of the uterus. The effects of this are, instantly to constringe the mouths of the vessels, to stop bleeding, to excite uterine contraction, and to corrugate the tissues and narrow the os uteri. When this state is induced there is safety. This is all the more important, because at the moment of reduction there is likely to be considerable collapse and uterine inertia; or the styptic may be applied by swabbing by means of a pledget of lint carried on a probang.

If uterine action be present, especially if the cervix and os are constringing the inverted part, the difficulty is greater, and it is no longer judicious to commence by pushing in the fundus. As Dr. M'Clintock has well shown,* to do this is to double the inflexion of the uterine walls, and thus to double the thickness of the mass that has to pass through the os. He advocates the method practised by Montgomery, which consists in regarding the inversion as a hernia, and *in replacing that part first which came down last.* The tumour must be grasped in its circumference near the constricting os; firmly compressing it towards the centre, and at the same time pushing it upwards, forwards, *and to one side.* The pressure must be steadily kept up, as it is sustained pressure that wears out the resistance of the os. After a time the os is felt to relax, the part nearest is pushed through, and then, generally suddenly, the body and fundus spring through. Two things facilitate this operation: chloroform and a semi-prone position of the patient.

If the opportunity of reducing within a few hours be lost, the difficulty increases through advancing involution of the uterus and contraction of the os. But you are not to be discouraged by the teaching, only till recently held, that after a few hours inversion of the uterus is all but irreducible. Reduction is simply a question of time: keep up pressure long enough, and the constricting os must yield. True, the necessary pressure may be more than you can sustain with your fingers; but we have other means. Dr. Tyler Smith † reduced an inverted uterus of twelve years'

* "Diseases of Women," 1863.

† "Med.-Chir. Trans.," 1858.

standing by maintaining pressure upon the tumour, and thus upon the os, by an air-pessary. The pressure was kept up for more than a week. Other cases of chronic inversion have since been treated on the same principle with perfect success by Drs. Charles West, Bockenthal, Birnbaum, and Schröder, and Messrs. Pridgin Teale, Hakes, and Lawson Tait. Sustained elastic pressure, properly applied, is free from danger; and when we consider that the alternative commonly is death by hæmorrhage and exhaustion, or reduction by violent taxis or amputation, both dangerous, often fatal in themselves, we cannot but regard this plan as a great gain in obstetric surgery. Still, there may be cases that will resist even sustained elastic pressure. Then you have another resource, which I believe I was the first to carry to a successful termination.* It is this: make a slight incision on either side of the cervix uteri, so as to relax the constriction. The proceeding is as follows :—Draw down the uterine tumour by means of a loop of tape slung round the body, so as to put the neck of the tumour upon the stretch; then with a bistoury make a longitudinal incision about half an inch long and a quarter of an inch deep, on either side, into the constricting os; then re-apply the elastic pressure. Next day, try the taxis, and re-apply the elastic pressure, if necessary. Elastic pressure alone, or aided by this operation, will, I am convinced, overcome every case of inversion, except where fixed by inflammatory adhesions. Amputation and taxis by direct force should be absolutely condemned and discarded from practice.

Courty ("Maladies de l'Utérus," 1866) describes a

* See "Medico-Chir. Trans." for 1869.

peculiar method of applying the taxis. In a case of ten months' duration, he succeeded in the following manner:—The uterus was dragged outside the vulva by Museux's vulsellum; then the index and middle finger of the right hand were passed into the rectum, and hooked forward over the neck of the uterus; then the uterus was seized with the left hand, and pressed back into the vagina, still holding the neck hooked down, the fundus of the uterus was turned so as to look forwards to the pubes, the neck turned to the sacrum. The fingers in the rectum separating, rest firmly in the angular sinuses formed by the utero-sacral ligaments; then the thumb and index of the left hand pressing on the pedicle of the tumour, gradually increase the depth of the utero-cervical grooves. The two hands acting thus in concert, the uterus was reduced without violence in a few minutes. He had failed with the air-pessary; the patient could not bear it. It appears to be the safest and most powerful kind of taxis.

The plans last described of course apply chiefly to chronic inversion. Let us endeavour to define the terms "recent" and "chronic" inversion of the uterus. Where is the line to be drawn between the two states? When does recent inversion merge into chronicity? I would distinguish the cases in this way: inversion is recent so long as the physiological process of involution of the uterine tissues is going on. When this process is complete, and the uterus has returned to its ordinary condition, the inversion is chronic. During the first condition, the tissues are more yielding, and taxis under chloroform may be enough to effect reduction. When involution is complete, the tissues are more rigid, and sustained elastic pressure should always be employed.

We may distinguish six modes of treating chronic inversion of the uterus. 1. We may amputate by the simple ligature, on Gooch's plan, tightening the ligature until it sloughs through, or we may perform the operation at once by the écraseur. 2. We may excise by the knife. 3. We may combine the ligature and excision. 4. We may attempt to reduce by forcible taxis. 5. We may keep up pressure on the tumour by a solid body. 6. We may apply sustained elastic pressure.

In appreciating the relative merits of these plans, we must not forget that the highest success attainable by the three first methods is achieved at the cost of mutilation; the woman is unsexed; whilst the penalty of failure too often is death. The forcible taxis, if successful, restores the woman to integrity; but the too frequent penalty, again, of failure is death. Considerable force is requisite to reduce; the necessary force cannot always be restricted within safe bounds; the uterus or vagina has been repeatedly lacerated.

Sustained solid pressure has been successful; but it is not easily borne, and it demands great care lest mischief result. The use of sustained elastic pressure seems free from serious drawback; it has been attended with a very great success; and if it fails, the patient is no worse; she is then in a condition to admit of resort to my operation of incising the constricted os.

I must not leave this subject without pointing out how the *diagnosis* is established. In recent inversion, the womb covered by placenta has been dragged upon so forcibly as even to tear it away from the body;

and in chronic inversion the uterus has been amputated in mistake for a polypus. If the proper means of examination are omitted, nothing is more easy than to fall into error; if they be used, the diagnosis is certain.

In recent inversion there ought to be no ambiguity, if you observe the good obstetric rule of following down the uterus during and after the expulsion of the placenta. The hand upon the fundus feels the uterus retreat into the pelvis. If, simultaneously with the appearance of a large mass in the vagina or externally, you find the uterine globe disappear from behind the symphysis, you have, combined with the symptoms of shock, strong presumptive evidence that the uterus has become inverted. You may make this sure by pressing the tumour back a little up the vagina with one hand, whilst you press in the flaccid abdominal wall with your other hand behind the symphysis; you may thus feel the funnel formed by the inversion, and having the tumour between your two hands, you may accurately determine its nature.

The recently-inverted uterus has a remarkable property which distinguishes it from every other tumour. Contractility is a vital act that no polypus possesses. If the tumour, then, hardens and relaxes alternately, we know it is the uterus.

Chronic inversion is more apt to deceive. The general symptoms and history may even mislead you. The observance of the following rules will save you from error.

First.—Never put a ligature or écraseur on a supposed polypus under chloroform. The true polypus is not sensitive; the uterus is acutely so, when compressed. The moment the ligature or wire is being tightened on the neck of the uterus, intense pain is felt. Let this be a warning to take off the ligature instantly.

Fig. 93.

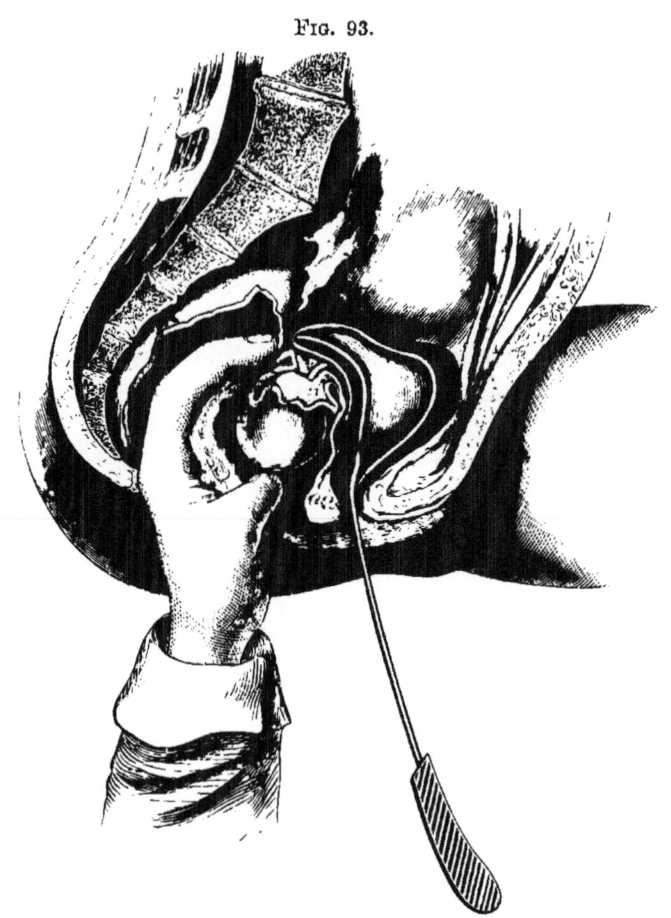

Secondly.—Properly this should, of course, be done in the first instance. Pass one or two fingers into the vagina to the root of the tumour; then press down the fingers of your other hand behind the symphysis. You will be able to make the fingers of the two hands meet where the uterus ought to intervene; you will also be able to push the finger from outside into the funnel of the inverted uterus.

Thirdly.—Pass a finger into the uterus so as to get the point above the root of the tumour; pass a sound into the bladder; turn the point of the sound backwards, so as to meet the finger in the rectum. You thus prove that the uterus is absent from its normal place, and that it must be the tumour in the vagina between your finger and the sound. This is illustrated in Fig. 93.

In polypus, besides, you may commonly pass a finger or the sound two or three inches beyond the stalk of the tumour.

In connection with the history of this very serious accident, it is proper to bear in mind that, in a few cases, inversion has existed for years without causing much distress. Dr. Woodman relates one,* which occurred at the London Hospital whilst I was obstetric physician there.

* "Obstetrical Transactions," vol. ix.

LECTURE XXIV.

HÆMORRHAGE AFTER THE REMOVAL OF THE PLACENTA—TWO SOURCES: THE PLACENTAL SITE; THE CERVIX UTERI—NATURAL AGENTS IN ARRESTING HÆMORRHAGE—SYMPTOMS, DIAGNOSIS, AND PROGNOSIS OF HÆMORRHAGE FROM INERTIA—ARTIFICIAL MEANS OF ARRESTING HÆMORRHAGE—MEANS DESIGNED TO CAUSE UTERINE CONTRACTION: PASSING THE HAND INTO THE UTERUS, ERGOT, TURPENTINE, COLD, KNEADING THE UTERUS, PLUGGING THE UTERUS, COMPRESSION OF THE AORTA, COMPRESSION OF THE UTERUS, BINDER AND COMPRESS—INDICATIONS HOW FAR TO TRUST THE FOREGOING AGENTS—THE DANGERS ATTENDING THEM—MEANS DESIGNED TO CLOSE THE BLEEDING VESSELS: BY COAGULA, THE PERCHLORIDE OF IRON—RESTORATIVE MEANS: OPIUM, CORDIALS, SALINES, REST—TRANSFUSION—SECONDARY HÆMORRHAGE.

HÆMORRHAGE after the removal of the placenta may arise from *two* sources. The *first* is *from the gaping vessels on the placental site.*

The *second source is from lesion of the cervix* or other part of the uterine structure. In the case of severe rupture of the uterus, this source of hæmorrhage is obvious enough; but minor lacerations of the cervix, especially after forcible delivery, although far more common, are seldom recognised. Contraction of the uterus is less effective in arresting hæmorrhage of this kind than that from the placental site; indeed, it persists when the uterus is well contracted. I have no doubt that laceration of the cervix is the true explanation of that form of hæmorrhage which Gooch described as due to an over-distended circulation, driving blood through the contracted uterus.

If, then, we find oozing or trickling of bright blood going on after labour, and with a well-contracted uterus, we may suspect this injury to be the cause. The remedy is to apply a powerful styptic, as perchloride of iron, direct to the bleeding surface.

What are *the means which Nature employs to arrest uterine hæmorrhage?*

1. The first, and the most efficient, is *active contraction of the muscular wall of the uterus.* This constricts, with the force of a ligature, the mouths of the arteries and veins on the placental site. So long as firm contraction holds, no blood can escape. To obtain this firm contraction is the great end of the obstetric practitioner. Active contraction settles into what may be called passive or tonic contraction, by which the volume of the uterus is permanently reduced. When this is effected, the patient is secure against a *return* of hæmorrhage.

2. The *uterine arteries have a certain retractile property.* Shrinking inwards, their mouths become narrowed, and the formation of thrombi is favoured.

3. The veins or sinuses of the uterus running obliquely or in strata in the uterine wall, and opening

obliquely on the surface, are most favourably disposed for closure by the approximation of their walls, and the valve-like arrangement where the sinuses pass from one stratum to another. Even moderate tonic contraction of the uterus will so close the uterine sinuses as to stop bleeding, provided the circulation is not unduly excited.

4. If the stream of blood through the uterine vessels be stopped for a short time, and diverted into the systemic circulation, so that there is temporary rest in the uterus, the opportunity is given for *the formation of clots, thrombi, in the vessels.* Under great losses of blood, the property of coagulating does not seem to be much impaired. No doubt it is owing to this property that many women are rescued, to all appearance from imminent death, after the most profuse and uncontrollable floodings. Under syncope, or a state approaching to it, the heart beats so feebly that the circulation is almost suspended; there is suspension of circulation in the uterus; and if ever so slight tonic contraction of the uterus goes on, the vessels get plugged by coagula. The probability of this event should encourage us never to despair of a case of hæmorrhage.

The *symptoms, diagnosis, and prognosis of hæmorrhage from atony of the uterus.* The first warning generally is the complaint of the patient that she feels blood flowing from her. Never disregard this. Examine the linen and the parts immediately. You will often see a thin stream of florid blood trickling down across the nates. This may seem insignificant in amount; but there should be none; and this "thin red line" is too often the indication of a greater loss, which is filling and stretching the uterus.

You feel the uterus, and find it has risen above the symphysis, perhaps above the umbilicus, that it is flaccid, or presents irregular hard prominences which shift their position. On compressing its fundus firmly, perhaps blood and clots will be forced out of the vulva. If the uterus is not brought to contract by the usual means, you pass your hand into the cavity, and feel that it is full of blood partly clotted, you feel the enlarged cavity, you feel the flaccid flabby walls. When the inertia is complete, it is sometimes difficult, by external manipulation, to make out the uterus at all. You miss the hard globe; and this negative sign is all. When the uterus has reached its full measure of distension, spasmodic contraction is sometimes excited, and a furious rush of blood is poured out. Often again, emotion, the dread of flooding, determines blood to the uterus, and a large quantity of blood is poured forth in a gush. Alternate contractions and relaxations, the uterus getting smaller, then larger, are certain signs of hæmorrhage from atony. These are the *local signs.*

The *general signs* are scarcely less marked or urgent. When hæmorrhage is copious enough to affect the system, a feeling of faintness, sometimes passing into actual syncope, comes on; irrepressible anxiety, a sense of fear, depression, are early symptoms; a degree of shock, of collapse, is conspicuous; the face is pale and cold; the whole surface is cold; the pulse is almost or quite imperceptible; the heart-beat is feeble and frequent; there is an indescribable sense of oppression on the chest; the patient calls out for air, will have the windows open, insists upon sitting up, sometimes would even get out of bed; she tosses her arms about, says she is sinking through the bed, is more or less delirious; her perception of external objects is dulled, or her appreciation of them is distorted; partial blindness, double vision, sometimes complete amaurosis set in, the pupils dilate, the iris seems paralyzed; she ceases to recognise at times the people about her; she complains of intense headache, noises in the ears, sometimes is manifestly deaf; she can hardly swallow, unless the fluid given be poured into the back of the mouth. So great is the loss of nervous force, that every organ, every tissue seems paralyzed; the uterus refuses to act under any stimulant; perhaps the sphincters relax, fæces and urine being voided. She rejects help, by word or sign entreats to be let alone; she would willingly die undisturbed.

From these symptoms, desperate as they look, the patient may recover. If the bleeding stop for awhile, slowly there is gathered up a little nerve-force; life that seemed ready to flit, holds its seat, with feeble and doubtful grasp, it is true, but gradually strengthening, if no fresh loss or accident occur.

If these signs are followed by marked collapse, contracting features, by short gasps or sobbing inspirations, which indicate that the chest-walls, unable to expand, make but an imperfect attempt to take in air, then quickly collapse, by convulsions, the case is indeed desperate. All power to respond to any remedy is exhausted. To persist in applying remedies now, is to harass the last moments of the patient in vain.

The favourable signs are, returning warmth and moisture of the surface, disposition to swallow, steady pulsation at the wrist, evidence of contraction of the uterus, a feeling of hopefulness and courage, a clearer perception of surrounding circumstances, a more accurate and steadier judgment.

Internal hæmorrhage is promoted by any cause that obstructs the escape of the blood externally. Thus, obliquity, or bagging of the uterus to the side, very likely to occur when the patient lies on one side, or even tending to the prone position with the pelvis raised, will form a depending sac in which blood will readily accumulate. Another condition promoting internal hæmorrhage is retroflexion of the uterus. This also is not very uncommon after labour with inertia. It was noticed by Burns. I have seen it several times myself, the hæmorrhage ceasing when the fundus was lifted into its proper position. Retroflexion is even more common as a cause of secondary hæmorrhage. The best way of restoring the uterus in the primary cases is to pass a hand into the cavity, and by it to lift the fundus forwards.

One or two *general rules* should be observed, whatever the particular method we may choose to rely upon to restrain the hæmorrhage.

1. *Place the patient on her back.*—The value of this rule is very great. Gravitation helps to let the uterus sink into the pelvis, instead of bagging over, as when the patient is on her side; the face is open to observation, to access of air, to administration of stimulants and food; the chest-walls can expand for better respiration; the uterus and aorta are under more easy control by the hand of the physician.

2. *Pass the catheter.*—A full bladder diverts the nervous force from the uterus, inducing irregular contractions, and impedes manipulation on the abdomen.

The indications in practice are drawn from our knowledge of the sources and natural modes of arrest of hæmorrhage, and observation of the symptoms.

In hæmorrhage after the removal of the placenta, we have, *first*, a class of remedies whose power is limited by one imperative condition. These *remedies act by exciting contraction of the uterine muscular fibre.* To effect this, there must exist a certain degree of nervous force, able to respond to centric or peripheral irritation. All the remedies commonly relied upon postulate this condition. *Ergot*, *compression of the uterus*, *cold*, all depend upon their power of inducing uterine contraction. If the nervous exhaustion be so great that irritability is lost, those agents are useless or injurious. The patient may indeed rally after syncope, but it may be truly said that at this point the art of the physician fails.

It fails unless, *secondly*, he has the courage to call to his aid another hæmostatic power, which acts even when to evoke contractility is impossible. When this is gone, he may still stop bleeding by the direct application of powerful styptics to the bleeding surface. The most useful of these agents is the *perchloride of iron*, the obstetric application of which I shall presently describe.

Passing the hand into the uterus to clear out the cavity. Although this proceeding is deprecated by many, and looked upon by most men as hazardous, I am satisfied that it is the first thing to be done whenever there is hæmorrhage with an enlarged uterus, and a suspicion that clots or other substances are retained in the cavity. If the enlarged uterus refuses to contract on external pressure, so as to expel its contents completely, I think there is no rule in obstetrics more imperative than to pass in the hand. If the patient is lying on her back, you can support the uterus externally whilst you introduce the other, without admitting much air into the cavity; you gain certain knowledge as to what is in the uterus; if there is nothing, the hand will act as a stimulus to contraction; if there are placental remains or clots, you have the opportunity of removing at once one of the most certain causes of flooding, primary and secondary, and of averting a very frequent cause of puerperal disease. In many cases this operation is followed by immediate success, little else being wanted. Often have I, when called in to a bad case of puerperal fever, wished that I could be fully assured there was nothing in the uterus. I perfectly agree with Collins and others, who insist upon the importance of this proceeding. I do not hesitate to repeat it two or three times if the uterus fills again; and in this latter case, the hand is ready to carry up the uterine-tube for the injection of perchloride of iron.

It is needless to insist that the operation should be performed gently. The shirt-sleeve should be rolled up to the shoulder; the back of the hand and all the arm be well oiled; the hand should be carefully directed to the axes of the pelvis.

Ergot, we know, possesses the property of causing uterine contraction. But how often does it fail to arrest uterine hæmorrhage? The caution I would urge in respect to ergot is this: if the first dose be not quickly followed by contraction, trust it no longer; do not lose all-precious time in repeating it. If it fail to act at once, it is probably because the nervous power is too much exhausted to respond to the excitation. I am sure I have seen the administration of ergot, when the system was much prostrated by loss of blood, cause further depression, doing harm instead of good; and if it does no harm, it is inert. In states of great depression, no absorption goes on in the stomach. Ergot, brandy, beef-tea, all alike simply load the stomach until rejected by vomiting.

The same may be said of *turpentine*, also, if the stomach will tolerate it, a very efficient hæmostatic. *Cold* is very much relied upon. How does it act? Producing a kind of shock, it excites reflex action, inducing contraction of the uterine muscles. The essential condition, therefore, of its action is, that there be sufficient nervous power to respond to the excitation. If this power be deficient, then the shock of cold only adds to the general depression; no contraction follows, or it is only momentary, the uterus quickly relaxing again. Nor does the harm cease here. The continuous application of ice and cold water will often cause congestion of the internal organs, and even lead to pleurisy, broncho-pneumonia, or peritonitis. I have seen cases of these forms of puerperal fever which I have no doubt were due to the patients being deluged in water, and being left for an hour or more chilled in wet clothes, because extreme prostration made it dangerous to change. Velpeau bears testimony to the same effect.

The rule, then, in the application of cold should be,

not to trust to it, if it be not quickly successful in evoking uterine contraction. The best form of using it, if the case is slight, if the nervous energy is good, is to apply the cold hand, or a lump of ice, or a plate taken out of ice-water, to the abdomen or back of the neck. Taking a draught of cold water into the stomach will sometimes excite uterine action.

The douche method, practised by Gooch and extolled by Collins, of pouring a stream of cold water from a height upon the abdomen, is the most certain to evoke contraction, if any degree of contractility remain. It is open to the objection that it swamps the bed, and exposes the patient to the danger of subsequent chill and inflammation. A very efficient plan, and free from the foregoing objection, is to flap the abdomen smartly with the corner of a wet towel. A good plan is to inject cold water into the rectum. Cold is more effectual if applied internally. Levret was, I believe, the first who used ice in this way. "Perfect," says Levret, "hit upon a very odd and ingenious expedient; he introduced a piece of ice into the uterus, which being struck with a sudden chill, immediately contracted and put a stop to the hæmorrhage." Of late years it has been a frequent practice to inject cold water into the uterus. I know it is often effectual. It has the advantage of washing out clots, as well as of exciting contraction; but I am not sure that it is free from danger. Levret's plan of placing a lump of ice in the uterus is the safest and best mode of applying cold. But I must repeat the warning, that unless there exist in the patient the power of reaction, cold will do harm instead of good. Unless it answer quickly, give it up.

Kneading the uterus, or compressing it firmly with the hand, is a means of exciting the uterus to contract much trusted to, and often useful. Even gentle friction or compression will often cause the uterus to contract and expel the placenta. A similar force applied when the uterus remains flaccid, or contracting spasmodically, will commonly induce firm, equable contraction. In states of greater exhaustion, with profuse flooding, even firm grasping is apt to fail. Whilst a strong hand is powerfully compressing the uterus, the hæmorrhage is certainly checked. But who can long keep up the requisite grasp? As soon as the tired hand relaxes, the hæmorrhage returns. Another hand takes the place of the first, and is in its turn exhausted; and so on, until the condition of the operator is almost as pitiable as that of the patient.

In resorting to this manœuvre, it is all-important to economise your own strength. You may do this greatly by placing the patient on her back near the side of the bed, so that you can stand well over her, and aid direct manual compression by the weight of your body. It is true that this manœuvre is often rewarded by success. If the hæmorrhage can be restrained for a few minutes, the strength will rally, and tonic contraction may return.

On the other hand, when there is deficient power, the uterus relaxes again and again; each renewal of flooding makes it more difficult to excite permanent contraction. Clots form again in the cavity, excite spasmodic action, and compel repeated introduction of the hand to remove them. Alternate relaxation and contraction of the uterus acts thus: expanding, it sucks up more blood from the aorta and vena cava;

contracting, this will escape again on the uterine surface; and so the uterus goes on relaxing and contracting, drawing blood from the system and discharging it into the uterine cavity. Thus the uterus acts like a Higginson's syringe pumping blood out of the body.

Not only is kneading uncertain, most painful to the patient, and exhausting to the physician, but it entails a special danger. This severe handling of the uterus, attended by bruising of the tissues, is liable to cause metritis. I have seen cases of metritic puerperal fever which I could only assign to this cause.

The compression of the abdominal aorta is a plan that deserves attention. It is practised in two ways. Ploucquet, who was the first to insist upon the method, compressed the aorta by the hand in the uterus. Baudelocque and Ulsamer compressed it through the flaccid abdominal walls. This is the plan most generally followed. It is recommended by Chailly, Caseaux, Jacquemier, and others. By it you may arrest the flooding, gaining a respite and time for the preparation of other means. It is performed in the following manner:—The patient lies on her back near the edge of the bed, the thighs drawn up. The physician stands at the right side; he presses the three middle fingers of his left hand gently into the abdominal wall near the umbilicus, curving them so as to fall obliquely upon the aorta, compressing it equally with all three fingers. The aorta is thus fixed upon the left side of the spinal column, avoiding the vena cava. The right hand can be used to aid the left, by supporting it. Compression kept up for a minute will often control the bleeding, giving time for the reproduction of nervous power and the contraction of the uterus. Faye relies mainly upon this method; Kiwisch objects to it, that compression on the aorta will simply compel the blood to go round by the spermatic arteries, and that the compression of the vena cava being unavoidable, the blood gets to the uterus by the numerous anastomoses of the pelvic circulation. The good resulting, he says, is due to the compression of the uterus. Boër and Hohl take similar objections. I have occasionally derived advantage from it; and look upon it as a momentary resource.

A form of pressure sometimes preferable to outward grasping of the uterus, is one recommended by Dr. G. Hamilton.* The fingers of the right hand are placed *under* the uterus, which the relaxed state of the parts usually allows, then, with the other hand upon the uterus externally, the organ is firmly compressed between the two hands. The cavity is closed by the anterior and posterior surfaces being flattened together. This manœuvre involves less violence, and is, for the time at least of its application, quite as effectual as external kneading.

Another form of applying pressure is by means of *the binder*, with or without a compress. I am a decided advocate for the binder after labour. That it is a source of comfort to the patient, almost every woman testifies; that it is a guarantee against hæmorrhage by exciting contraction of the uterus and preventing accumulation of blood in the uterine cavity, scarcely any practitioner of experience doubts. It has another advantage: in severe flooding from inertia, the bulky

* "Edinburgh Medical Journal," 1861.

flaccid uterus falls towards the diaphragm and forwards, encountering no support or resistance from the distended and paralyzed abdominal walls; this falling forwards, creating a vacuum, draws air into the vagina and cavity of the uterus, supplying the conditions for decomposition of clots, and thus favouring septicæmia. If the hand be introduced into the uterus, as it is likely to be in these emergencies, to remove placental remains or clots, air may often be *felt and heard* rushing in. Another urgent reason is this: just as, after tapping an ovarian cyst, the sudden release of the abdominal vessels from pressure diminishing the tension under which the heart acts, disposes to syncope, so does the sudden emptying of the uterus dispose to syncope. A well-applied bandage is a security against these possibly fatal accidents.

Various special obstetric binders and compresses have been devised. A compress acting with a spring like a hernia-pad, contrived by Mr. Toulmin, of Clapton, has several times given me great satisfaction, by relieving the tired hand. But these means more properly come into use to support the uterus when you feel that the hæmorrhage is fairly subdued; they are not to be depended upon as principal agents in arresting flooding.

Plugging has been practised in various ways. Paul, of Ægina, placed a sponge soaked in vinegar in the uterus. In later times, Leroux was a warm advocate of this method. Rouget (1810) advised the introduction of a sheep's or pig's bladder into the uterus, and then distending it. Diday (1850) used a vulcanized caoutchouc bladder in this way. If there is nervous force enough to respond to the irritation provoked by the presence of these foreign bodies, the plan may be useful. But as excitants to contraction they are inferior to other means; and, as plugs, I am sure they cannot be trusted.

Now, let us assume that the means which act by inducing uterine contraction have failed. The hæmorrhage goes on. Must we abandon the patient to a too probable death? Is there no other condition for stopping the stream of blood from the open vessels than contraction? There are styptics powerful enough. Why should we not use them? If they will act precisely when other means fail, is it not unreasonable to reject their aid?

This is the place to discuss the application of the perchloride of iron. This remedy is not mentioned in English or French text-books; it is still regarded by most practitioners as a dangerous innovation, not sufficiently reflecting that hæmorrhage is more dangerous still.

As I am mainly responsible for its introduction into obstetric practice, at least in this country, it is desirable that I should give a more connected account of the matter than has hitherto been done:

The styptic virtue of the perchloride and other salts of iron has long been known. These salts have been applied to bleeding surfaces by swabs by many surgeons. The idea of applying styptics to the bleeding uterus is also not new. But until recent times there prevailed an unreasonable dread of touching the inner surface of the uterus with styptics; yet from time to time they have been used with a success that might have encouraged us to look upon them with more confidence. Perfect says he stuffed the vagina with tow and oxycrate (vinegar and water). Hoffmann succeeded in stopping a profuse hæmorrhage by passing pledgets of lint dipped in a solution of "colcothar of vitriol"—a mixture of sulphate and oxide of iron—as high into the vagina as possible.

It appears from Hohl that the injection of perchloride of iron was first used by D'Outrepont; but reference to the subject in those of his works which I have read has escaped me. It is strongly praised by Kiwisch,* who used a solution of two drachms of the muriate of iron in eight ounces of water.

The first idea of using this means in uterine hæmorrhage was suggested to my mind on reading that perchloride of iron had been injected into an aneurismal sac. If the salt could thus be thrown almost into the stream of the circulation with safety, *à fortiori*, the open mouths of the sinuses on a free surface might be bathed with it; and surely the often irresistible and desperate course of uterine hæmorrhage was motive enough to try even a doubtful remedy. In the Lettsomian Lectures on Placenta Prævia, delivered by me in 1857,† I recommended this practice. In a lecture on "The Obstetric Bag," published in the *Lancet*, in 1862, I again recommended that perchloride of iron should be carried for the purpose of arresting hæmorrhage. Again, at the Obstetrical Society's meeting, in February, 1865, I made the following remarks, which were soon afterwards reported in the medical journals.‡ The subject under discussion was puerperal

* "Beiträge zur Geburtskunde," 1846.

† "Lancet," 1857, vol. ii., and "The Physiology and Treatment of Placenta Prævia," 1858.

‡ "Obstetrical Transactions," 1866, vol. vii.

fever. "As a means of preventing the loss of blood—as hæmorrhage undoubtedly predisposed to puerperal fever—I have found nothing of equal efficacy to the injection of a solution of perchloride of iron into the uterus after clearing out the cavity of placental remains and clots. I have used this plan for several years, and in a large number of cases after labour and abortion, and have always had reason to congratulate myself upon the result. The perchloride had the further advantage of being antiseptic. It instantly coagulated the blood in the mouths of the uterine vessels." As many of the cases referred to had occurred in consultation, of course my practice had been observed by many professional brethren. Now, confirmation is not wanting. Dr. W. Read, in the "Library of Medicine," 1861, quotes Dr. Von Schreier, of Hamburg, as having stopped hæmorrhage after abortion by a swab soaked in a weak solution. Dr. Alfred Hegar, in a memoir on the treatment of placental remains in 1862, says he had used the perchloride in various forms. Dr. Noeggerath, in 1862, had used it in metrorrhagia. Scanzoni, in his systematic work on "Obstetrics" (edition 1867), recommends it to be used in the most desperate cases, and quotes D'Outrepont and Kiwisch as advising it. In April, 1869, Mr. Norris, of South Petherton, published a paper in the *British Medical Journal*, which had been read in 1865, relating four cases in which he had injected the salt to arrest post-partum hæmorrhage. He has since narrated further cases. At the British Medical Association meeting, at Leeds, in August, 1869, Dr. Braxton Hicks related a series of cases in which he had successfully followed this method. Isolated cases in the hands of other practitioners have come to my knowledge.

The Mode of using the Perchloride of Iron.

An important preliminary step is to clear the uterus of placental remains and clots, so that the fluid, when injected, may come into immediate contact with the walls of the cavity. At one time it was my habit to wash out the uterus with iced water first. I now prefer not to lose time in doing this. Besides, at the period when the perchloride is especially indicated, the exhaustion is generally so great that the injection of iced water is ill borne. I am inclined to think that, under the circumstances, the injection of iced water is more hazardous than the injection of perchloride of iron.

You have the Higginson's syringe adapted with an uterine tube eight or nine inches long. Into a deep basin or shallow jug, pour a mixture of four ounces of the Liquor Ferri Perchloridi Fortior of the British Pharmacopœia and twelve ounces of water. The suction-tube of the syringe should reach to the bottom of the vessel. Pump through the delivery-tube two or three times to expel air, and insure the filling of the apparatus with the fluid before passing the uterine tube into the uterus. This, guided by the fingers of the left hand in the os uteri, should be passed quite up to the fundus. Then inject slowly and steadily. You will feel the fluid come back into the vagina mixed with coagula, caused by the styptic action of the fluid. The hæmostatic effect of the iron is produced in three ways: first, there is its direct action in coagulating the blood in the mouths of the vessels; secondly, it acts as a powerful astringent on the inner membrane of the uterus, strongly corrugating the surface, and thus constringing the mouths of the vessels; thirdly, it often provokes some amount of contractile action of the muscular wall. An inestimable benefit gained by the perchloride is, that it acts precisely in those cases of powerless labour where there is no contractile energy to be depended upon; it therefore comes to our aid exactly in the more severe cases, where other remedies fail, rescuing women whose position would be otherwise hopeless. Like the cervical detachment of the placenta, like the hydrostatic dilator, it is a new power in obstetric practice.

It is rare to find any renewal of hæmorrhage after the injection has been made as described. If there should be any return, it may be repeated.

Reasoning from general knowledge, it may be apprehended that the injection of perchloride of iron into the uterus is not free from danger. Examining my clinical experience, I do not find this apprehension verified.

May not some of the fluid penetrate into the circulation and cause thrombi in the blood-vessels or heart? This risk is met by the property the iron possesses of instantly coagulating the blood it comes in contact with at the mouths of the vessels. The thrombi then formed protect the circulation beyond. Thrombi, indeed, always, or nearly so, form in the mouths of the uterine vessels under ordinary circumstances. These artificially-produced thrombi break down in a few days, and under the contraction and involution of the uterus, get extruded into the cavity. For some days the *débris* from these thrombi, small coagula formed by free blood, and some serous oozing, come away. The discharge is black; it is apt to stain linen; and the nurse should be warned of this, lest she should take alarm at the unusual appearance. The discharge sometimes becomes a little offensive. This may be corrected by washing the vagina with Condy's fluid. But the perchloride of iron is a valuable antiseptic; and I believe this property is useful as a preservative against the septicæmia to which women who have gone through a labour with placenta prævia are so prone.

Several disastrous accidents have happened from the injection of minute quantities of a solution of iron into nævi. It is right to bear these in mind. Mr. R. B. Carter* refers to a case where immediate death followed the injection of five minims into a nævus on the nose of an infant eleven weeks old. Another case is recorded by Mr. N. Crisp, of Swallowfield. Examination showed that the point of the syringe had penetrated the transverse facial vein, and that the blood in the right cavity of the heart had been immediately coagulated. Dr. Aveling, now of Rochester, has informed me of another case. It is impossible not to entertain some misgiving upon this point. But these cases suggest that there is some special danger attending the injection of perchloride of iron into nævi. It will be observed that an injecting force is used; that in one case the point of the syringe actually penetrated a vein; and that the blood in the venous cavities of the heart was coagulated. Did the perchloride reach the heart, and there cause coagulation? This is doubtful. The

* "Medical Times and Gazette," 1864.

process, more probably, was as follows:—The moment the perchloride touched the blood in the facial vein, it created a small clot or thrombus. This thrombus was carried to the heart, and there served as a nucleus for further coagulation.

May not air be carried into the uterine sinuses, and thence into the heart? Possibly. The only case known to me in which symptoms causing a suspicion of this accident arose, was one of abortion, in which an injection was made by means of a clumsy india-rubber bottle. Symptoms like those described as following on air entering the circulation and death ensued. In the face of these catastrophes, we cannot regard the application of the perchloride to the uterus without some misgiving. But there are reassuring considerations. The small, rigid uterus, with an imperfectly dilated os, as in abortion, is different from the large, flaccid uterus, with widely-expanded os, after labour. In abortion, I have already insisted that it is better to swab than to inject. It is further probable that the injection of nævi offers no very close analogy to the washing the inner surface of the uterus. The veins about the face and neck present peculiar facilities to the entry of injected air or fluid. Most of the recorded accidents in this region have occurred during operations where the veins come within the range of the aspiration or suction-force of the chest.

Granting that there is reason for apprehension that the perchloride *may* do mischief, it must not be forgotten that the continued loss of blood *must* be hurtful, probably fatal, and that the other common means for checking hæmorrhage are, as we have seen, not free from danger. The choice is between a certain good alloyed with a possible risk, and a certain evil, ending probably in death. The courage of timidity is scarcely less singular than is its process of reasoning. A woman is dying of hæmorrhage: it dares to let her die, rather than use an unaccustomed remedy.

The injection of the perchloride to arrest hæmorrhage has been followed in my practice:—

1st. In many cases in which flooding persisted in spite of the use of ordinary means, by instant arrest of the bleeding and steady recovery. In a large proportion of these cases, the patients have borne children subsequently.

2ndly. In several cases where profuse bleeding had taken place, it was also instantly arrested; phlegmasia dolens appeared. All recovered well in the end.

3rdly. In several cases where the patients were already *in extremis* from loss of blood, the injection stopped further bleeding; but death ensued. In these cases I was unable to see that the injection in any way contributed to the fatal issue. The patients were moribund; the remedy was used too late.

The commentary which I would offer upon these three classes of cases is, that, of the first class, some women would clearly have died but for the remedy, whilst others were certainly spared an amount of blood which secured them against the severe secondary effects of flooding.

That the supervention of phlegmasia dolens may possibly, in some cases, have been due to the extension of the thrombi formed in the mouths of the uterine sinuses by the action of the perchloride; but that, on the other hand, we cannot regard this thrombotic process as altogether injurious. It is undoubtedly often a conservative process, cutting off from the general circulation noxious matter generated on the surface of the uterus. And, again, it must be borne in mind that phlegmasia dolens is by no means uncommon after severe flooding, where the iron injection is not used; and this is especially true of placenta prævia. Finally, some of these patients would probably have died. It was surely better to run the hazard of phlegmasia dolens.

That, in the last class of cases, those in which death ensued after the injection, the cause being collapse from loss of blood, recovery was not to be expected from anything less than the supply of that which was lost. Here the remedy is transfusion. The practical lesson to be deduced from these unfortunate cases is, not to defer resort to the injection too long. It will, indeed, at almost any stage, stop the bleeding, but it cannot restore the blood which has been lost; it will, indeed, often give the system time and opportunity to rally from the shock it has sustained; but it cannot recall the life that is ebbing away. Although, therefore, the injection of perchloride of iron will, not seldom, rescue women from death by flooding after other means have failed, it is important to use it betimes, before the stage of utter collapse, whilst there is still inherent in the system the power of reaction.

This end will be secured, if the warning I have given, as to the extent to which ergot, compression, and cold should be trusted, be regarded. If these agents be abandoned the moment we observe that the expected response in uterine contraction is not given, we may then count upon the best results from injecting perchloride of iron. One more emphatic reason I would plead in favour of early resort to this treatment is, that it is not enough to save bare life; we must save all the blood we can. For this reason, and having acquired confidence in the efficacy and safety of this crowning remedy, I never now lose time in trying other means that may fail, and which, having failed even partially, leave the patient the worse by the quantity of blood lost. The growth of confidence in a new remedy is often slow; but I do not doubt that this practice will come to be generally adopted.

We have, then, three stages of hæmorrhage to deal with:—

First, there is hæmorrhage with active contractility of the uterus. Here the diastaltic function may be relied upon; excitants of contraction find their application.

Secondly, there is the stage beyond the first, when contractility is seriously impaired, or even lost. Here excitants of contraction are useless. Our reliance must be upon the direct application of styptics to the bleeding surface.

Thirdly, there is the stage beyond the two first, when not only contractility, but all vital force is spent —when no remedy holds out a hope unless it be transfusion; and when even this will probably be too late.

A few words upon the *restorative treatment.* The great physiological principle of *rest* must govern all treatment. The system drained of blood, and therefore barely nourished enough to keep the ordinary functions going on the most reduced scale compatible with life, can ill tolerate any unnecessary call. The first effort must be directed to produce reaction from the stage of collapse. This is especially necessary if the patient have been much prostrated by the persistent use of cold. Since cold is so universally trusted to, and is pushed so far, it requires some firmness to insist upon resort to its opposite, warmth. But unless you would see your patient sink under collapse, you must restore the circulation. To this end, hot bottles should be applied to the feet; wet, cold linen removed, if possible, without unduly disturbing the patient; if not, warm, dry clothes must be wrapped around her; gentle friction of the hands and feet may be used. Cordials and stimulants must be given in small quantities, *warm.* It is the common practice to give everything cold, under the fear that warmth in any form is likely to cause a renewal of the bleeding. But we are now dealing with a state of vital depression greatly caused by cold. You will not cure by making the patient colder. Cordials are, no doubt, increased in efficacy if taken warm. What are the best cordials? The best of all, undoubtedly, is cognac. Give it at first in small sips, diluted with equal parts of hot water. A good mull of claret, and cinnamon or nutmeg, is often useful. Watchfulness is required, lest over-anxiety ply the weak, all but paralyzed stomach too freely. The power of imbibition is very feeble; if you pour in any quantity, it simply accumulates, loads the stomach, until it is forced to reject the whole by vomiting.

It is at this stage, or when the system is rallying, that opium is so valuable. Opium is, in my opinion, decidedly contra-indicated during the flooding. It then tends to relax the uterus. But when the object is to support the system, to allay nervous irritability, there is no remedy like opium. It may be given in doses of thirty or forty drops of laudanum or Battley's Solution, every two or three hours. If the stomach will not bear it, it may be given in an enema. The bowel is also our resource in the administration of cordials and nourishment. An enema of beef-tea and brandy is sometimes extremely useful. It may be worth while to try the subcutaneous injection of $\frac{1}{8}$ to $\frac{1}{4}$ grain of morphia; but this must be done cautiously.

When a little strength has been gathered up, you must think of nutrition. "Little and often" is a good maxim. What you give, if it be absorbed, is soon used up; but if you give much it is lost by vomiting; therefore, the supplies must be frequent and moderate.

When reaction has set in, the heart beats painfully, the temples throb, agonizing headache follows. Then salines, of which the best is the fresh-made acetate of ammonia, are most serviceable.

The opium can be combined with it. Nothing tranquillizes the hæmorrhagic excitation like salines; and it is more than probable that salines supply a pabulum the circulation wants. Salines continue to be useful for many days during what has been aptly called the "hæmorrhagic fever." Ice will now be useful; it quenches thirst, and acts as a sedative. Do not be in a hurry to give iron. If given prematurely, it parches the mucous membranes, increases headache, and fever. In short, it cannot be borne at this stage; it comes in usefully at a more advanced period of convalescence.

In cases of great exhaustion from hæmorrhage, do not suffer the patient to rise on any pretext; linen can be changed without quitting the recumbent posture; and it is better to use the catheter two or three times a day than to incur the risk of sitting up. In fine, all depends upon constant attention to the smallest details, upon good, noiseless, gentle nursing. Guard your patient against all excitement, even from the effort of talking. Bar the hall-door against officious, sympathizing friends. I am confident that more than half the mischief I have seen in childbed might have been traced to bad nursing, to the neglect of these precautions.

Transfusion.—When the patient has sunk to the last extremity from loss of blood, the hope that seldom fails to linger in the human breast under severest trials is still justified by the remarkable recoveries that have followed the operation of transfusion.

This hope will be strengthened by the reflection that the recoveries—resuscitations they may fairly be called—have occurred under the most desperate circumstances. The operation appears terrible to the bystanders; the conditions of success are not yet established; there are practical difficulties attending it; but there seems no reason to doubt that every difficulty may be overcome, and that we shall not long have to regret that the operation occupies an equivocal position amongst the resources of obstetric medicine.

Two great questions to settle are, first, the form in which the renovating fluid shall be prepared; secondly, the apparatus to be used in the proceeding. It is obviously essential that the fluid shall be in a condition to answer the great end of rallying the system, that it shall enter as an harmonious constituent into the body of the patient. It is also essential that the apparatus shall be simple, and of convenient manipulation.

Blood has been used in two different states: first, just as it flows from the vein of the giver; secondly, after it has been deprived of its fibrin. Pure blood has been used in two different ways: namely, by immediate transmission from the vein of the giver to the vein of the receiver; and by mediate transmission, that is, by first collecting the blood in a vessel, and then injecting it into the vein of the receiver by a syringe. I shall presently submit grounds for using, in part, at least, an artificially-prepared fluid.

I must limit myself here to a condensed account of what will serve as a guide in practice. Let us first try to discover what is *the best renovating fluid.* It seems most natural to conclude that pure blood would be the best. Many cases are on record in which the transfusion of pure blood has been attended with complete success. One objection to it is its tendency to coagulate. This property has often baffled the operator, by plugging the apparatus or the vein near the point of injection; and sometimes by the blood clotting in the heart. The blood of many animals begins to coagulate almost immediately on escaping from the vein. Human blood sometimes begins to coagulate in a minute, and within a few minutes, four or five, there is no security against this event. Can coagulation be in any way averted? Oré shows that cold, and preventing contact of air, retard coagulation. Müller, Dieffenbach, Bischoff, and others, overcome coagulation by depriving the blood of its fibrin. In about fifteen cases, the injection of defibrinated blood has failed. In one case, related by Dr. Polli, and in one by Nussbaum, it succeeded. It is established by experiments that the fibrin is not essential to success; it is also established that the globules are enough (Oré); but Panum contends that the serum contains a vivifying principle. Dr. Richardson, in his admirable prize essay on the "Cause of Coagulation of the Blood," proved that ammonia prevented fibrillation, and he used it for the purpose in transfusion. Dr. Hicks * used phosphate of soda for the same purpose. This seems better than defibrination.

Lastly, the numerous observations of Dr. Little, Mr. L. S. Little, and others, prove that persons *in extremis* from cholera-collapse may recover by the injection of saline alcoholic fluid alone.

The choice of the renovating fluid will, to a great extent, determine the choice of the apparatus for injecting. In many of the experiments on animals, immediate transfusion has been effected by a tube having a canula at each end, so as to carry the blood direct from the giver to the receiver.

On the other hand, almost all those who have practised the transfusion either of pure or defibrinated blood on the human subject, have used some form of injecting apparatus, previously collecting the blood in a reservoir, generally a funnel communicating directly with the pipe of the syringe, so that the blood is propelled partly by gravitation. An excellent instrument of this kind is Mr. Higginson's, who, by it, saved two lives out of seven. (*See* "Liverpool Med. Chir.

* "Guy's Reports," 1869.

Journal," 1857.) To this there is no piston; gravitation is aided by squeezing the barrel, which is made of vulcanized india-rubber. If a mediate injecting apparatus be determined upon, that of Dr. Little, figured and described in the "London Hospital Reports," vol. iii., 1866, may be used. But a better one still has been shown me by that accomplished physiologist, Dr. Richardson. He also looks favourably upon immediate transfusion from vein to vein, using a simple elastic tube with a suitable canula at either end. But he recommends mediate transfusion as preferable. He uses a glass syringe, holding about eight ounces, to which is attached an elastic delivery tube. The syringe has a piston with a jointed rod. Where the rod is united to the piston, this is perforated, so that on pulling up the rod fluid will run through, getting below. On pushing the rod down, this opening is closed, and the fluid is propelled. A little water is first put in the syringe. The blood is received direct from the giver into the syringe, the jointed rod and perforated piston admitting of this without any difficulty. Coagulation is prevented by adding about three drops of ammonia to every ounce of blood. This enables the operator to proceed leisurely and without fear of obstruction.

If it be determined to use pure blood, the plan of immediate transfusion advocated by Dr. Aveling* is worthy of attention for the following cogent reasons:—

The chances of coagulation are small, because the blood glides through the pipe, and comes in contact only with a thin coating of coagulated blood, and

* "Obstetrical Transactions," vol. vi., 1865.

because it is removed from the living vessels for only a few seconds during its transit from the vein of the giver to that of the receiver, being never exposed to the air. His apparatus is remarkably simple. It may be described as a small Higginson's syringe without valves, and armed at each end with a silver canula to enter the vessels. It is, in truth, a continuous tube with a dilatation in the middle. This dilated part holds two drachms. It may be used to propel the blood, if flowing slowly. By pinching the tube on either side of the expanded portion, valves are dispensed with. If a bit of glass tube be interpolated, the operator can see if the blood is flowing. In animals, Dr. Aveling says, the apparatus answers perfectly. It has been doubted whether a sufficient supply could be drawn by it from the human giver. In support of Aveling's apparatus may be cited the numerous experiments of Richardson and Oré. Oré, after many trials and modifications, finished by contriving an apparatus essentially similar to Aveling's. It is, however, furnished with valves and a stopcock, which Aveling shows to be an unnecessary complication. Air may be got rid of by first pumping water through it, or, better still, a saline solution.

In using defibrinated blood, the blood is first received into a basin (not less than eight ounces should be collected); then it is stirred round rapidly with a stick, to let the fibrin cohere and be removed; it should then be filtered through fine muslin, to stop any small coagula which might be dispersed in the fluid. These minute clots might otherwise act as emboli, and prove injurious.

Of those who inject saline fluids, Dr. Little uses

the gravitation injecting syringe described; Mr. Little uses an apparatus in which hydraulic pressure gives the propelling force; Dr. Woodman and Mr. Heckford used an ordinary Higginson's syringe. One of Dr. Woodman's three cases made a good recovery.

On reviewing all the facts connected with the problem to be solved, we may hold the following propositions to be established:—

1. The life of a person on the point of dying from loss of blood may be restored by injecting into his veins blood taken from another person.

2. The blood used for the purpose may be either whole or defibrinated.

3. A person seemingly about to die from cholera-collapse may be restored by the injection of a saline alcoholic fluid.

4. That a remarkably small quantity of new blood is required; sometimes two ounces has been enough; four to six ounces need rarely be exceeded.

5. That in using saline alcoholic solutions, a greater quantity, that is, from one to three pints, is required; also that re-injection is sometimes required.

My personal experience of transfusion is limited. In a case in which I was consulted by Dr. Fowler, Mr. Hutchinson injected by means of a syringe about three or four ounces of blood into a woman *in extremis* from hæmorrhage. She rallied obviously, but died in about six hours.

I should now practise, and recommend to others, the following plans:—

1st. If suitable blood can be procured, I would prefer Richardson's plan, or immediate transfusion by Aveling's apparatus.

The operation.—1. *To prepare the apparatus.*—Pump through it Mr. Little's solution, composed of—

Chloride of sodium	. .	60 grains.
Chloride of potassium	. .	6 ,,
Phosphate of soda	. .	3 ,,
Carbonate of soda	. .	20 ,,
Distilled water	. . .	20 ounces.

If these ingredients cannot be procured, a suitable saline may be improvised by dissolving a teaspoonful of common salt and half the quantity of carbonate of soda in a pint of water, at the temperature of 98° to 100° Fahr., or a solution of phosphate of soda alone, or water with a few drops of ammonia in it.

The apparatus, after pumping through it some of the saline solution a few times, should be kept in the solution full.

2. *To prepare the patient.*—Expose a vein at the bend of the elbow, and pass a probe under it.

3. *To procure the blood.*—Open a vein at the bend of the elbow of the giver; insert the nozzle of the tube which is full of saline solution; then open by a longitudinal slit the exposed vein of the patient, and complete the communication by inserting the nozzle of the other end of the tube.

Then compress the inflated portion of the apparatus gently, to propel its saline contents into the vein of the patient. When allowed to expand, the tube on the patient's side of the apparatus must be pinched to close it; it will then take in blood from the giver's side; this done, pinch the giver's side of the tube, and propel gently onwards towards the patient. This process should be continued until signs of amendment are observed in the patient, or until about four to six ounces of blood have been injected.

In this mode of proceeding we have in the saline fluid an additional security against fibrillation; at the same time, additional security against the entry of air; and, further, what is almost certainly an advantage, a quantity of saline fluid injected as well as blood. A reservoir containing saline fluid might be so adapted to the tube as to give a further supply to mix with the blood as desired; or we might throw into the barrel or expanded portion of the tube a drop or two of ammonia by means of a hair-trocar or a subcutaneous injecting syringe, from time to time, so as to ensure liquidity of the blood.

But it will not rarely happen that to procure blood is impossible, or entails too much loss of time. In such cases I would strenuously urge the expediency of giving the sinking woman the chance of recovery promised by saline alcoholic injections. If such injections will restore a patient dying from a disease, which to extreme draining of the circulating fluid adds a dire poisonous influence, it may fairly be expected that they will act not less effectually where there is nothing but the loss of blood to contend with.

The method of saline injections has the incontestable advantages of being always available at short notice; of avoiding all danger of failure from coagulation; of enabling us to use simple forms of apparatus; and of being a more simple proceeding.

If, then, it be determined to employ saline injections, we have first *to prepare the fluid.* This may be either Mr. Little's fluid already described, adding two drachms of pure alcohol to the pint, or a similar solution, omitting the chloride of potassium. Three or four pints of this may be prepared in a basin, and kept at a temperature of 98° to 100°.

The apparatus may be Mr. Little's, figured in the "London Hospital Reports," 1866.

With this apparatus, the whole operation can be performed by one person. The vessel holds forty ounces; a lamp underneath and a thermometer in it regulate the temperature; from the tap near the bottom proceeds a thick india-rubber tube four feet long, with a silver nozzle at the end. When this instrument was placed at the bedside, about on a level with the patient's head, and the nozzle inserted into a vein at the bend of the elbow, its contents flowed into the vein in about ten minutes. The instrument might be made to contain eighty ounces. In Mr. Little's more successful cases, eighty ounces were introduced at a time, generally in between twenty minutes and half-an-hour.

Such an apparatus cannot conveniently enter into the obstetric bag, but it might very well be a part of the armamentarium of lying-in hospitals.

But a small Higginson's syringe armed with an appropriate nozzle, as used by Dr. Woodman and Mr. Heckford, should be carried for emergencies of this kind; or Aveling's little instrument might serve for the saline injections as well as for immediate transfusion. Dr. Woodman informs me that he examined the blood-globules in a patient who had undergone saline injection; he found them quite unaltered. Pure water, we know, will affect the globules.

The indications for the operation.—The quantity of blood lost is no criterion. We must look to the state of the system. Absence of pulse, sinking of the features, gasping for air, jactitation, unwillingness to be disturbed with a view to treatment, are signs of threatening dissolution. It should be our endeavour to act before these are far advanced. We shall come by-and-by, I have little doubt, to regard transfusion and venous injection as much less formidable operations than they now seem; we shall then take courage to act earlier, before the case is desperate, and the results will be proportionally better.

As to re-injection.—If the patient flag again, showing symptoms of sinking, the operation should be repeated.

Transfusion has been performed before delivery in cases of placenta prævia, where the patient was too prostrated to be able to survive delivery unless previously rallied. Many patients perish under the attempt to deliver them when in a state of extreme depression. If rallied first, they may bear artificial delivery, and recover.

In cases of hæmorrhage during and after labour, it has been apprehended that the new blood transfused might again escape on the uterine surface. This source of failure may, I think, be overcome by the local application of perchloride of iron. The two methods of arrest by styptics, and of replenishment by transfusion, should be thoughtfully studied together. In most cases, the timely use of the styptic will render resort to transfusion unnecessary; but in some we should be prepared to follow up the styptic by transfusion.

I would further urge the expediency of extending the application of transfusion of blood, or of the injection of saline fluid, to other conditions of great emergency as well as to hæmorrhage. In two conditions, the indication is especially strong. I mean puerperal convulsions and extreme prostration from uncontrollable vomiting in pregnancy. In convulsions with albuminuria, the blood is undoubtedly poisoned; most men of experience in this disease still recognise that venesection is beneficial; that the benefit thus derived is partly due to the abstraction of a portion of the poison, it seems reasonable to believe; and there seems to be ground for hope that the injection of healthy blood or saline fluids, to replace the poisoned blood abstracted, may be useful in diluting the poison still circulating and in rallying the system from the prostration that threatens to be fatal. I believe one successful case of this kind has been published.

With regard to the application in cases of extreme exhaustion from obstinate vomiting, I have even less hesitation. Here there is undoubtedly not only a high degree of anæmia, but the little blood circulating is degraded by admixture with the products of diseased action. The source of nutrition is stopped up; there is no other door open for the admission of fresh supplies than through the veins.

After-treatment.—When transfusion has been performed, great care is still required to economise and to increase the little store of heat and strength the system may retain. Take care that warm dry linen surround the patient's body; apply hot bottles to the feet, legs, and sides of the body; when opportunity offers, throw an enema of a pint of warm gruel or beef-tea, containing an ounce or two of brandy, into the rectum. When she can swallow, give small quantities of hot brandy-and-water.

Is it not strange that in these days, when blood is looked upon as so precious, when venesection is almost gone out of use, when the dread of losing a pint of blood by an operation is too often an obstacle to relief, that restoring this vital fluid to a patient dying for the want of it, should be still struggling for a settled place in obstetric medicine?

I think it ought to be an accepted aphorism in medicine, *that no one should be permitted to die of hæmorrhage.* We should never rest until the means of preventing dangerous losses of blood, and of restoring those who are in danger from such losses, shall be brought to perfection.

Those who may wish for complete historical information on this deeply-interesting subject, and for detailed experiments and observations, should consult especially Dr. Little's Hunterian Oration, 1852; Blundell's Memoir, "Med. Chir. Trans.," 1818; Dr. Waller's Memoir, "Obstetr. Trans.," 1860; Prof. Ed. Martin, Berlin, 1859; Prof. Panum, "Embolie, Transfusion, und Blutmenge," Berlin, 1864; Dr. Oré's "Études sur la Transfusion du Sang," Paris, 1868; Dr. Routh, *Medical Times*, vol. xx.; and Dr. Richardson's "Essay on the Cause of the Coagulation of the Blood."

Secondary Puerperal Hæmorrhage.

When the immediate perils of childbirth seem to have been weathered; when the patient has been deemed safe from hæmorrhage; at any rate, at a period more or less remote from labour, varying from a few hours to two or three weeks, this terrible complication may yet make its appearance. If we have taken due care in the conduct of the labour, and especially in the management of the placenta and any primary hæmorrhage, we shall rarely experience the mortification of seeing secondary hæmorrhage. This will be evident if we examine the principal ascertained causes of this catastrophe. They are :—

A. *Local.*—1. A portion of placenta or membranes may have remained *in utero.*

2. Clots of blood may have formed and been retained.

3. Laceration or abrasion of the cervix, vagina, or perinæum, or a vesico-vaginal or recto-vaginal fistula.

4. Hæmatocele or thrombus of the cervix, vagina, vulva, or perinæum.

5. Chronic hypertrophy, congestion or ulceration of the cervix uteri.

6. General relaxation of the uterine tissues.

7. Fibroid tumours and polypi.

8. Inversion of the uterus.

9. Retroflexion of the uterus.

B. — *Constitutional or remote* conditions causing disturbance of the vascular system.

1. Emotions.

2. Sexual intercourse.

3. Heart disease, including imperfect involution.

4. Liver disease.

5. General debility of tissue, malnutrition of nervous system, and irritable heart from anæmia.

A. 1. *Retention of placenta.*—This is perhaps the most common cause. It is that which I have most frequently met with; it is that which may be surely avoided. Since it has become a recognised practice not to allow the placenta to remain more than an hour, it must be very rare to find secondary hæmorrhage occurring from retention of the entire placenta, or even the greater part of it. But I have been called in to such cases. In these the flooding has usually come on within twenty-four hours of labour. The explanation has been that rigidity of the cervix uteri rendered it impossible to get the placenta away. I have removed the whole placenta several days, even a week after labour. In all these cases there was not only flooding—indeed, flooding was not always the urgent symptom—but there were unmistakeable marks of septicæmia; a pallid straw-coloured complexion, with hectic flush; rigors; quick pulse, running over 100, to 120 or 130; thirst; distressing perspirations alternating with hot skin; vomiting; general prostration. The local signs were: some distension of the abdomen; tenderness on pressure over the uterus, which could be felt rising perhaps as high as the umbilicus; on examination by vagina, the uterus was felt enlarged; the cervix more or less open, at least admitting one or two fingers, and on withdrawing the finger, it was found smeared with muco-purulent sanguineous fluid generally of peculiarly offensive odour.

These symptoms are enough to justify, even to demand, exploration of the cavity of the uterus. The tenderness of the uterus and abdomen and the partial closure of the cervix render the introduction of the hand, or even a finger, a very distressing operation. You may sometimes acquire information to guide further proceedings by passing in the uterine sound. This will give you the exact measure of the cavity; and you may be able by it to make out the presence of a foreign body in the cavity. If the sound goes four inches or more beyond the os, you may conclude that the uterus is of abnormal size, and that it contains something solid. If you perceive any fluid issuing from it, or feel any substance in it, you must make up your mind to pass your finger fairly into the cavity, so as to explore fully, and to bring away whatever is in it.

It will be next to impossible to remove an entire placenta without passing your hand into the uterus; and this is an operation of no small difficulty. If the cervix be so contracted as not to admit more than a finger, it will be wise to introduce a faggot of three or four large laminaria tents, and leave the patient for a few hours to let them expand. The next step may be to introduce the medium or largest sized caoutchouc bag, and distend with water until there is room to pass three or four fingers. When you can do this, by steady pressure, supporting and pushing the fundus down upon the inside hand by the hand outside, you will succeed in getting your left hand into the cavity. This done, take care to grasp well the whole mass of the placenta and withdraw it. When you have got it out, you will be glad to get rid of it as quickly as possible. There is perhaps no stench more offensive than that of a placenta that has decomposed *in utero*.

This being so, what must be the condition of the uterus itself? It clearly wants disinfecting. It should be washed out with chlorine water, Condy's fluid, weak solution of carbolic acid, creosote, or perchloride of iron; and if none of these are at hand, with tepid water. The uterine tube should be carried fairly into the uterus. The operation should be repeated two or three times a day for some days. An excellent plan is to irrigate the parts by means of such an apparatus as that contrived by Dr. Rasch. The fluid is carried into the uterus by a syphon, and runs out by a discharge-tube, which empties itself into a vessel by the side of the bed.

More common is the retention of a portion of placenta. This may be either adherent or loose. It is more likely to have adhered at the time of the removal of the bulk of the placenta. In this case the uterus remains large; there is constantly tenderness on pressure; often there is an offensive discharge attending or alternating with hæmorrhage.

The same proceedings must be adopted as in the case of retention of the whole placenta. But there may be great difficulty in separating all the adherent portions from the surface of the uterus. You must take pains to break off all that will readily come, carefully avoiding digging your nails into the substance of the uterus. It often answers, as in the case of early abortion, to break up any adherent masses, if you cannot detach them. It is far better to be content with this than to err by excess of pertinacity. If a portion project at all prominently in a polypoid form, I strongly advise removing it by the wire-écraseur, which does it cleanly, and without injuring the uterus. Having done your best in removing *débris*, wash out with disinfecting fluid.

2. A *blood-clot* may form in the uterine cavity from imperfect contraction after labour; it grows by receiving accretions on its surface, and being com-

pressed by the uterus, it becomes condensed into a mass having the shape of the cavity. This acts as a foreign body. It excites spasmodic action; it attracts blood to the uterus; and the congestion is relieved by oozing of blood and serum from the uterine wall. Some of this escapes externally, but not always in sufficient quantity to arrest attention as "flooding." The napkins are sodden with a serous fluid streaked with blood, and often somewhat offensive to the smell.

Sometimes, when the uterus remains flaccid, the clot is not moulded and compressed, but more blood collects, and forms a number of dark, loose clots, mixed with fluid blood.

In either case the uterus must be emptied; and this is generally done by passing the fingers or hand into the cavity. Here, again, irrigation by disinfectants is advisable.

3. *Laceration or abrasion of the cervix* is more frequent than is commonly suspected. Hæmorrhage from this cause is sometimes protracted and serious in quantity. We arrive at the diagnosis by a method of exclusion. If we find the body of the uterus firmly contracted, of size corresponding to the date after delivery—a point accurately determinable by the introduction of the uterine sound—our attention is fixed upon the state of the cervix. A rent here may then be felt, and even seen by the speculum. The blood is usually florid. The treatment is to apply a solution of perchloride of iron to the wound by a swab, carried through the speculum, or on pledgets of lint, in the form of a compress or plug.

Laceration of the Vagina or Perinæum.

Any part of the vagina may be torn during labour. Laceration of the upper part is, commonly, an extension of rupture of the lower part of the uterus; and symptoms indicating the gravity of the injury will be apparent at the time of labour. Laceration of the middle part of the vagina is rare; but laceration of the lower part, merging in rent of the perinæum, is common. This may become a source of secondary bleeding. Digital and visual examination will reveal it. If not discovered for some days, the better course will be to stop the bleeding by compresses steeped in perchloride of iron, if necessary, and to leave restoration to future consideration.

A little bleeding may also continue from vesico-vaginal or recto-vaginal rents or fistulæ; but the quantity is rarely enough to cause anxiety as hæmorrhage.

4. Sometimes a collection of blood forms gradually after labour under the mucous membrane of the cervix uteri, some part of the vagina, or of the perinæum or vulva. This is the *thrombus or hœmatocele* of the part concerned. It is formed in one of two ways—the more common is from friction during the passage of the head. The head, advancing with difficulty, carries the mucous membrane with which it is in contact down with it, sliding it along on the moveable submucous stratum, and even tearing it partly away. In this process, numerous blood-vessels are necessarily torn. These pour out blood sometimes quickly; and the thrombus may be fully formed during or immediately after labour. Sometimes, however, the

blood collects very slowly, so that four or five days may elapse before the tumour attains its full size. It may then burst, and a hæmorrhage more or less profuse will be observed.

The other way in which these thrombi are caused is through the intense congestion arising below the head, leading to extravasation from obstruction to the return of blood. This cause chiefly produces the thrombi of the vulva and perinæum. Those of the cervix, I believe, are mostly caused by friction, as above described.

It is admitted as a general rule that these thrombi should not be opened at once; but after a few days, when they have attained their full size, they may be punctured, if they do not burst. Then *compression* is the best remedy; it may be applied by compresses or plugging.

During their formation, a bladder filled with ice applied inside the vagina may be useful in checking the effusion.

5. *Chronic hypertrophy, congestion or ulceration*, will be detected by touch and by speculum. The treatment will be to touch the parts lightly with the solid nitrate of silver every third day. Lotions of lead or sulphate of zinc are also useful.

6. *General relaxation of the uterine* tissues is mostly associated with systemic debility and malnutrition. Constitutional treatment is here especially serviceable. Iron, strychnine, phosphoric acid, cinchona, are particularly valuable.

7. The complication of *fibroid tumours and polypi* has been discussed in the last lecture. If they be first discovered some days after labour, the treatment

is still the same as that recommended when found at the time of labour. In the case of fibroid tumours, we can only try to restrain hæmorrhage by the topical application of perchloride of iron. Polypi should be removed by the wire-écraseur.

8. *Inversion* has been discussed in the preceding lecture.

9. *Retroflexion of the uterus* is, in my experience, a frequent cause of secondary hæmorrhage. The displacement no doubt occurs soon after the labour, the heavy fundus falling backwards, whilst the tissues are still in a relaxed state. In some cases, I ascertained that there had been chronic retroflexion in the non-pregnant state. It is highly probable that there has frequently been antecedent retroflexion, and also, that occurring for the first time after a labour, the condition becomes permanent. When it occurs, the free return of blood from the body of the uterus is impeded by the flexion at the neck, the involution of the fundus and body is arrested, these parts become congested and relaxed, the body is found bulky, and the pressure upon the pelvic organs no doubt also favours local hyperæmia.

The diagnosis is made out by the finger in the vagina feeling the rounded mass of the fundus of the uterus behind the os uteri, bulging downwards and forwards the posterior and upper part of the vagina; by the finger in the rectum, which determines the rounded mass of the fundus even more accurately; by the finger in the vagina passing up in front of the os uteri to meet the hand pressed down from above the symphysis, revealing the absence of the uterus between them; and still more absolutely, by the uterine sound, the point of which must be turned back to enter the body of the uterus.

The treatment is first to restore the uterus. This may be done by the sound at once, aided or not by the pressure of the finger in the vagina or rectum below on the fundus. It may be maintained in position by an air pessary placed in the fundus of the vagina, or better still, by a large Hodge's pessary.

If bleeding continue after restoration, the interior of the uterus should be swabbed with perchloride of iron. The constringing effect of this application, by lessening the bulk of the fundus, tends still further to correct the retroflexion. At a later stage, the insertion of a five-grain stick of sulphate of zinc inside the uterus every third day will much conduce to bring about a healthy condition.

B. The management of secondary hæmorrhage depending upon the constitutional or remote causes must obviously consist in avoiding or lessening the influence of those causes, and in pursuing the treatment indicated for the cure or mitigation of the diseases disposing to uterine hæmorrhage.

INDEX OF AUTHORITIES.

INDEX OF SUBJECTS.

LONDON: BENJAMIN PARDON AND SON, PRINTERS, PATERNOSTER ROW

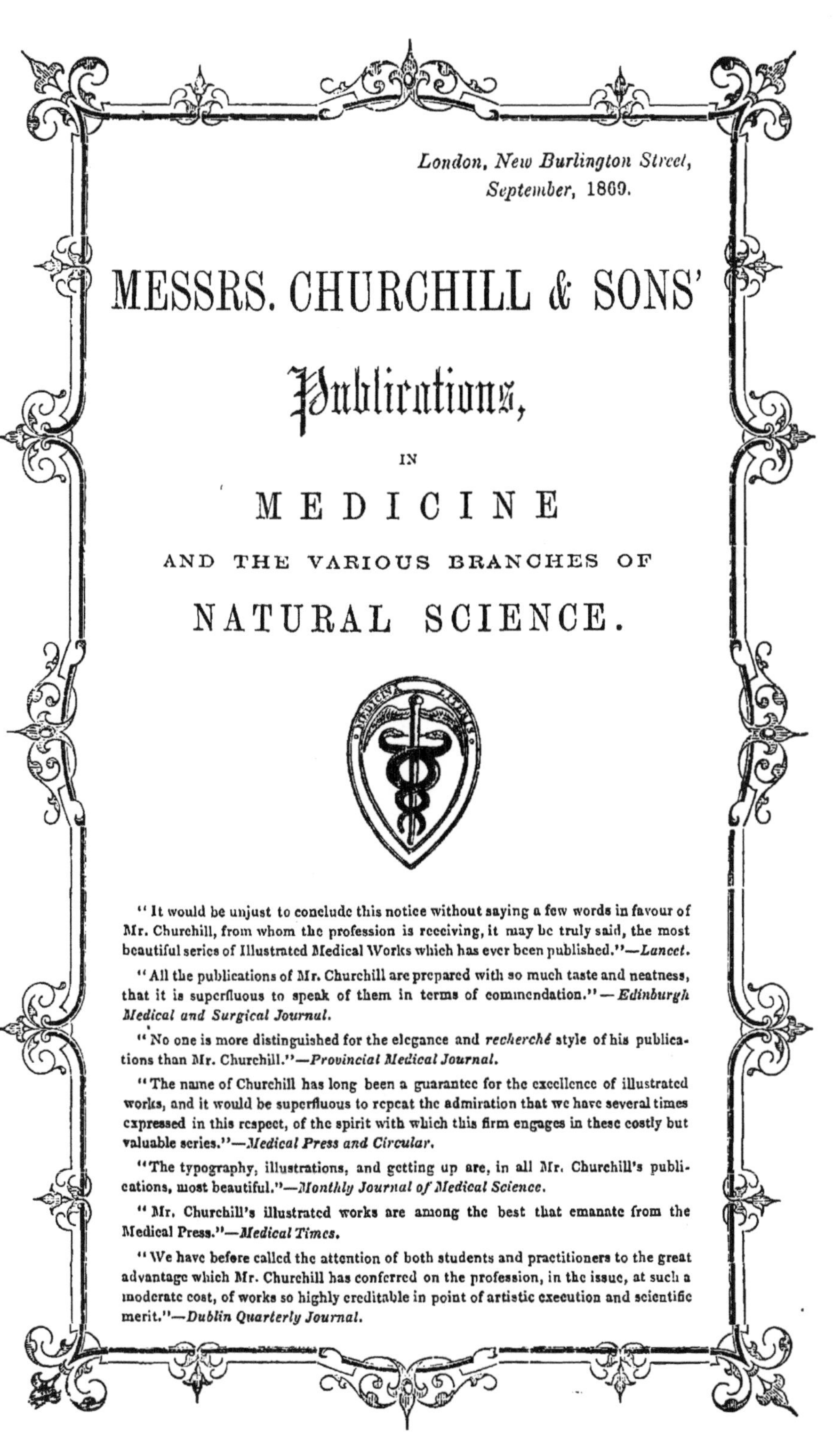

London, New Burlington Street,
September, 1869.

MESSRS. CHURCHILL & SONS'

Publications,

IN

MEDICINE

AND THE VARIOUS BRANCHES OF

NATURAL SCIENCE.

"It would be unjust to conclude this notice without saying a few words in favour of Mr. Churchill, from whom the profession is receiving, it may be truly said, the most beautiful series of Illustrated Medical Works which has ever been published."—*Lancet.*

"All the publications of Mr. Churchill are prepared with so much taste and neatness, that it is superfluous to speak of them in terms of commendation."—*Edinburgh Medical and Surgical Journal.*

"No one is more distinguished for the elegance and *recherché* style of his publications than Mr. Churchill."—*Provincial Medical Journal.*

"The name of Churchill has long been a guarantee for the excellence of illustrated works, and it would be superfluous to repeat the admiration that we have several times expressed in this respect, of the spirit with which this firm engages in these costly but valuable series."—*Medical Press and Circular.*

"The typography, illustrations, and getting up are, in all Mr. Churchill's publications, most beautiful."—*Monthly Journal of Medical Science.*

"Mr. Churchill's illustrated works are among the best that emanate from the Medical Press."—*Medical Times.*

"We have before called the attention of both students and practitioners to the great advantage which Mr. Churchill has conferred on the profession, in the issue, at such a moderate cost, of works so highly creditable in point of artistic execution and scientific merit."—*Dublin Quarterly Journal.*

A CLASSIFIED INDEX

TO

MESSRS. CHURCHILL & SONS' CATALOGUE.

MEDICINE—*continued.*

MICROSCOPE.

MISCELLANEOUS.

MISCELLANEOUS—*cont^d.*

NERVOUS DISORDERS AND INDIGESTION.

OBSTETRICS.

OPHTHALMOLOGY.

OPHTHALMOLOGY—*cont^d.*

PHYSIOLOGY.

PSYCHOLOGY.

PULMONARY and CHEST DISEASES, &c.

RENAL and URINARY DISEASES.

SCIENCE.

SURGERY.

SURGERY—*continued.*

VETERINARY MEDICINE.

MR. ACTON, M.R.C.S.

I.

A PRACTICAL TREATISE ON DISEASES OF THE URINARY AND GENERATIVE ORGANS IN BOTH SEXES. Third Edition. 8vo. cloth, £1. 1s. With Plates, £1. 11s. 6d. The Plates alone, limp cloth, 10s. 6d.

II.

THE FUNCTIONS AND DISORDERS OF THE REPRODUC-TIVE ORGANS IN CHILDHOOD, YOUTH, ADULT AGE, AND ADVANCED LIFE, considered in their Physiological, Social, and Moral Relations. Fourth Edition. 8vo. cloth, 10s. 6d.

III.

PROSTITUTION: Considered in its Moral, Social, and Sanitary Bearings, with a View to its Amelioration and Regulation. 8vo. cloth, 10s. 6d.

DR. ADAMS, A.M.

A TREATISE ON RHEUMATIC GOUT; OR, CHRONIC RHEUMATIC ARTHRITIS. 8vo. cloth, with a Quarto Atlas of Plates, 21s.

MR. WILLIAM ADAMS, F.R.C.S.

I.

ON THE PATHOLOGY AND TREATMENT OF LATERAL AND OTHER FORMS OF CURVATURE OF THE SPINE. With Plates. 8vo. cloth, 10s. 6d.

II.

CLUBFOOT: its Causes, Pathology, and Treatment. Jacksonian Prize Essay for 1864. With 100 Engravings. 8vo. cloth, 12s.

III.

ON THE REPARATIVE PROCESS IN HUMAN TENDONS AFTER SUBCUTANEOUS DIVISION FOR THE CURE OF DEFORMITIES. With Plates. 8vo. cloth, 6s.

IV.

SKETCH OF THE PRINCIPLES AND PRACTICE OF SUBCUTANEOUS SURGERY. 8vo. cloth, 2s. 6d.

DR. WILLIAM ADDISON, F.R.S.

I.

CELL THERAPEUTICS. 8vo. cloth, 4s.

II.

ON HEALTHY AND DISEASED STRUCTURE, AND THE TRUE PRINCIPLES OF TREATMENT FOR THE CURE OF DISEASE, ESPECIALLY CONSUMPTION AND SCROFULA, founded on MICROSCOPICAL ANALYSIS. 8vo. cloth, 12s.

DR. ALDIS.

AN INTRODUCTION TO HOSPITAL PRACTICE IN VARIOUS COMPLAINTS; with Remarks on their Pathology and Treatment. 8vo. cloth, 5s. 6d.

DR. SOMERVILLE SCOTT ALISON, M.D.EDIN., F.R.C.P.

THE PHYSICAL EXAMINATION OF THE CHEST IN PUL-MONARY CONSUMPTION, AND ITS INTERCURRENT DISEASES. With Engravings. 8vo. cloth, 12s.

DR. ALTHAUS, M.D., M.R.C.P.

ON EPILEPSY, HYSTERIA, AND ATAXY. Cr. 8vo. cloth, 4*s.*

THE ANATOMICAL REMEMBRANCER; OR, COMPLETE POCKET ANATOMIST. Sixth Edition, carefully Revised. 32mo. cloth, 3*s.* 6*d.*

DR. MCCALL ANDERSON, M.D.

I.

THE PARASITIC AFFECTIONS OF THE SKIN. Second Edition. With Engravings. 8vo. cloth, 7*s.* 6*d.*

II.

ECZEMA. Second Edition. 8vo. cloth, 6*s.*

III.

PSORIASIS AND LEPRA. With Chromo-lithograph. 8vo. cloth, 5*s.*

DR. ANDREW ANDERSON, M.D.

TEN LECTURES INTRODUCTORY TO THE STUDY OF FEVER. Post 8vo. cloth, 5*s.*

DR. ARLIDGE.

ON THE STATE OF LUNACY AND THE LEGAL PROVISION FOR THE INSANE; with Observations on the Construction and Organisation of Asylums. 8vo. cloth, 7*s.*

DR. ALEXANDER ARMSTRONG, R.N.

OBSERVATIONS ON NAVAL HYGIENE AND SCURVY. More particularly as the latter appeared during a Polar Voyage. 8vo. cloth, 5*s.*

MR. T. J. ASHTON.

I.

ON THE DISEASES, INJURIES, AND MALFORMATIONS OF THE RECTUM AND ANUS. Fourth Edition. 8vo. cloth, 8*s.*

II.

PROLAPSUS, FISTULA IN ANO, AND HÆMORRHOIDAL AFFECTIONS; their Pathology and Treatment. Second Edition. Post 8vo. cloth 2*s.* 6*d.*

MR. THOS. J. AUSTIN, M.R.C.S.ENG.

A PRACTICAL ACCOUNT OF GENERAL PARALYSIS: Its Mental and Physical Symptoms, Statistics, Causes, Seat, and Treatment. 8vo. cloth, 6*s.*

DR. THOMAS BALLARD, M.D.

A NEW AND RATIONAL EXPLANATION OF THE DISEASES PECULIAR TO INFANTS AND MOTHERS; with obvious Suggestions for their Prevention and Cure. Post 8vo. cloth, 4*s.* 6*d.*

MR. BEASLEY.

I.

THE BOOK OF PRESCRIPTIONS; containing 3000 Prescriptions. Collected from the Practice of the most eminent Physicians and Surgeons, English and Foreign. Third Edition. 18mo. cloth, 6*s.*

II.

THE DRUGGIST'S GENERAL RECEIPT-BOOK: comprising a copious Veterinary Formulary and Table of Veterinary Materia Medica; Patent and Proprietary Medicines, Druggists' Nostrums, &c.; Perfumery, Skin Cosmetics, Hair Cosmetics, and Teeth Cosmetics; Beverages, Dietetic Articles, and Condiments; Trade Chemicals, Miscellaneous Preparations and Compounds used in the Arts, &c.; with useful Memoranda and Tables. Sixth Edition. 18mo. cloth, 6*s.*

III.

THE POCKET FORMULARY AND SYNOPSIS OF THE BRITISH AND FOREIGN PHARMACOPŒIAS; comprising standard and approved Formulæ for the Preparations and Compounds employed in Medical Practice. Eighth Edition, corrected and enlarged. 18mo. cloth, 6*s.*

DR. HENRY BENNET.

I.

A PRACTICAL TREATISE ON UTERINE DISEASES. Fourth Edition, revised, with Additions. 8vo. cloth, 16*s.*

II.

WINTER IN THE SOUTH OF EUROPE; OR, MENTONE, THE RIVIERA, CORSICA, SICILY, AND BIARRITZ, AS WINTER CLIMATES. Third Edition, with numerous Plates, Maps, and Wood Engravings. Post 8vo. cloth, 10*s.* 6*d.*

PROFESSOR BENTLEY, F.L.S.

A MANUAL OF BOTANY. With nearly 1,200 Engravings on Wood. Fcap. 8vo. cloth, 12*s.* 6*d.*

DR. BERNAYS.

NOTES FOR STUDENTS IN CHEMISTRY; being a Syllabus compiled from the Manuals of Miller, Fownes, Berzelius, Gerhardt, Gorup-Besanez, &c. Fourth Edition. Fcap. 8vo. cloth, 3*s.*

MR. HENRY HEATHER BIGG.

ORTHOPRAXY: a complete Guide to the Modern Treatment of Deformities by Mechanical Appliances. With 300 Engravings. Second Edition. Post 8vo. cloth, 10*s.*

DR. S. B. BIRCH, M.D., M.R.C.P.

I.

OXYGEN: ITS ACTION, USE, AND VALUE IN THE TREATMENT OF VARIOUS DISEASES OTHERWISE INCURABLE OR VERY INTRACTABLE. Second Edition. Post 8vo. cloth, 3*s.* 6*d.*

II.

CONSTIPATED BOWELS: the Various Causes and the Different Means of Cure. Third Edition. Post 8vo. cloth, 3*s.* 6*d.*

DR. GOLDING BIRD, F.R.S.

URINARY DEPOSITS; THEIR DIAGNOSIS, PATHOLOGY, AND THERAPEUTICAL INDICATIONS. With Engravings. Fifth Edition. Edited by E. LLOYD BIRKETT, M.D. Post 8vo. cloth, 10s. 6d.

MR. BISHOP, F.R.S.

I.

ON DEFORMITIES OF THE HUMAN BODY, their Pathology and Treatment. With Engravings on Wood. 8vo. cloth, 10s.

II.

ON ARTICULATE SOUNDS, AND ON THE CAUSES AND CURE OF IMPEDIMENTS OF SPEECH. 8vo. cloth, 4s.

MR. BLAINE.

OUTLINES OF THE VETERINARY ART; OR, A TREATISE ON THE ANATOMY, PHYSIOLOGY, AND DISEASES OF THE HORSE, NEAT CATTLE, AND SHEEP. Seventh Edition. By Charles Steel, M.R.C.V.S.L. With Plates. 8vo. cloth, 18s.

MR. BLOXAM.

I.

CHEMISTRY, INORGANIC AND ORGANIC; with Experiments and a Comparison of Equivalent and Molecular Formulæ. With 276 Engravings on Wood. 8vo. cloth, 16s.

II.

LABORATORY TEACHING; OR PROGRESSIVE EXERCISES IN PRACTICAL CHEMISTRY. With 89 Engravings. Crown, 8vo. cloth, 5s. 6d.

DR. BOURGUIGNON.

ON THE CATTLE PLAGUE; OR, CONTAGIOUS TYPHUS IN HORNED CATTLE: its History, Origin, Description, and Treatment. Post 8vo. 5s.

MR. JOHN E. BOWMAN, & MR. C. L. BLOXAM.

I.

PRACTICAL CHEMISTRY, including Analysis. With numerous Illustrations on Wood. Fifth Edition. Foolscap 8vo. cloth, 6s. 6d.

II.

MEDICAL CHEMISTRY; with Illustrations on Wood. Fourth Edition, carefully revised. Fcap. 8vo. cloth, 6s. 6d.

DR. BRAIDWOOD, M.D. EDIN.

ON PYÆMIA, OR SUPPURATIVE FEVER: the Astley Cooper Prize Essay for 1868. With 12 Plates. 8vo. cloth, 10s. 6d.

DR. JAMES BRIGHT.

ON DISEASES OF THE HEART, LUNGS, & AIR PASSAGES; with a Review of the several Climates recommended in these Affections. Third Edition. Post 8vo. cloth, 9s.

DR. BRINTON, F.R.S.

I.

THE DISEASES OF THE STOMACH, with an Introduction on its Anatomy and Physiology; being Lectures delivered at St. Thomas's Hospital. Second Edition. 8vo. cloth, 10*s.* 6*d.*

II.

INTESTINAL OBSTRUCTION. Edited by Dr. Buzzard. Post 8vo. cloth, 5*s.*

MR. BERNARD E. BRODHURST, F.R.C.S.

I.

CURVATURES OF THE SPINE: their Causes, Symptoms, Pathology, and Treatment. Second Edition. Roy. 8vo. cloth, with Engravings, 7*s.* 6*d.*

II.

ON THE NATURE AND TREATMENT OF CLUBFOOT AND ANALOGOUS DISTORTIONS involving the TIBIO-TARSAL ARTICULATION. With Engravings on Wood. 8vo. cloth, 4*s.* 6*d.*

III.

PRACTICAL OBSERVATIONS ON THE DISEASES OF THE JOINTS INVOLVING ANCHYLOSIS, and on the TREATMENT for the RESTORATION of MOTION. Third Edition, much enlarged, 8vo. cloth, 4*s.* 6*d.*

MR BROOKE, M.A., M B., F.R.S.

ELEMENTS OF NATURAL PHILOSOPHY. Based on the Work of the late Dr. Golding Bird. Sixth Edition. With 700 Engravings. Fcap. 8vo. cloth, 12*s.* 6*d.*

DR. T. L. BRUNTON, B.Sc., M.B.

ON DIGITALIS. With some Observations on the Urine. Fcap. 8vo. cloth, 4*s.* 6*d.*

MR. THOMAS BRYANT, F.R.C.S.

I.

ON THE DISEASES AND INJURIES OF THE JOINTS. CLINICAL AND PATHOLOGICAL OBSERVATIONS. Post 8vo. cloth, 7*s.* 6*d.*

II.

CLINICAL SURGERY. Parts I. to VII. 8vo., 3*s.* 6*d.* each.

DR. BUCKLE, M.D., L.R.C.P.LOND.

VITAL AND ECONOMICAL STATISTICS OF THE HOSPITALS, INFIRMARIES, &c., OF ENGLAND AND WALES. Royal 8vo. 5*s.*

DR. JOHN CHARLES BUCKNILL, F.R.S., & DR. DANIEL H. TUKE.

A MANUAL OF PSYCHOLOGICAL MEDICINE: containing the History, Nosology, Description, Statistics, Diagnosis, Pathology, and Treatment of Insanity. Second Edition. 8vo. cloth, 15*s.*

DR. BUDD, F.R.S.

I.

ON DISEASES OF THE LIVER.

Illustrated with Coloured Plates and Engravings on Wood. Third Edition. 8vo. cloth, 16*s.*

II.

ON THE ORGANIC DISEASES AND FUNCTIONAL DISORDERS OF THE STOMACH. 8vo. cloth, 9*s.*

MR. CALLENDER, F.R.C.S.

FEMORAL RUPTURE: Anatomy of the Parts concerned. With Plates. 8vo. cloth, 4*s.*

DR. JOHN M. CAMPLIN, F.L.S.

ON DIABETES, AND ITS SUCCESSFUL TREATMENT. Third Edition, by Dr. Glover. Fcap. 8vo. cloth, 3*s.* 6*d.*

MR. ROBERT B. CARTER, M.R.C.S.

I.

ON THE INFLUENCE OF EDUCATION AND TRAINING IN PREVENTING DISEASES OF THE NERVOUS SYSTEM. Fcap. 8vo., 6*s.*

II.

THE PATHOLOGY AND TREATMENT OF HYSTERIA. Post 8vo. cloth, 4*s.* 6*d.*

DR. CARPENTER, F.R.S.

I.

PRINCIPLES OF HUMAN PHYSIOLOGY. With nearly 300 Illustrations on Steel and Wood. Seventh Edition. Edited by Mr. Henry Power. 8vo. cloth, 28*s.*

II.

A MANUAL OF PHYSIOLOGY. With 252 Illustrations on Steel and Wood. Fourth Edition. Fcap. 8vo. cloth, 12*s.* 6*d.*

III.

THE MICROSCOPE AND ITS REVELATIONS. With more than 400 Engravings on Steel and Wood. Fourth Edition. Fcap. 8vo. cloth, 12*s.* 6*d.*

MR. JOSEPH PEEL CATLOW, M.R.C.S.

ON THE PRINCIPLES OF ÆSTHETIC MEDICINE; or the Natural Use of Sensation and Desire in the Maintenance of Health and the Treatment of Disease. 8vo. cloth, 9*s.*

DR. CHAMBERS.

I.

LECTURES, CHIEFLY CLINICAL. Fourth Edition. 8vo. cloth, 14*s.*

II.

THE INDIGESTIONS OR DISEASES OF THE DIGESTIVE ORGANS FUNCTIONALLY TREATED. Second Edition. 8vo. cloth, 10*s.* 6*d.*

III.

SOME OF THE EFFECTS OF THE CLIMATE OF ITALY. Crown 8vo. cloth, 4*s.* 6*d.*

DR. CHANCE, M.B.

VIRCHOW'S CELLULAR PATHOLOGY, AS BASED UPON PHYSIOLOGICAL AND PATHOLOGICAL HISTOLOGY. With 144 Engravings on Wood. 8vo. cloth, 16*s.*

MR. H. T. CHAPMAN, F.R.C.S.

I.

THE TREATMENT OF OBSTINATE ULCERS AND CUTANEOUS ERUPTIONS OF THE LEG WITHOUT CONFINEMENT. Third Edition. Post 8vo. cloth, 3*s.* 6*d.*

II.

VARICOSE VEINS: their Nature, Consequences, and Treatment, Palliative and Curative. Second Edition. Post 8vo. cloth, 3*s.* 6*d.*

MR. PYE HENRY CHAVASSE, F.R.C.S.

I.

ADVICE TO A MOTHER ON THE MANAGEMENT OF HER CHILDREN. Ninth Edition. Foolscap 8vo., 2*s.* 6*d.*

II.

COUNSEL TO A MOTHER: being a Continuation and the Completion of "Advice to a Mother." Fcap. 8vo. 2*s.* 6*d.*

III.

ADVICE TO A WIFE ON THE MANAGEMENT OF HER OWN HEALTH. With an Introductory Chapter, especially addressed to a Young Wife. Eighth Edition. Fcap. 8vo., 2*s.* 6*d.*

MR. LE GROS CLARK, F.R.C.S.

OUTLINES OF SURGERY; being an Epitome of the Lectures on the Principles and the Practice of Surgery delivered at St. Thomas's Hospital. Fcap. 8vo. cloth, 5*s.*

MR. JOHN CLAY, M.R.C.S.

KIWISCH ON DISEASES OF THE OVARIES: Translated, by permission, from the last German Edition of his Clinical Lectures on the Special Pathology and Treatment of the Diseases of Women. With Notes, and an Appendix on the Operation of Ovariotomy. Royal 12mo. cloth, 16*s.*

DR. COCKLE, M.D.

ON INTRA-THORACIC CANCER. 8vo. 6*s.* 6*d.*

MR. COLLIS, M.B.DUB., F.R.C.S.I.

THE DIAGNOSIS AND TREATMENT OF CANCER AND THE TUMOURS ANALOGOUS TO IT. With coloured Plates. 8vo. cloth, 14*s.*

MR. COOLEY.

COMPREHENSIVE SUPPLEMENT TO THE PHARMACOPŒIAS.

THE CYCLOPÆDIA OF PRACTICAL RECEIPTS, PROCESSES, AND COLLATERAL INFORMATION IN THE ARTS, MANUFACTURES, PROFESSIONS, AND TRADES, INCLUDING MEDICINE, PHARMACY, AND DOMESTIC ECONOMY; designed as a General Book of Reference for the Manufacturer, Tradesman, Amateur, and Heads of Families. Fourth and greatly enlarged Edition, 8vo. cloth, 28*s.*

MR. W. WHITE COOPER.

I.

ON WOUNDS AND INJURIES OF THE EYE. Illustrated by 17 Coloured Figures and 41 Woodcuts. 8vo. cloth, 12*s.*

II.

ON NEAR SIGHT, AGED SIGHT, IMPAIRED VISION, AND THE MEANS OF ASSISTING SIGHT. With 31 Illustrations on Wood. Second Edition. Fcap. 8vo. cloth, 7*s.* 6*d.*

SIR ASTLEY COOPER, BART., F.R.S.

ON THE STRUCTURE AND DISEASES OF THE TESTIS. With 24 Plates. Second Edition. Royal 4to., 20*s.*

MR. COOPER.

A DICTIONARY OF PRACTICAL SURGERY AND ENCYCLOPÆDIA OF SURGICAL SCIENCE. New Edition, brought down to the present time. By SAMUEL A. LANE, F.R.C.S., assisted by various eminent Surgeons. Vol. I., 8vo. cloth, £1. 5*s.*

MR. HOLMES COOTE, F.R.C.S.

A REPORT ON SOME IMPORTANT POINTS IN THE TREATMENT OF SYPHILIS. 8vo. cloth, 5*s.*

DR. COTTON.

PHTHISIS AND THE STETHOSCOPE; OR, THE PHYSICAL SIGNS OF CONSUMPTION. Third Edition. Foolscap 8vo. cloth, 3*s.*

MR. COULSON.

ON DISEASES OF THE BLADDER AND PROSTATE GLAND. New Edition, revised. *In Preparation.*

MR. WALTER COULSON, F.R.C.S.

I.

A TREATISE ON SYPHILIS. 8vo. cloth, 10*s.*

II.

STONE IN THE BLADDER: With Special Reference to its Prevention, Early Symptoms, and Treatment by Lithotrity. 8vo. cloth, 6*s.*

MR. WILLIAM CRAIG, L.F.P.S.

ON THE INFLUENCE OF VARIATIONS OF ELECTRIC TENSION AS THE REMOTE CAUSE OF EPIDEMIC AND OTHER DISEASES. 8vo. cloth, 10*s.*

MR. CURLING, F.R.S.

I.

OBSERVATIONS ON DISEASES OF THE RECTUM. Third Edition. 8vo. cloth, 7*s.* 6*d.*

II.

A PRACTICAL TREATISE ON DISEASES OF THE TESTIS, SPERMATIC CORD, AND SCROTUM. Third Edition, with Engravings. 8vo. cloth, 16*s.*

DR. WILLIAM DALE, M.D.LOND.

A COMPENDIUM OF PRACTICAL MEDICINE AND MORBID ANATOMY. With Plates, 12mo. cloth, 7*s.*

DR. DALRYMPLE, M.R.C.P., F.R.C.S.

THE CLIMATE OF EGYPT: METEOROLOGICAL AND MEDICAL OBSERVATIONS, with Practical Hints for Invalid Travellers. Post 8vo. cloth, 4*s.*

MR. JOHN DALRYMPLE, F.R.S., F.R.C.S.

PATHOLOGY OF THE HUMAN EYE. Complete in Nine Fasciculi: imperial 4to., 20*s.* each; half-bound morocco, gilt tops, 9*l.* 15*s.*

DR. HERBERT DAVIES.

ON THE PHYSICAL DIAGNOSIS OF DISEASES OF THE LUNGS AND HEART. Second Edition. Post 8vo. cloth, 8*s.*

DR. DAVEY.

I.

THE GANGLIONIC NERVOUS SYSTEM: its Structure, Functions, and Diseases. 8vo. cloth, 9*s.*

II.

ON THE NATURE AND PROXIMATE CAUSE OF INSANITY. Post 8vo. cloth, 3*s.*

DR. HENRY DAY, M.D., M.R.C.P.

CLINICAL HISTORIES; with Comments. 8vo. cloth, 7*s.* 6*d.*

MR. DIXON.

A GUIDE TO THE PRACTICAL STUDY OF DISEASES OF THE EYE. Third Edition. Post 8vo. cloth, 9*s.*

DR. DOBELL.

I.

DEMONSTRATIONS OF DISEASES IN THE CHEST, AND THEIR PHYSICAL DIAGNOSIS. With Coloured Plates. 8vo. cloth, 12*s.* 6*d.*

II.

LECTURES ON THE GERMS AND VESTIGES OF DISEASE, and on the Prevention of the Invasion and Fatality of Disease by Periodical Examinations. 8vo. cloth, 6*s.* 6*d.*

III.

ON TUBERCULOSIS: ITS NATURE, CAUSE, AND TREATMENT; with Notes on Pancreatic Juice. Second Edition. Crown 8vo. cloth, 3*s.* 6*d.*

IV.

LECTURES ON WINTER COUGH (CATARRH, BRONCHITIS, EMPHYSEMA, ASTHMA); with an Appendix on some Principles of Diet in Disease. Post 8vo. cloth, 5*s.* 6*d.*

V.

LECTURES ON THE TRUE FIRST STAGE OF CONSUMPTION. Crown 8vo. cloth, 3*s.* 6*d.*

DR. TOOGOOD DOWNING.

NEURALGIA: its various Forms, Pathology, and Treatment. THE JACKSONIAN PRIZE ESSAY FOR 1850. 8vo. cloth, 10*s.* 6*d.*

DR. DRUITT, F.R.C.S.

THE SURGEON'S VADE-MECUM; with numerous Engravings on Wood. Ninth Edition. Foolscap 8vo. cloth, 12*s.* 6*d.*

MR. DUNN, F.R.C.S.

PSYCHOLOGY—PHYSIOLOGICAL, 4*s.*; MEDICAL, 3*s.*

MR. ERNEST EDWARDS, B.A.

PHOTOGRAPHS OF EMINENT MEDICAL MEN, with brief Analytical Notices of their Works. Vols. I. and II. (24 Portraits), 4to. cloth, 24*s.* each.

SIR JAMES EYRE, M.D.

I.

THE STOMACH AND ITS DIFFICULTIES. Sixth Edition, by Mr. BEALE. Fcap. 8vo., 2*s.* 6*d.*

II.

PRACTICAL REMARKS ON SOME EXHAUSTING DISEASES. Second Edition. Post 8vo. cloth, 4*s.* 6*d.*

DR. FAYRER, M.D., F.R.C.S., C.S.I.

CLINICAL SURGERY IN INDIA. With Engravings. 8vo. cloth, 16*s.*

DR. FENWICK.

THE MORBID STATES OF THE STOMACH AND DUODENUM, AND THEIR RELATIONS TO THE DISEASES OF OTHER ORGANS. With 10 Plates. 8vo. cloth, 12*s.*

SIR WILLIAM FERGUSSON, BART., F.R.S.

I.

A SYSTEM OF PRACTICAL SURGERY; with numerous Illustrations on Wood. Fourth Edition. Fcap. 8vo. cloth, 12*s.* 6*d.*

II.

LECTURES ON THE PROGRESS OF ANATOMY AND SURGERY DURING THE PRESENT CENTURY. With numerous Engravings. 8vo. cloth, 10*s.* 6*d.*

SIR JOHN FIFE, F.R.C.S. AND MR. URQUHART.

MANUAL OF THE TURKISH BATH. Heat a Mode of Cure and a Source of Strength for Men and Animals. With Engravings. Post 8vo. cloth, 5*s.*

MR. FLOWER, F.R.S., F.R.C.S.

DIAGRAMS OF THE NERVES OF THE HUMAN BODY, exhibiting their Origin, Divisions, and Connexions, with their Distribution to the various Regions of the Cutaneous Surface, and to all the Muscles. Folio, containing Six Plates, 14*s.*

MR. FLUX.

THE LAW TO REGULATE THE SALE OF POISONS WITHIN GREAT BRITAIN. Crown 8vo. cloth, 2*s.* 6*d.*

MR. FOWNES, PH.D., F.R.S.

I.

A MANUAL OF CHEMISTRY; with 187 Illustrations on Wood. Tenth Edition. Fcap. 8vo. cloth, 14*s.*
Edited by H. BENCE JONES, M.D., F.R.S., and HENRY WATTS, B.A., F.R.S.

II.

CHEMISTRY, AS EXEMPLIFYING THE WISDOM AND BENEFICENCE OF GOD. Second Edition. Fcap. 8vo. cloth, 4*s.* 6*d.*

III.

INTRODUCTION TO QUALITATIVE ANALYSIS. Post 8vo. cloth, 2*s.*

DR. D. J. T. FRANCIS.

CHANGE OF CLIMATE; considered as a Remedy in Dyspeptic, Pulmonary, and other Chronic Affections; with an Account of the most Eligible Places of Residence for Invalids, at different Seasons of the Year. Post 8vo. cloth, 8*s.* 6*d.*

DR. FULLER.

I.

ON DISEASES OF THE LUNGS AND AIR PASSAGES. Second Edition. 8vo. cloth, 12*s.* 6*d.*

II.

ON DISEASES OF THE HEART AND GREAT VESSELS. 8vo. cloth, 7*s.* 6*d.*

III.

ON RHEUMATISM, RHEUMATIC GOUT, AND SCIATICA: their Pathology, Symptoms, and Treatment. Third Edition. 8vo. cloth, 12*s.* 6*d.*

PROFESSOR FRESENIUS.

A SYSTEM OF INSTRUCTION IN CHEMICAL ANALYSIS,
Edited by ARTHUR VACHER.
QUALITATIVE. Seventh Edition. 8vo. cloth, 9*s.*
QUANTITATIVE. Fourth Edition. 8vo. cloth, 18*s.*

MR. GALLOWAY.

I.

THE FIRST STEP IN CHEMISTRY. With numerous Engravings. Fourth Edition. Fcap. 8vo. cloth, 6*s.* 6*d.*

II.

A KEY TO THE EXERCISES CONTAINED IN ABOVE. Fcap. 8vo., 2*s.* 6*d.*

III.

THE SECOND STEP IN CHEMISTRY; or, the Student's Guide to the Higher Branches of the Science. With Engravings. 8vo. cloth, 10*s.*

IV.

A MANUAL OF QUALITATIVE ANALYSIS. Fourth Edition. Post 8vo. cloth, 6*s.* 6*d.*

V.

CHEMICAL TABLES. On Five Large Sheets, for School and Lecture Rooms. Second Edition. 4*s.* 6*d.*

MR. J. SAMPSON GAMGEE.

HISTORY OF A SUCCESSFUL CASE OF AMPUTATION AT THE HIP-JOINT (the limb 48-in. in circumference, 99 pounds weight). With 4 Photographs. 4to cloth, 10*s.* 6*d.*

MR. F. J. GANT, F.R.C.S.

I.

THE PRINCIPLES OF SURGERY: Clinical, Medical, and Operative. With Engravings. 8vo. cloth, 18*s.*

II.

THE IRRITABLE BLADDER: its Causes and Curative Treatment. Second Edition, enlarged. Crown 8vo. cloth, 5*s.*

MR. GAY, F.R.C.S.

ON VARICOSE DISEASE OF THE LOWER EXTREMITIES. LETTSOMIAN LECTURES. With Plates. 8vo. cloth, 5*s.*

SIR DUNCAN GIBB, BART., M.D.

I.

ON DISEASES OF THE THROAT AND WINDPIPE, as reflected by the Laryngoscope. Second Edition. With 116 Engravings. Post 8vo. cloth, 10*s.* 6*d.*

II.

THE LARYNGOSCOPE IN DISEASES OF THE THROAT, with a Chapter on RHINOSCOPY. Third Edition, with Engravings. Crown 8vo., cloth, 5*s.*

MRS. GODFREY.

ON THE NATURE, PREVENTION, TREATMENT, AND CURE OF SPINAL CURVATURES and DEFORMITIES of the CHEST and LIMBS, without ARTIFICIAL SUPPORTS or any MECHANICAL APPLIANCES. Third Edition, Revised and Enlarged. 8vo. cloth 5*s.*

DR. GORDON, M.D., C.B.

I.

ARMY HYGIENE. 8vo. cloth, 20*s.*

II.

CHINA, FROM A MEDICAL POINT OF VIEW; IN 1860 AND 1861; With a Chapter on Nagasaki as a Sanatarium. 8vo. cloth, 10*s.* 6*d.*

DR. GAIRDNER.

ON GOUT; its History, its Causes, and its Cure. Fourth Edition. Post 8vo. cloth, 8s. 6*d*.

DR. GRANVILLE, F.R.S.

I.

THE MINERAL SPRINGS OF VICHY: their Efficacy in the Treatment of Gout, Indigestion, Gravel, &c. 8vo. cloth, 3s.

II.

ON SUDDEN DEATH. Post 8vo., 2*s*. 6*d*.

DR. GRAVES M.D., F.R.S.

STUDIES IN PHYSIOLOGY AND MEDICINE. Edited by Dr. Stokes. With Portrait and Memoir. 8vo. cloth, 14*s*.

MR. GRIFFITHS.

CHEMISTRY OF THE FOUR SEASONS—Spring, Summer, Autumn, Winter. Illustrated with Engravings on Wood. Second Edition. Foolscap 8vo. cloth, 7*s*. 6*d*.

DR. GULLY.

THE SIMPLE TREATMENT OF DISEASE; deduced from the Methods of Expectancy and Revulsion. 18mo. cloth, 4*s*.

DR. GUY AND DR. JOHN HARLEY.

HOOPER'S PHYSICIAN'S VADE-MECUM; OR, MANUAL OF THE PRINCIPLES AND PRACTICE OF PHYSIC. Seventh Edition. With Engravings. Foolscap 8vo. cloth, 12*s*. 6*d*.

GUY'S HOSPITAL REPORTS. Third Series. Vol. XIV., 8vo. 7*s*. 6*d*.

DR. HABERSHON, F.R.C.P.

I.

ON DISEASES OF THE ABDOMEN, comprising those of the Stomach and other Parts of the Alimentary Canal, Œsophagus, Stomach, Cæcum, Intestines, and Peritoneum. Second Edition, with Plates. 8vo. cloth, 14*s*.

II.

ON THE INJURIOUS EFFECTS OF MERCURY IN THE TREATMENT OF DISEASE. Post 8vo. cloth, 3*s*. 6*d*.

DR. C. RADCLYFFE HALL.

TORQUAY IN ITS MEDICAL ASPECT AS A RESORT FOR PULMONARY INVALIDS. Post 8vo. cloth, 5*s*.

DR. MARSHALL HALL, F.R.S.

I.

PRONE AND POSTURAL RESPIRATION IN DROWNING AND OTHER FORMS OF APNŒA OR SUSPENDED RESPIRATION. Post 8vo. cloth. 5*s*.

II.

PRACTICAL OBSERVATIONS AND SUGGESTIONS IN MEDICINE. Second Series. Post 8vo. cloth, 8*s*. 6*d*.

MR. HARDWICH.

A MANUAL OF PHOTOGRAPHIC CHEMISTRY. With Engravings. Seventh Edition. Foolscap 8vo. cloth, 7*s*. 6*d*.

DR. J. BOWER HARRISON, M.D., M.R.C.P.

I.

LETTERS TO A YOUNG PRACTITIONER ON THE DISEASES OF CHILDREN. Foolscap 8vo. cloth, 3*s.*

II.

ON THE CONTAMINATION OF WATER BY THE POISON OF LEAD, and its Effects on the Human Body. Foolscap 8vo. cloth, 3*s.* 6*d.*

DR. HARTWIG.

I.

ON SEA BATHING AND SEA AIR. Second Edition. Fcap. 8vo., 2*s.* 6*d.*

II.

ON THE PHYSICAL EDUCATION OF CHILDREN. Fcap. 8vo., 2*s.* 6*d.*

DR. A. H. HASSALL.

THE URINE, IN HEALTH AND DISEASE; being an Explanation of the Composition of the Urine, and of the Pathology and Treatment of Urinary and Renal Disorders. Second Edition. With 79 Engravings (23 Coloured). Post 8vo. cloth, 12*s.* 6*d.*

MR. ALFRED HAVILAND, M.R.C.S.

CLIMATE, WEATHER, AND DISEASE; being a Sketch of the Opinions of the most celebrated Ancient and Modern Writers with regard to the Influence of Climate and Weather in producing Disease. With Four coloured Engravings. 8vo. cloth, 7*s.*

MR. W. HAYCOCK, M.R.C.V.S.

HORSES; HOW THEY OUGHT TO BE SHOD: being a plain and practical Treatise on the Principles and Practice of the Farrier's Art. With 14 Plates. Cloth, 7*s.* 6*d.*

DR. HEADLAND, M.D., F.R.C.P.

I.

ON THE ACTION OF MEDICINES IN THE SYSTEM. Fourth Edition. 8vo. cloth, 14*s.*

II.

A MEDICAL HANDBOOK; comprehending such Information on Medical and Sanitary Subjects as is desirable in Educated Persons. Second Thousand. Foolscap 8vo. cloth, 5*s.*

DR. HEALE.

I.

A TREATISE ON THE PHYSIOLOGICAL ANATOMY OF THE LUNGS. With Engravings. 8vo. cloth, 8*s.*

II.

A TREATISE ON VITAL CAUSES. 8vo. cloth, 9*s.*

MR. CHRISTOPHER HEATH, F.R.C.S.

I.

PRACTICAL ANATOMY: a Manual of Dissections. With numerous Engravings. Fcap. 8vo. cloth, 10*s.* 6*d.*

II.

A MANUAL OF MINOR SURGERY AND BANDAGING, FOR THE USE OF HOUSE-SURGEONS, DRESSERS, AND JUNIOR PRACTITIONERS. With Illustrations. Third Edition. Fcap. 8vo. cloth, 5*s.*

III.

INJURIES AND DISEASES OF THE JAWS. JACKSONIAN PRIZE ESSAY. With Engravings. 8vo. cloth, 12*s.*

MR. HIGGINBOTTOM, F.R.S., F.R.C.S.E.

A PRACTICAL ESSAY ON THE USE OF THE NITRATE OF SILVER IN THE TREATMENT OF INFLAMMATION, WOUNDS, AND ULCERS. Third Edition, 8vo. cloth, 6*s.*

DR. HINDS.

THE HARMONIES OF PHYSICAL SCIENCE IN RELATION TO THE HIGHER SENTIMENTS; with Observations on Medical Studies, and on the Moral and Scientific Relations of Medical Life. Post 8vo. cloth, 4*s.*

MR. J. A. HINGESTON, M.R.C.S.

TOPICS OF THE DAY, MEDICAL, SOCIAL, AND SCIENTIFIC. Crown 8vo. cloth, 7*s.* 6*d.*

DR. HODGES.

THE NATURE, PATHOLOGY, AND TREATMENT OF PUERPERAL CONVULSIONS. Crown 8vo. cloth, 3*s.*

DR. DECIMUS HODGSON.

THE PROSTATE GLAND, AND ITS ENLARGEMENT IN OLD AGE. With 12 Plates. Royal 8vo. cloth, 6*s.*

MR. JABEZ HOGG.

A MANUAL OF OPHTHALMOSCOPIC SURGERY; being a Practical Treatise on the Use of the Ophthalmoscope in Diseases of the Eye. Third Edition. With Coloured Plates. 8vo. cloth, 10*s.* 6*d.*

MR. LUTHER HOLDEN, F.R.C.S.

I.

HUMAN OSTEOLOGY: with Plates, showing the Attachments of the Muscles. Third Edition. 8vo. cloth, 16*s.*

II.

A MANUAL OF THE DISSECTION OF THE HUMAN BODY. With Engravings on Wood. Third Edition. 8vo. cloth, 16*s.*

MR. BARNARD HOLT, F.R.C.S.

ON THE IMMEDIATE TREATMENT OF STRICTURE OF THE URETHRA. Third Edition, Enlarged. 8vo. cloth, 6*s.*

SIR CHARLES HOOD, M.D.

SUGGESTIONS FOR THE FUTURE PROVISION OF CRIMINAL LUNATICS. 8vo. cloth, *5s. 6d.*

DR. P. HOOD.

THE SUCCESSFUL TREATMENT OF SCARLET FEVER; also, OBSERVATIONS ON THE PATHOLOGY AND TREATMENT OF CROWING INSPIRATIONS OF INFANTS. Post 8vo. cloth, *5s.*

MR. JOHN HORSLEY.

A CATECHISM OF CHEMICAL PHILOSOPHY; being a Familiar Exposition of the Principles of Chemistry and Physics. With Engravings on Wood. Designed for the Use of Schools and Private Teachers. Post 8vo. cloth, *6s. 6d.*

DR. JAMES A. HORTON, M.D.

PHYSICAL AND MEDICAL CLIMATE AND METEOROLOGY OF THE WEST COAST OF AFRICA. 8vo. cloth, *10s.*

MR. LUKE HOWARD, F.R.S.

ESSAY ON THE MODIFICATIONS OF CLOUDS. Third Edition, by W. D. and E. HOWARD. With 6 Lithographic Plates, from Pictures by Kenyon. 4to. cloth, *10s. 6d.*

DR. HAMILTON HOWE, M.D.

A THEORETICAL INQUIRY INTO THE PHYSICAL CAUSE OF EPIDEMIC DISEASES. Accompanied with Tables. 8vo. cloth, *7s.*

DR. HUFELAND.

THE ART OF PROLONGING LIFE. Second Edition. Edited by ERASMUS WILSON, F.R.S. Foolscap 8vo., *2s. 6d.*

MR. W. CURTIS HUGMAN, F.R.C.S.

ON HIP-JOINT DISEASE; with reference especially to Treatment by Mechanical Means for the Relief of Contraction and Deformity of the Affected Limb. With Plates. Re-issue, enlarged. 8vo. cloth, *3s. 6d.*

MR. HULKE, F.R.C.S.

A PRACTICAL TREATISE ON THE USE OF THE OPHTHALMOSCOPE. Being the Jacksonian Prize Essay for 1859. Royal 8vo. cloth, *8s.*

DR. HENRY HUNT.

ON HEARTBURN AND INDIGESTION. 8vo. cloth, *5s.*

MR. G. Y. HUNTER, M.R.C.S.

BODY AND MIND: the Nervous System and its Derangements. Fcap. 8vo. cloth, *3s. 6d.*

MR. JONATHAN HUTCHINSON, F.R.C.S.

A CLINICAL MEMOIR ON CERTAIN DISEASES OF THE

EYE AND EAR, CONSEQUENT ON INHERITED SYPHILIS; with an appended Chapter of Commentaries on the Transmission of Syphilis from Parent to Offspring, and its more remote Consequences. With Plates and Woodcuts, 8vo. cloth, 9*s.*

PROF. HUXLEY, LL.D., F.R.S.

INTRODUCTION TO THE CLASSIFICATION OF ANIMALS.

With Engravings. 8vo. cloth, 6*s.*

DR. INMAN, M.R.C.P.

I.

ON MYALGIA: ITS NATURE, CAUSES, AND TREATMENT;

being a Treatise on Painful and other Affections of the Muscular System. Second Edition. 8vo. cloth, 9*s.*

II.

FOUNDATION FOR A NEW THEORY AND PRACTICE

OF MEDICINE. Second Edition. Crown 8vo. cloth, 10*s.*

DR. JAGO, M.D.OXON., A.B.CANTAB.

ENTOPTICS, WITH ITS USES IN PHYSIOLOGY AND

MEDICINE. With 54 Engravings. Crown 8vo. cloth, 5*s.*

DR. PROSSER JAMES, M.D.

SORE-THROAT: ITS NATURE, VARIETIES, AND TREAT-

MENT; including the Use of the LARYNGOSCOPE as an Aid to Diagnosis. Second Edition, with numerous Engravings. Post 8vo. cloth, 5*s.*

DR. JENCKEN, M.D., M.R.C.P.

THE CHOLERA: ITS ORIGIN, IDIOSYNCRACY, AND

TREATMENT. Fcap. 8vo. cloth, 2*s.* 6*d.*

DR. HANDFIELD JONES, M.B., F.R.C.P.

CLINICAL OBSERVATIONS ON FUNCTIONAL NERVOUS

DISORDERS. Post 8vo. cloth, 10*s.* 6*d.*

DR. H. BENCE JONES, M.D., F.R.S.

I.

LECTURES ON SOME OF THE APPLICATIONS OF

CHEMISTRY AND MECHANICS TO PATHOLOGY AND THERAPEUTICS. 8vo. cloth, 12*s.*

II.

CROONIAN LECTURES ON MATTER AND FORCE.

Fcap. 8vo. cloth, 5*s.*

DR. HANDFIELD JONES, F.R.S., & DR. EDWARD H. SIEVEKING.

A MANUAL OF PATHOLOGICAL ANATOMY.

Illustrated with numerous Engravings on Wood. Foolscap 8vo. cloth, 12*s.* 6*d.*

DR. JAMES JONES, M.D., M.R.C.P.

ON THE USE OF PERCHLORIDE OF IRON AND OTHER CHALYBEATE SALTS IN THE TREATMENT OF CONSUMPTION. Crown 8vo. cloth, 3*s.* 6*d.*

MR. WHARTON JONES, F.R.S.

I.

A MANUAL OF THE PRINCIPLES AND PRACTICE OF OPHTHALMIC MEDICINE AND SURGERY; with Nine Coloured Plates and 173 Wood Engravings. Third Edition, thoroughly revised. Foolscap 8vo. cloth, 12*s.* 6*d.*

II.

THE WISDOM AND BENEFICENCE OF THE ALMIGHTY, AS DISPLAYED IN THE SENSE OF VISION. Actonian Prize Essay. With Illustrations on Steel and Wood. Foolscap 8vo. cloth, 4*s.* 6*d.*

III.

DEFECTS OF SIGHT AND HEARING: their Nature, Causes, Prevention, and General Management. Second Edition, with Engravings. Fcap. 8vo. 2*s.* 6*d.*

IV.

A CATECHISM OF THE MEDICINE AND SURGERY OF THE EYE AND EAR. For the Clinical Use of Hospital Students. Fcap. 8vo. 2*s.* 6*d.*

V.

A CATECHISM OF THE PHYSIOLOGY AND PHILOSOPHY OF BODY, SENSE, AND MIND. For Use in Schools and Colleges. Fcap. 8vo., 2*s.* 6*d.*

DR. LAENNEC.

A MANUAL OF AUSCULTATION AND PERCUSSION. Translated and Edited by J. B. SHARPE, M.R.C.S. 3*s.*

DR. LANE, M.A.

HYDROPATHY; OR, HYGIENIC MEDICINE. An Explanatory Essay. Second Edition. Post 8vo. cloth, 5*s.*

SIR WM. LAWRENCE, BART., F.R.S

I.

LECTURES ON SURGERY. 8vo. cloth, 16*s.*

II.

A TREATISE ON RUPTURES. The Fifth Edition, considerably enlarged. 8vo. cloth, 16*s.*

DR. LEARED, M.R.C.P.

IMPERFECT DIGESTION: ITS CAUSES AND TREATMENT. Fourth Edition. Foolscap 8vo. cloth, 4*s.*

MR. HENRY LEE, F.R.C.S.

I.

ON SYPHILIS. Second Edition. With Coloured Plates. 8vo. cloth, 10*s.*

II.

ON DISEASES OF TH VEINS, HÆMORRHOIDAL TUMOURS, AND OTHER AFFECTIONS OF THE RECTUM. Second Edition. 8vo. cloth, 8*s.*

DR. EDWIN LEE.

I.

THE EFFECT OF CLIMATE ON TUBERCULOUS DISEASE, with Notices of the chief Foreign Places of Winter Resort. Small 8vo. cloth, 4*s.* 6*d.*

II.

THE WATERING PLACES OF ENGLAND, CONSIDERED with Reference to their Medical Topography. Fourth Edition. Fcap. 8vo. cloth, 7*s.* 6*d.*

III.

THE BATHS OF FRANCE. Fourth Edition. Fcap. 8vo. cloth, 4*s.* 6*d.*

IV.

THE BATHS OF GERMANY. Fourth Edition. Post 8vo. cloth, 7*s.*

V.

THE BATHS OF SWITZERLAND. 12mo. cloth, 3*s.* 6*d.*

VI.

HOMŒOPATHY AND HYDROPATHY IMPARTIALLY APPRECIATED. Fourth Edition. Post 8vo. cloth, 3*s.*

DR. ROBERT LEE, F.R.S.

I.

CONSULTATIONS IN MIDWIFERY. Foolscap 8vo. cloth, 4*s.* 6*d.*

II.

A TREATISE ON THE SPECULUM; with Three Hundred Cases. 8vo. cloth, 4*s.* 6*d.*

III.

CLINICAL REPORTS OF OVARIAN AND UTERINE DISEASES, with Commentaries. Foolscap 8vo. cloth, 6*s.* 6*d.*

IV.

CLINICAL MIDWIFERY: comprising the Histories of 545 Cases of Difficult, Preternatural, and Complicated Labour, with Commentaries. Second Edition. Foolscap 8vo. cloth, 5*s.*

DR. LEISHMAN, M.D., F.F.P.S.

THE MECHANISM OF PARTURITION: An Essay, Historical and Critical. With Engravings. 8vo. cloth, 5*s.*

MR. F. HARWOOD LESCHER.

THE ELEMENTS OF PHARMACY. 8vo. cloth, 7*s.* 6*d.*

MR. LISTON, F.R.S.

PRACTICAL SURGERY. Fourth Edition. 8vo. cloth, 22*s.*

MR. H. W. LOBB, L.S.A., M.R.C.S.E.

ON SOME OF THE MORE OBSCURE FORMS OF NERVOUS AFFECTIONS, THEIR PATHOLOGY AND TREATMENT. Re-issue, with the Chapter on Galvanism entirely Re-written. With Engravings. 8vo. cloth, 8*s.*

DR. LOGAN, M.D., M.R.C.P.LOND.

ON OBSTINATE DISEASES OF THE SKIN. Fcap.8vo.cloth,2*s.*6*d.*

LONDON HOSPITAL.

CLINICAL LECTURES AND REPORTS BY THE MEDICAL AND SURGICAL STAFF. With Illustrations. Vols. I. to IV. 8vo. cloth, 7*s.* 6*d.*

LONDON MEDICAL SOCIETY OF OBSERVATION.

WHAT TO OBSERVE AT THE BED-SIDE, AND AFTER DEATH. Published by Authority. Second Edition. Foolscap 8vo. cloth, 4*s.* 6*d.*

MR. HENRY LOWNDES, M.R.C.S.

AN ESSAY ON THE MAINTENANCE OF HEALTH. Fcap. 8vo. cloth, 2*s.* 6*d.*

MR. M'CLELLAND, F.L.S., F.G.S.

THE MEDICAL TOPOGRAPHY, OR CLIMATE AND SOILS, OF BENGAL AND THE N. W. PROVINCES. Post 8vo. cloth, 4*s.* 6*d.*

DR. MACLACHLAN, M.D., F.R.C.P.L.

THE DISEASES AND INFIRMITIES OF ADVANCED LIFE. 8vo. cloth, 16*s.*

DR. A. C. MACLEOD, M.R.C.P.LOND.

ACHOLIC DISEASES; comprising Jaundice, Diarrhœa, Dysentery, and Cholera. Post 8vo. cloth, 5*s.* 6*d.*

DR. GEORGE H. B. MACLEOD, F.R.C.S.E.

I.

OUTLINES OF SURGICAL DIAGNOSIS. 8vo. cloth, 12*s.* 6*d.*

II.

NOTES ON THE SURGERY OF THE CRIMEAN WAR; with REMARKS on GUN-SHOT WOUNDS. 8vo. cloth, 10*s.* 6*d.*

DR. WM. MACLEOD, F.R.C.P.EDIN.

THE THEORY OF THE TREATMENT OF DISEASE ADOPTED AT BEN RHYDDING. Fcap. 8vo. cloth, 2*s.* 6*d.*

MR. JOSEPH MACLISE, F.R.C.S.

I.

SURGICAL ANATOMY. A Series of Dissections, illustrating the Principal Regions of the Human Body. Second Edition, folio, cloth, £3. 12*s.*; half-morocco, £4. 4*s.*

II.

ON DISLOCATIONS AND FRACTURES. This Work is Uniform with "Surgical Anatomy;" folio, cloth, £2. 10*s.*; half-morocco, £2. 17*s.*

MR. MACNAMARA.

I.

A MANUAL OF THE DISEASES OF THE EYE. With Coloured Plates. Fcap. 8vo. cloth, 12*s.* 6*d.*

II.

ON DISEASES OF THE EYE; referring principally to those Affections requiring the aid of the Ophthalmoscope for their Diagnosis. With coloured plates. 8vo. cloth, 10*s.* 6*d.*

DR. MC NICOLL, M.R.C.P.

A HAND-BOOK FOR SOUTHPORT, MEDICAL & GENERAL; with Copious Notices of the Natural History of the District. Second Edition. Post 8vo. cloth, 3*s.* 6*d.*

DR. MARCET, F.R.S.

ON CHRONIC ALCOHOLIC INTOXICATION; with an INQUIRY INTO THE INFLUENCE OF THE ABUSE OF ALCOHOL AS A PREDISPOSING CAUSE OF DISEASE. Second Edition, much enlarged. Foolscap 8vo. cloth, 4*s.* 6*d.*

DR. J. MACPHERSON, M.D.

CHOLERA IN ITS HOME; with a Sketch of the Pathology and Treatment of the Disease. Crown 8vo. cloth, 5s.

DR. MARKHAM.

I.

DISEASES OF THE HEART: THEIR PATHOLOGY, DIAGNOSIS, AND TREATMENT. Second Edition. Post 8vo. cloth, 6s.

II.

SKODA ON AUSCULTATION AND PERCUSSION. Post 8vo. cloth, 6s.

III.

BLEEDING AND CHANGE IN TYPE OF DISEASES. Gulstonian Lectures for 1864. Crown 8vo. 2s. 6d.

DR. ALEXANDER MARSDEN, M.D., F.R.C.S.

A NEW AND SUCCESSFUL MODE OF TREATING CERTAIN FORMS OF CANCER; to which is prefixed a Practical and Systematic Description of all the Varieties of this Disease. With Coloured Plates. 8vo. cloth, 6s. 6d.

SIR RANALD MARTIN, K.C.B., F.R.S.

INFLUENCE OF TROPICAL CLIMATES IN PRODUCING THE ACUTE ENDEMIC DISEASES OF EUROPEANS; including Practical Observations on their Chronic Sequelæ under the Influences of the Climate of Europe. Second Edition, much enlarged. 8vo. cloth, 20s.

DR. P. MARTYN, M.D.LOND.

HOOPING-COUGH; ITS PATHOLOGY AND TREATMENT. With Engravings. 8vo. cloth, 2s. 6d.

MR. C. F. MAUNDER, F.R.C.S.

OPERATIVE SURGERY. With 158 Engravings. Post 8vo. 6s.

DR. MAYNE, M.D., LL.D.

I.

AN EXPOSITORY LEXICON OF THE TERMS, ANCIENT AND MODERN, IN MEDICAL AND GENERAL SCIENCE. 8vo. cloth, £2. 10s.

II.

A MEDICAL VOCABULARY; or, an Explanation of all Names, Synonymes, Terms, and Phrases used in Medicine and the relative branches of Medical Science. Third Edition. Fcap. 8vo. cloth, 8s. 6d.

DR. MERYON, M.D., F.R.C.P.

PATHOLOGICAL AND PRACTICAL RESEARCHES ON THE VARIOUS FORMS OF PARALYSIS. 8vo. cloth, 6s.

DR. W. J. MOORE, M.D.

I.

HEALTH IN THE TROPICS; or, Sanitary Art applied to Europeans in India. 8vo. cloth, 9s.

II.

A MANUAL OF THE DISEASES OF INDIA. Fcap. 8vo. cloth, 5s.

DR. JAMES MORRIS, M.D.LOND.

I.

GERMINAL MATTER AND THE CONTACT THEORY: An Essay on the Morbid Poisons. Second Edition. Crown 8vo. cloth, 4s. 6d.

II.

IRRITABILITY: Popular and Practical Sketches of Common Morbid States and Conditions bordering on Disease; with Hints for Management, Alleviation, and Cure. Crown 8vo. cloth, 4s. 6d.

PROFESSOR MULDER, UTRECHT.

THE CHEMISTRY OF WINE. Edited by H. BENCE JONES, M.D., F.R.S. Fcap. 8vo. cloth, 6s.

DR. W. MURRAY, M.D., M.R.C.P.

EMOTIONAL DISORDERS OF THE SYMPATHETIC SYSTEM OF NERVES. Crown 8vo. cloth, 3s. 6d.

DR. MUSHET, M.B., M.R.C.P.

ON APOPLEXY, AND ALLIED AFFECTIONS OF THE BRAIN. 8vo. cloth, 7s.

MR. NAYLER, F.R.C.S.

ON THE DISEASES OF THE SKIN. With Plates. 8vo. cloth, 10s. 6d.

DR. BIRKBECK NEVINS.

THE PRESCRIBER'S ANALYSIS OF THE BRITISH PHARMACOPEIA of 1867. 32mo. cloth, 3s. 6d.

DR. THOS. NICHOLSON, M.D.

ON YELLOW FEVER; comprising the History of that Disease as it appeared in the Island of Antigua. Fcap. 8vo. cloth, 2s. 6d.

DR. NOAD, PH.D., F.R.S.

THE INDUCTION COIL, being a Popular Explanation of the Electrical Principles on which it is constructed. Third Edition. With Engravings. Fcap. 8vo. cloth, 3s.

DR. NOBLE.

THE HUMAN MIND IN ITS RELATIONS WITH THE BRAIN AND NERVOUS SYSTEM. Post 8vo. cloth, 4s. 6d.

MR. NUNNELEY, F.R.C.S.E.

I.

ON THE ORGANS OF VISION: THEIR ANATOMY AND PHYSIOLOGY. With Plates, 8vo. cloth, 15s.

II.

A TREATISE ON THE NATURE, CAUSES, AND TREATMENT OF ERYSIPELAS. 8vo. cloth, 10s. 6d.

DR. OPPERT, M.D.

I.

HOSPITALS, INFIRMARIES, AND DISPENSARIES; their Construction, Interior Arrangement, and Management, with Descriptions of existing Institutions. With 58 Engravings. Royal 8vo. cloth, 10s. 6d.

II.

VISCERAL AND HEREDITARY SYPHILIS. 8vo. cloth, 5s.

MR. LANGSTON PARKER.

THE MODERN TREATMENT OF SYPHILITIC DISEASES, both Primary and Secondary; comprising the Treatment of Constitutional and Confirmed Syphilis, by a safe and successful Method. Fourth Edition, 8vo. cloth, 10s.

DR. PARKES, F.R.S., F.R.C.P.

I.

A MANUAL OF PRACTICAL HYGIENE; intended especially for the Medical Officers of the Army. With Plates and Woodcuts. 3rd Edition, 8vo. cloth, 16s.

II.

THE URINE: ITS COMPOSITION IN HEALTH AND DISEASE, AND UNDER THE ACTION OF REMEDIES. 8vo. cloth, 12*s.*

DR. PARKIN, M.D., F.R.C.S.

I.

THE ANTIDOTAL TREATMENT AND PREVENTION OF THE EPIDEMIC CHOLERA. Third Edition. 8vo. cloth, 7*s.* 6*d.*

II.

THE CAUSATION AND PREVENTION OF DISEASE; with the Laws regulating the Extrication of Malaria from the Surface, and its Diffusion in the surrounding Air. 8vo. cloth, 5*s.*

MR. JAMES PART, F.R.C.S.

THE MEDICAL AND SURGICAL POCKET CASE BOOK, for the Registration of important Cases in Private Practice, and to assist the Student of Hospital Practice. Second Edition. 2*s.* 6*d.*

DR. PATTERSON, M.D.

EGYPT AND THE NILE AS A WINTER RESORT FOR PULMONARY AND OTHER INVALIDS. Fcap. 8vo. cloth, 3*s.*

DR. PAVY, M.D., F.R.S., F.R.C.P.

I.

DIABETES: RESEARCHES ON ITS NATURE AND TREATMENT. Second Edition. With Engravings. 8vo. cloth, 10*s.*

II.

DIGESTION: ITS DISORDERS AND THEIR TREATMENT. Second Edition. 8vo. cloth, 8*s.* 6*d.*

DR. PEACOCK, M.D., F.R.C.P.

I.

ON MALFORMATIONS OF THE HUMAN HEART. With Original Cases and Illustrations. Second Edition. With 8 Plates. 8vo. cloth, 10*s.*

II.

ON SOME OF THE CAUSES AND EFFECTS OF VALVULAR DISEASE OF THE HEART. With Engravings. 8vo. cloth, 5*s.*

DR W. H. PEARSE, M.D.EDIN.

NOTES ON HEALTH IN CALCUTTA AND BRITISH EMIGRANT SHIPS, including Ventilation, Diet, and Disease. Fcap. 8vo. 2*s.*

DR. PEREIRA, F.R.S.

SELECTA E PRÆSCRIPTIS. Fifteenth Edition. 24mo. cloth, 5*s.*

DR. PICKFORD.

HYGIENE; or, Health as Depending upon the Conditions of the Atmosphere, Food and Drinks, Motion and Rest, Sleep and Wakefulness, Secretions, Excretions, and Retentions, Mental Emotions, Clothing, Bathing, &c. Vol. I. 8vo. cloth, 9*s.*

PROFESSOR PIRRIE, F.R.S.E.

THE PRINCIPLES AND PRACTICE OF SURGERY. With numerous Engravings on Wood. Second Edition. 8vo. cloth, 24*s.*

PROFESSOR PIRRIE & DR. KEITH.

ACUPRESSURE: an excellent Method of arresting Surgical Hæmorrhage and of accelerating the healing of Wounds. With Engravings. 8vo. cloth, 5*s.*

DR. PIRRIE. M.D.

ON HAY ASTHMA, AND THE AFFECTION TERMED HAY FEVER. Fcap. 8vo. cloth, 2*s.* 6*d.*

PROFESSORS PLATTNER & MUSPRATT.

THE USE OF THE BLOWPIPE IN THE EXAMINATION OF MINERALS, ORES, AND OTHER METALLIC COMBINATIONS. Illustrated by numerous Engravings on Wood. Third Edition. 8vo. cloth, 10*s.* 6*d.*

MR. HENRY POWER, F.R.C.S., M.B.LOND.

ILLUSTRATIONS OF SOME OF THE PRINCIPAL DISEASES OF THE EYE: With an Account of their Symptoms, Pathology and Treatment. Twelve Coloured Plates. 8vo. cloth, 20*s.*

DR. HENRY F. A. PRATT, M.D., M.R.C.P.

I.

THE GENEALOGY OF CREATION, newly Translated from the Unpointed Hebrew Text of the Book of Genesis, showing the General Scientific Accuracy of the Cosmogony of Moses and the Philosophy of Creation. 8vo. cloth, 14*s.*

II.

ON ECCENTRIC AND CENTRIC FORCE: A New Theory of Projection. With Engravings. 8vo. cloth, 10*s.*

III.

ON ORBITAL MOTION: The Outlines of a System of Physical Astronomy. With Diagrams. 8vo. cloth, 7*s.* 6*d.*

IV.

ASTRONOMICAL INVESTIGATIONS. The Cosmical Relations of the Revolution of the Lunar Apsides. Oceanic Tides. With Engravings. 8vo. cloth, 5*s.*

V.

THE ORACLES OF GOD: An Attempt at a Re-interpretation. Part I. The Revealed Cosmos. 8vo. cloth, 10*s.*

THE PRESCRIBER'S PHARMACOPŒIA; containing all the Medicines in the British Pharmacopœia, arranged in Classes according to their Action, with their Composition and Doses. By a Practising Physician. Fifth Edition. 32mo. cloth, 2*s.* 6*d.*; roan tuck (for the pocket), 3*s.* 6*d.*

DR. JOHN ROWLISON PRETTY.

AIDS DURING LABOUR, including the Administration of Chloroform, the Management of Placenta and Post-partum Hæmorrhage. Fcap. 8vo. cloth, 4*s.* 6*d.*

MR. P. C. PRICE, F.R.C.S.

AN ESSAY ON EXCISION OF THE KNEE-JOINT. With Coloured Plates. With Memoir of the Author and Notes by Henry Smith, F.R.C.S. Royal 8vo. cloth, 14*s.*

MR. LAKE PRICE.

PHOTOGRAPHIC MANIPULATION: A Manual treating of the Practice of the Art, and its various Applications to Nature. With numerous Engravings. Second Edition. Crown 8vo. cloth, 6s. 6d.

DR. PRIESTLEY.

LECTURES ON THE DEVELOPMENT OF THE GRAVID UTERUS. 8vo. cloth, 5s. 6d.

MR. RAINEY.

ON THE MODE OF FORMATION OF SHELLS OF ANIMALS, OF BONE, AND OF SEVERAL OTHER STRUCTURES, by a Process of Molecular Coalescence, Demonstrable in certain Artificially-formed Products. Fcap. 8vo. cloth, 4s. 6d.

MR. ROBERT RAMSAY AND MR. J. OAKLEY COLES.

DEFORMITIES OF THE MOUTH, CONGENITAL AND ACCIDENTAL: Their Mechanical Treatment. With Illustrations. 8vo. cloth, 5s.

DR. F. H. RAMSBOTHAM.

THE PRINCIPLES AND PRACTICE OF OBSTETRIC MEDICINE AND SURGERY. Illustrated with One Hundred and Twenty Plates on Steel and Wood; forming one thick handsome volume. Fifth Edition. 8vo. cloth, 22s.

DR. READE, M.B.T.C.D., L.R.C.S.I.

SYPHILITIC AFFECTIONS OF THE NERVOUS SYSTEM, AND A CASE OF SYMMETRICAL MUSCULAR ATROPHY; with other Contributions to the Pathology of the Spinal Marrow. Post 8vo. cloth, 5s.

PROFESSOR REDWOOD, PH. D.

A SUPPLEMENT TO THE PHARMACOPŒIA: A concise but comprehensive Dispensatory, and Manual of Facts and Formulæ, for the use of Practitioners in Medicine and Pharmacy. Third Edition. 8vo. cloth, 22s.

DR. DU BOIS REYMOND.

ANIMAL ELECTRICITY; Edited by H. Bence Jones, M.D., F.R.S. With Fifty Engravings on Wood. Foolscap 8vo. cloth, 6s.

DR. REYNOLDS, M.D.LOND., F.R.S.

I.

EPILEPSY: ITS SYMPTOMS, TREATMENT, AND RELATION TO OTHER CHRONIC CONVULSIVE DISEASES. 8vo. cloth, 10s.

II.

THE DIAGNOSIS OF DISEASES OF THE BRAIN, SPINAL CORD, AND THEIR APPENDAGES. 8vo. cloth, 8s.

DR. B. W. RICHARDSON, F.R.S.

ON THE CAUSE OF THE COAGULATION OF THE BLOOD. Being the Astley Cooper Prize Essay for 1856. With a Practical Appendix. 8vo. cloth, 16s.

DR. RITCHIE, M.D.

ON OVARIAN PHYSIOLOGY AND PATHOLOGY. With Engravings. 8vo. cloth, 6*s.*

DR. WILLIAM ROBERTS, M.D., F.R.C.P.

AN ESSAY ON WASTING PALSY; being a Systematic Treatise on the Disease hitherto described as ATROPHIE MUSCULAIRE PROGRESSIVE. With Four Plates. 8vo. cloth, 5*s.*

DR. ROUTH.

INFANT FEEDING, AND ITS INFLUENCE ON LIFE; Or, the Causes and Prevention of Infant Mortality. Second Edition. Fcap. 8vo. cloth, 6*s.*

DR. W. H. ROBERTSON.

I.

THE NATURE AND TREATMENT OF GOUT. 8vo. cloth, 10*s.* 6*d.*

II.

A TREATISE ON DIET AND REGIMEN. Fourth Edition. 2 vols. 12*s.* post 8vo. cloth.

DR. ROWE.

NERVOUS DISEASES, LIVER AND STOMACH COMPLAINTS, LOW SPIRITS, INDIGESTION, GOUT, ASTHMA, AND DISORDERS PRODUCED BY TROPICAL CLIMATES. With Cases. Sixteenth Edition. Fcap. 8vo. 2*s.* 6*d.*

DR. ROYLE, F.R.S., AND DR. HEADLAND, M.D.

A MANUAL OF MATERIA MEDICA AND THERAPEUTICS. With numerous Engravings on Wood. Fifth Edition. Fcap. 8vo. cloth, 12*s.* 6*d.*

DR. RYAN, M.D.

INFANTICIDE: ITS LAW, PREVALENCE, PREVENTION, AND HISTORY. 8vo. cloth, 5*s.*

ST. BARTHOLOMEW'S HOSPITAL.

A DESCRIPTIVE CATALOGUE OF THE ANATOMICAL MUSEUM. Vol. I. (1846), Vol. II. (1851), Vol. III. (1862), 8vo. cloth, 5*s.* each.

ST. GEORGE'S HOSPITAL REPORTS. Vols. I. to III. 8vo. 7*s.* 6*d.*

MR. T. P. SALT, BIRMINGHAM.

ON DEFORMITIES AND DEBILITIES OF THE LOWER EXTREMITIES AND THE MECHANICAL TREATMENT EMPLOYED IN THE PROMOTION OF THEIR CURE. With Plates. 8vo. cloth, 15*s.*

DR. SALTER, F.R.S.

ASTHMA. Second Edition. 8vo. cloth, 10*s.*

DR. SANKEY, M.D.LOND.

LECTURES ON MENTAL DISEASES. 8vo. cloth, 8*s.*

DR. SANSOM, M.D.LOND.

I.

CHLOROFORM: ITS ACTION AND ADMINISTRATION. A Handbook. With Engravings. Crown 8vo. cloth, 5*s.*

II.

THE ARREST AND PREVENTION OF CHOLERA; being a Guide to the Antiseptic Treatment. Fcap. 8vo. cloth, 2*s.* 6*d.*

MR. SAVORY.

A COMPENDIUM OF DOMESTIC MEDICINE, AND COMPANION TO THE MEDICINE CHEST; intended as a Source of Easy Reference for Clergymen, and for Families residing at a Distance from Professional Assistance. Seventh Edition. 12mo. cloth, 5*s.*

DR. SCHACHT.

THE MICROSCOPE, AND ITS APPLICATION TO VEGETABLE ANATOMY AND PHYSIOLOGY. Edited by FREDERICK CURREY, M.A. Fcap. 8vo. cloth, 6*s.*

DR. SCORESBY-JACKSON, M.D., F.R.S.E.

MEDICAL CLIMATOLOGY; or, a Topographical and Meteorological Description of the Localities resorted to in Winter and Summer by Invalids of various classes both at Home and Abroad. With an Isothermal Chart. Post 8vo. cloth, 12*s.*

DR. SEMPLE.

ON COUGH: its Causes, Varieties, and Treatment. With some practical Remarks on the Use of the Stethoscope as an aid to Diagnosis. Post 8vo. cloth, 4*s.* 6*d.*

DR. SEYMOUR.

I.

ILLUSTRATIONS OF SOME OF THE PRINCIPAL DISEASES OF THE OVARIA: their Symptoms and Treatment; to which are prefixed Observations on the Structure and Functions of those parts in the Human Being and in Animals. On India paper. Folio, 16*s.*

II.

THE NATURE AND TREATMENT OF DROPSY; considered especially in reference to the Diseases of the Internal Organs of the Body, which most commonly produce it. 8vo. 5*s.*

DR. SHAPTER, M.D., F.R.C.P.

THE CLIMATE OF THE SOUTH OF DEVON, AND ITS INFLUENCE UPON HEALTH. Second Edition, with Maps. 8vo. cloth, 10*s.* 6*d.*

MR. SHAW, M.R.C.S.

THE MEDICAL REMEMBRANCER; OR, BOOK OF EMERGENCIES. Fifth Edition. Edited, with Additions, by JONATHAN HUTCHINSON, F.R.C.S. 32mo. cloth, 2*s.* 6*d.*

DR. SHEA, M.D., B.A.

A MANUAL OF ANIMAL PHYSIOLOGY With an Appendix of Questions for the B.A. London and other Examinations. With Engravings. Foolscap 8vo. cloth, 5*s.* 6*d.*

DR. SHRIMPTON.

CHOLERA: ITS SEAT, NATURE, AND TREATMENT. With Engravings. 8vo. cloth, 4*s.* 6*d.*

MR. U. J. KAY-SHUTTLEWORTH.

FIRST PRINCIPLES OF MODERN CHEMISTRY: a Manual of Inorganic Chemistry. Crown 8vo. cloth, 4*s.* 6*d.*

DR. SIBSON, F.R.S.

MEDICAL ANATOMY. With coloured Plates. Imperial folio. Complete in Seven Fasciculi. 5*s.* each.

DR. E. H. SIEVEKING.

ON EPILEPSY AND EPILEPTIFORM SEIZURES: their Causes, Pathology, and Treatment. Second Edition. Post 8vo. cloth, 10*s.* 6*d.*

DR. SIMMS.

A WINTER IN PARIS: being a few Experiences and Observations of French Medical and Sanitary Matters. Fcap. 8vo. cloth, 4*s.*

MR. SINCLAIR AND DR. JOHNSTON.

PRACTICAL MIDWIFERY: Comprising an Account of 13,748 Deliveries, which occurred in the Dublin Lying-in Hospital, during a period of Seven Years. 8vo. cloth, 10*s.*

DR. SIORDET, M.B.LOND., M.R.C.P.

MENTONE IN ITS MEDICAL ASPECT. Foolscap 8vo. cloth, 2*s.* 6*d.*

MR. ALFRED SMEE, F.R.S.

GENERAL DEBILITY AND DEFECTIVE NUTRITION; their Causes, Consequences, and Treatment. Second Edition. Fcap. 8vo. cloth, 3*s.* 6*d.*

DR. SMELLIE.

OBSTETRIC PLATES: being a Selection from the more Important and Practical Illustrations contained in the Original Work. With Anatomical and Practical Directions. 8vo. cloth, 5*s.*

MR. HENRY SMITH, F.R.C.S.

I.

ON STRICTURE OF THE URETHRA. 8vo. cloth, 7*s.* 6*d.*

II.

HÆMORRHOIDS AND PROLAPSUS OF THE RECTUM: Their Pathology and Treatment, with especial reference to the use of Nitric Acid. Third Edition. Fcap. 8vo. cloth, 3*s.*

III.

THE SURGERY OF THE RECTUM. Lettsomian Lectures. Second Edition. Fcap. 8vo. 3*s.* 6*d.*

DR. J. SMITH, M.D., F.R.C.S.EDIN.

HANDBOOK OF DENTAL ANATOMY AND SURGERY, FOR THE USE OF STUDENTS AND PRACTITIONERS. Fcap. 8vo. cloth, 3*s.* 6*d.*

DR. W. TYLER SMITH.

A MANUAL OF OBSTETRICS, THEORETICAL AND PRACTICAL. Illustrated with 186 Engravings. Fcap. 8vo. cloth, 12*s.* 6*d.*

DR. SNOW.

ON CHLOROFORM AND OTHER ANÆSTHETICS: THEIR ACTION AND ADMINISTRATION. Edited, with a Memoir of the Author, by Benjamin W. Richardson, M.D. 8vo. cloth, 10*s.* 6*d.*

MR. J. VOSE SOLOMON, F.R.C.S.

TENSION OF THE EYEBALL; GLAUCOMA: some Account of the Operations practised in the 19th Century. 8vo. cloth, 4*s.*

DR. STANHOPE TEMPLEMAN SPEER.

PATHOLOGICAL CHEMISTRY, IN ITS APPLICATION TO THE PRACTICE OF MEDICINE. Translated from the French of MM. BECQUEREL and RODIER. 8vo. cloth, reduced to 8*s.*

MR. J. K. SPENDER, M.D.LOND.

A MANUAL OF THE PATHOLOGY AND TREATMENT OF ULCERS AND CUTANEOUS DISEASES OF THE LOWER LIMBS. 8vo. cloth, 4*s.*

MR. BALMANNO SQUIRE, M.B.LOND.

I.

CLINICAL LECTURES ON SKIN DISEASES. Illustrated by Coloured Photographs from Life. Complete in 36 Numbers, price 1*s.* 6*d.* each. Nos. I.—XXX. are now ready.

II.

A MANUAL OF THE DISEASES OF THE SKIN. Illustrated by Coloured Plates of the Diseases, and by Woodcuts of the Parasites of the Skin. Post 8vo. cloth, 24*s.*

MR. PETER SQUIRE.

I.

A COMPANION TO THE BRITISH PHARMACOPŒIA. Seventh Edition. 8vo. cloth, 10*s.* 6*d.*

II.

THE PHARMACOPŒIAS OF THE LONDON HOSPITALS, arranged in Groups for easy Reference and Comparison. Second Edition. 18mo. cloth, 5*s.*

DR. STEGGALL.

I.

A MEDICAL MANUAL FOR APOTHECARIES' HALL AND OTHER MEDICAL BOARDS. Twelfth Edition. 12mo. cloth, 10*s.*

II.

A MANUAL FOR THE COLLEGE OF SURGEONS; intended for the Use of Candidates for Examination and Practitioners. Second Edition. 12mo. cloth, 10*s.*

III.

FIRST LINES FOR CHEMISTS AND DRUGGISTS PREPARING FOR EXAMINATION AT THE PHARMACEUTICAL SOCIETY. Third Edition. 18mo. cloth, 3*s.* 6*d.*

MR. STOWE, M.R.C.S.

A TOXICOLOGICAL CHART, exhibiting at one view the Symptoms, Treatment, and Mode of Detecting the various Poisons, Mineral, Vegetable, and Animal. To which are added, concise Directions for the Treatment of Suspended Animation. Twelfth Edition, revised. On Sheet, 2*s.*; mounted on Roller, 5*s.*

MR. FRANCIS SUTTON, F.C.S.

A SYSTEMATIC HANDBOOK OF VOLUMETRIC ANALYSIS; or, the Quantitative Estimation of Chemical Substances by Measure. With Engravings. Post 8vo. cloth, 7*s.* 6*d.*

DR. SWAYNE.

OBSTETRIC APHORISMS FOR THE USE OF STUDENTS COMMENCING MIDWIFERY PRACTICE. With Engravings on Wood. Fourth Edition. Fcap. 8vo. cloth, 3*s.* 6*d.*

MR. TAMPLIN, F.R.C.S.E.

LATERAL CURVATURE OF THE SPINE: its Causes, Nature, and Treatment. 8vo. cloth, 4*s.*

SIR ALEXANDER TAYLOR, M.D., F.R.S.E.

THE CLIMATE OF PAU; with a Description of the Watering Places of the Pyrenees, and of the Virtues of their respective Mineral Sources in Disease. Third Edition. Post 8vo. cloth, 7*s.*

DR. ALFRED S. TAYLOR, F.R.S.

I.

THE PRINCIPLES AND PRACTICE OF MEDICAL JURISPRUDENCE. With 176 Wood Engravings. 8vo. cloth, 28*s.*

II.

A MANUAL OF MEDICAL JURISPRUDENCE. Eighth Edition. With Engravings. Fcap. 8vo. cloth, 12*s.* 6*d.*

III.

ON POISONS, in relation to MEDICAL JURISPRUDENCE AND MEDICINE. Second Edition. Fcap. 8vo. cloth, 12*s.* 6*d.*

MR. TEALE.

ON AMPUTATION BY A LONG AND A SHORT RECTANGULAR FLAP. With Engravings on Wood. 8vo. cloth, 5*s.*

DR. THEOPHILUS THOMPSON, F.R.S.

CLINICAL LECTURES ON PULMONARY CONSUMPTION; with additional Chapters by E. SYMES THOMPSON, M.D. With Plates. 8vo. cloth, 7*s.* 6*d.*

DR. THOMAS.

THE MODERN PRACTICE OF PHYSIC; exhibiting the Symptoms, Causes, Morbid Appearances, and Treatment of the Diseases of all Climates. Eleventh Edition. Revised by ALGERNON FRAMPTON, M.D. 2 vols. 8vo. cloth, 28*s.*

SIR HENRY THOMPSON, F.R.C.S.

I.

STRICTURE OF THE URETHRA AND URINARY FISTULÆ; their Pathology and Treatment. Jacksonian Prize Essay. With Plates. Third Edition. 8vo. cloth, 10*s.*

II.

THE DISEASES OF THE PROSTATE; their Pathology and Treatment. With Plates. Third Edition. 8vo. cloth, 10*s.*

III.

PRACTICAL LITHOTOMY AND LITHOTRITY; or, An Inquiry into the best Modes of removing Stone from the Bladder. With numerous Engravings, 8vo. cloth, 9*s.*

IV.

CLINICAL LECTURES ON DISEASES OF THE URINARY ORGANS. With Engravings. Crown 8vo. cloth, 5*s.*

DR. THUDICHUM.

I.

A TREATISE ON THE PATHOLOGY OF THE URINE, Including a complete Guide to its Analysis. With Plates, 8vo. cloth, 14*s.*

II.

A TREATISE ON GALL STONES: their Chemistry, Pathology, and Treatment. With Coloured Plates. 8vo. cloth, 10*s.*

DR. TILT.

I.

ON UTERINE AND OVARIAN INFLAMMATION, AND ON THE PHYSIOLOGY AND DISEASES OF MENSTRUATION. Third Edition. 8vo. cloth, 12*s.*

II.

A HANDBOOK OF UTERINE THERAPEUTICS AND OF DISEASES OF WOMEN. Third Edition. Post 8vo. cloth, 10*s.*

III.

THE CHANGE OF LIFE IN HEALTH AND DISEASE: a Practical Treatise on the Nervous and other Affections incidental to Women at the Decline of Life. Second Edition. 8vo. cloth, 6*s.*

DR. GODWIN TIMMS.

CONSUMPTION: its True Nature and Successful Treatment. Re-issue, enlarged. Crown 8vo. cloth, 10*s.*

DR. ROBERT B. TODD, F.R.S.

I.

CLINICAL LECTURES ON THE PRACTICE OF MEDICINE. *New Edition, in one Volume, Edited by* DR. BEALE, *8vo. cloth,* 18*s.*

II.

ON CERTAIN DISEASES OF THE URINARY ORGANS, AND ON DROPSIES. Fcap. 8vo. cloth, 6*s.*

MR. TOMES, F.R.S.

A MANUAL OF DENTAL SURGERY. With 208 Engravings on Wood. Fcap. 8vo. cloth, 12*s.* 6*d.*

DR. TURNBULL.

I.

AN INQUIRY INTO THE CURABILITY OF CONSUMPTION, ITS PREVENTION, AND THE PROGRESS OF IMPROVEMENT IN THE TREATMENT. Third Edition. 8vo. cloth, 6*s.*

II.

A PRACTICAL TREATISE ON DISORDERS OF THE STOMACH with FERMENTATION; and on the Causes and Treatment of Indigestion, &c. 8vo. cloth, 6*s.*

DR. TWEEDIE, F.R.S.

CONTINUED FEVERS: THEIR DISTINCTIVE CHARACTERS, PATHOLOGY, AND TREATMENT. With Coloured Plates. 8vo. cloth, 12*s.*

VESTIGES OF THE NATURAL HISTORY OF CREATION. Eleventh Edition. Illustrated with 106 Engravings on Wood. 8vo. cloth, 7*s.* 6*d.*

DR. UNDERWOOD.

TREATISE ON THE DISEASES OF CHILDREN. Tenth Edition, with Additions and Corrections by HENRY DAVIES, M.D. 8vo. cloth, 15*s.*

MR. WADE, F.R.C.S.

STRICTURE OF THE URETHRA, ITS COMPLICATIONS AND EFFECTS; a Practical Treatise on the Nature and Treatment of those Affections. Fourth Edition. 8vo. cloth, 7*s.* 6*d.*

DR. WAHLTUCH, M.D.

A DICTIONARY OF MATERIA MEDICA AND THERAPEUTICS. 8vo. cloth, 15*s.*

DR. WALKER, M.B.LOND.

ON DIPHTHERIA AND DIPHTHERITIC DISEASES. Fcap. 8vo. cloth, 3*s.*

DR. WALLER.

ELEMENTS OF PRACTICAL MIDWIFERY; or, Companion to the Lying-in Room. Fourth Edition, with Plates. Fcap. cloth, 4*s.* 6*d.*

MR. HAYNES WALTON, F.R.C.S.

SURGICAL DISEASES OF THE EYE. With Engravings on Wood. Second Edition. 8vo. cloth, 14*s.*

DR. WARING, M.D., M.R.C.P.LOND.

I.

A MANUAL OF PRACTICAL THERAPEUTICS. Second Edition, Revised and Enlarged. Fcap. 8vo. cloth, 12*s.* 6*d.*

II.

THE TROPICAL RESIDENT AT HOME. Letters addressed to Europeans returning from India and the Colonies on Subjects connected with their Health and General Welfare. Crown 8vo. cloth, 5*s.*

DR. WATERS, F.R.C.P.

I.

DISEASES OF THE CHEST. CONTRIBUTIONS TO THEIR CLINICAL HISTORY, PATHOLOGY, AND TREATMENT. With Plates. 8vo. cloth, 12*s.* 6*d.*

II.

THE ANATOMY OF THE HUMAN LUNG. The Prize Essay to which the Fothergillian Gold Medal was awarded by the Medical Society of London. Post 8vo. cloth, 6*s.* 6*d.*

III.

RESEARCHES ON THE NATURE, PATHOLOGY, AND TREATMENT OF EMPHYSEMA OF THE LUNGS, AND ITS RELATIONS WITH OTHER DISEASES OF THE CHEST. With Engravings. 8vo. cloth, 5*s.*

DR. ALLAN WEBB, F.R.C.S.L.

THE SURGEON'S READY RULES FOR OPERATIONS IN SURGERY. Royal 8vo. cloth, 10*s.* 6*d.*

MR. SOELBERG WELLS.

I.

A TREATISE ON THE DISEASES OF THE EYE. With Coloured Plates and Wood Engravings. 8vo. cloth, 24*s*.

II.

ON LONG, SHORT, AND WEAK SIGHT, and their Treatment by the Scientific Use of Spectacles. Third Edition. With Plates. 8vo. cloth, 6*s*.

MR. T. SPENCER WELLS, F.R.C.S.

SCALE OF MEDICINES FOR MERCHANT VESSELS. With Observations on the Means of Preserving the Health of Seamen, &c. &c. Seventh Thousand. Fcap. 8vo. cloth, 3*s*. 6*d*.

DR. WEST.

LECTURES ON THE DISEASES OF WOMEN. Third Edition. 8vo. cloth, 16*s*.

MR. WHEELER.

HAND-BOOK OF ANATOMY FOR STUDENTS OF THE FINE ARTS. With Engravings on Wood. Fcap. 8vo., 2*s*. 6*d*.

DR. WHITEHEAD, F.R.C.S.

ON THE TRANSMISSION FROM PARENT TO OFFSPRING OF SOME FORMS OF DISEASE, AND OF MORBID TAINTS AND TENDENCIES. Second Edition. 8vo. cloth, 10*s*. 6*d*.

DR. WILLIAMS, F.R.S.

PRINCIPLES OF MEDICINE: An Elementary View of the Causes, Nature, Treatment, Diagnosis, and Prognosis, of Disease. With brief Remarks on Hygienics, or the Preservation of Health. The Third Edition. 8vo. cloth, 15*s*.

DR. WINSLOW, M.D., D.C.L.OXON.

OBSCURE DISEASES OF THE BRAIN AND MIND. Fourth Edition. Carefully Revised. Post 8vo. cloth, 10*s*. 6*d*.

DR. WISE, M.D., F.R.C.P.EDIN.

REVIEW OF THE HISTORY OF MEDICINE AMONG ASIATIC NATIONS. Two Vols. 8vo. cloth, 16*s*.

MR. ERASMUS WILSON, F.R.S.

I.

THE ANATOMIST'S VADE-MECUM: A SYSTEM OF HUMAN ANATOMY. With numerous Illustrations on Wood. Eighth Edition. Foolscap 8vo. cloth, 12*s.* 6*d.*

II.

ON DISEASES OF THE SKIN: A SYSTEM OF CUTANEOUS MEDICINE. Sixth Edition. 8vo. cloth, 18*s.*

THE SAME WORK; illustrated with finely executed Engravings on Steel, accurately coloured. 8vo. cloth, 36*s.*

III.

HEALTHY SKIN: A Treatise on the Management of the Skin and Hair in relation to Health. Seventh Edition. Foolscap 8vo. 2*s.* 6*d.*

IV.

PORTRAITS OF DISEASES OF THE SKIN. Folio. Fasciculi I. to XII., completing the Work. 20*s.* each. The Entire Work, half morocco, £13.

V.

THE STUDENT'S BOOK OF CUTANEOUS MEDICINE AND DISEASES OF THE SKIN. Post 8vo. cloth, 8*s.* 6*d.*

VI.

ON SYPHILIS, CONSTITUTIONAL AND HEREDITARY; AND ON SYPHILITIC ERUPTIONS. With Four Coloured Plates. 8vo. cloth, 16*s.*

VII.

A THREE WEEKS' SCAMPER THROUGH THE SPAS OF GERMANY AND BELGIUM, with an Appendix on the Nature and Uses of Mineral Waters. Post 8vo. cloth, 6*s.* 6*d.*

VIII.

THE EASTERN OR TURKISH BATH: its History, Revival in Britain, and Application to the Purposes of Health. Foolscap 8vo., 2*s.*

DR. G. C. WITTSTEIN.

PRACTICAL PHARMACEUTICAL CHEMISTRY: An Explanation of Chemical and Pharmaceutical Processes, with the Methods of Testing the Purity of the Preparations, deduced from Original Experiments. Translated from the Second German Edition, by STEPHEN DARBY. 18mo. cloth, 6*s.*

DR. WOLFE, M.D.

AN IMPROVED METHOD OF EXTRACTION OF CATARACT. With Results of 107 Operations. 8vo. cloth, 2*s.* 6*d.*

DR. HENRY G. WRIGHT.

I.

UTERINE DISORDERS: their Constitutional Influence and Treatment. 8vo. cloth, 7*s.* 6*d.*

II.

HEADACHES; their Causes and their Cure. Fourth Edition. Fcap. 8vo. 2*s.* 6*d.*

DR. YEARSLEY, M.D., M.R.C.S.

I.

DEAFNESS PRACTICALLY ILLUSTRATED; being an Exposition as to the Causes and Treatment of Diseases of the Ear. Sixth Edition. 8vo. cloth, 6*s.*

II.

ON THROAT AILMENTS, MORE ESPECIALLY IN THE ENLARGED TONSIL AND ELONGATED UVULA. Eighth Edition. 8vo. cloth, 5*s.*

CHURCHILL'S SERIES OF MANUALS.

Fcap. 8vo. cloth, 12*s.* 6*d.* each.

"We here give Mr. Churchill public thanks for the positive benefit conferred on the Medical Profession, by the series of beautiful and cheap Manuals which bear his imprint."—*British and Foreign Medical Review.*

AGGREGATE SALE, 154,000 COPIES.

ANATOMY. With numerous Engravings. Eighth Edition. By ERASMUS WILSON, F.R.C.S., F.R.S.

BOTANY. With numerous Engravings. By ROBERT BENTLEY, F.L.S., Professor of Botany, King's College, and to the Pharmaceutical Society.

CHEMISTRY. With numerous Engravings. Tenth Edition, 14*s.* By GEORGE FOWNES, F.R.S., H. BENCE JONES, M.D., F.R.S., and HENRY WATTS, B.A., F.R.S.

DENTAL SURGERY. With numerous Engravings. By JOHN TOMES, F.R.S.

EYE, DISEASES OF. With coloured Plates and Engravings on Wood. By C. MACNAMARA.

MATERIA MEDICA. With numerous Engravings. Fifth Edition. By J. FORBES ROYLE, M.D., F.R.S., and F. W. HEADLAND, M.D., F.L.S.

MEDICAL JURISPRUDENCE. With numerous Engravings. Eighth Edition. By ALFRED SWAINE TAYLOR, M.D., F.R.S.

PRACTICE OF MEDICINE. Second Edition. By G. HILARO BARLOW, M.D., M.A.

The **MICROSCOPE and its REVELATIONS.** With numerous Plates and Engravings. Fourth Edition. By W. B. CARPENTER, M.D., F.R.S.

NATURAL PHILOSOPHY. With numerous Engravings. Sixth Edition. By CHARLES BROOKE, M.B., M.A., F.R.S. *Based on the Work of the late Dr. Golding Bird.*

OBSTETRICS. With numerous Engravings. By W. TYLER SMITH, M.D., F.R.C.P.

OPHTHALMIC MEDICINE and SURGERY. With coloured Plates and Engravings on Wood. Third Edition. By T. WHARTON JONES, F.R.C.S., F.R.S.

PATHOLOGICAL ANATOMY. With numerous Engravings. By C. HANDFIELD JONES, M.B., F.R.S., and E. H. SIEVEKING, M.D., F.R.C.P.

PHYSIOLOGY. With numerous Engravings. Fourth Edition. By WILLIAM B. CARPENTER, M.D., F.R.S.

POISONS. Second Edition. By ALFRED SWAINE TAYLOR, M.D., F.R.S.

PRACTICAL ANATOMY. With numerous Engravings. (10*s.* 6*d.*) By CHRISTOPHER HEATH, F.R.C.S.

PRACTICAL SURGERY. With numerous Engravings. Fourth Edition. By Sir WILLIAM FERGUSSON, Bart., F.R.C.S., F.R.S.

THERAPEUTICS. Second Edition. By E. J. Waring, M.D., M.R.C.P.

Printed by W. BLANCHARD & SONS, 62, Millbank Street, Westminster.